Health Effects

of

Toxic Substances

Second Edition

M. J. Malachowski, PhD

Arleen F. Goldberg, *Contributing Author*

Government Institutes
a Division of ABS Group Inc.
Rockville, Maryland

Government Institutes, a Division of ABS Group Inc.
4 Research Place, Rockville, Maryland 20850, USA
Phone: (301) 921-2300
Fax: (301) 921-0373
Email: giinfo@govinst.com
Internet: http://www.govinst.com

03 02 01 00 99   5   4   3   2

Library of Congress Cataloging-in-Publication Data

Malachowski, M. J. (Michael J.)
        Health effects of toxic substances / by M. J. Malachowski. — 2nd ed.
        p.   cm.
        Includes bibliographical references and index.
        ISBN: 0-86587-649-5
        1. Industrial toxicology.   I. Title.
        [DNLM: 1. Hazardous Substances—adverse effects. 2. Occupational Exposure—adverse effects. 3. Occupational Exposure—prevention & control. 4. Environmental Monitoring.   WA 465 M236h   1999]
        RA1229.M35   1999
        615.9'02—dc21
        DNLM/DCL
        for Library of Congress   98-32315
                                                               CIP

Printed in the United States of America

Summary Contents

Expanded Contents

Foreword

In the 1970s, the San Francisco Bay Area was rapidly developing into a hotbed of radical environmental causes. With the passage of the Resource Conservation and Recovery Act of 1976 (RCRA), environmental activity gained significant validation. Vista Community College in Berkeley developed a variety of Environmental Certificate Programs in the late 1970s. One of the authors, Dr. Malachowski, was hired to teach the Solid and Hazardous Waste Management Certificate Program. The program offered its first courses in 1980. Among the courses taught were environmental toxicology and industrial toxicology. Courses catered to the developing environmental professionals and the California State and City of Berkeley Public Health Departments. These are both located in the City of Berkeley.

Due to district administrative cutbacks, the Vista programs were relocated to the Merritt Community College Environmental Studies Department in the late 1980s. The program was given new life with the advent of the State of California's Environmental Training Program and the formation of the Partnership of Environmental Technology Education (PETE). More than twenty community colleges in California banded together to offer six core certificate courses. The goal was to provide a seamless offering to students throughout the state. The program encouraged cooperation among participating colleges. In Northern California, the colleges banded together to form the Northern California Environmental Technology Consortium (NorCal ETC).

NorCal ETC coordinated with the University of California and the State University Systems to provide the training and retraining for displaced defense workers for base closure and cleanup. In the San Francisco Bay Area, these included the Presidio and Hunters Point/Treasure Island in San Francisco, Alameda Naval Air Station and the Oakland Army Base in the East Bay, and Mare Island. A number of special courses and programs were developed to meet the specific needs of the base closure events. Many different health and safety elements were compiled into the *Health Effects of Toxic Substances* text.

Hans Dieter, Ph.D.

Preface

Welcome to the second edition of *Health Effects of Toxic Substances*. The first edition was based upon a decade of lecture notes which were modularized for class participants over the years. In 1991, the State of California established the California Environmental Hazardous Materials Consortium. This effort, supported by a Community College Funds for Instructional Improvement (FII) Grant, developed a set of core curricula for an environmental training program. These courses formed the basis for the Partnership in Environmental Technology Education (PETE) program, which started with what is now known as Western PETE.

When I first started teaching Industrial Toxicology, Environmental Toxicology, and Health Effects, nothing was available in the way of textbooks at the community college level. In the 1970s and 80s, literature was sparse in these areas. My first courses, therefore, were almost exclusively lecture based, accompanied by numerous handouts. Over the years I put together modules of materials that I could distribute for specific lecture topics. Finally, with the development of the California curriculum, this was all combined into a single volume.

The first edition of *Health Effects of Toxic Substances* was a synthesis of my previous courses, the FII suggested curriculum, and the PETE needs. My desire was to produce a single product that I could provide to my class participants. This text would provide the information needed to assist them through the course. Because of the length of the semester, material was divided into sixteen chapters, one a week, with an extra week for exams. It is not designed as a reference text; a number of these exist. Rather this is a working document for the participant. The concept is to minimize the need to take class lecture notes, and instead, to have the lectures be a chance to listen and annotate the text.

It is assumed that the reader has had basic courses in science. It is highly recommended that human biology, or anatomy and physiology, and chemistry be taken before attempting this course. The necessary background material for understanding the *Health Effects of Toxic Substances* can be found in texts for those courses. This same background material will also aid in understanding useful references, such as Dreisbach and Robertson's *Handbook of Poisoning*.

Over the last four years I have received voluminous comments on the text. In response to these comments, the second edition has been significantly augmented. Much of the text was rewritten to facilitate clarity. As the field has emerged, marked interest has grown in some areas. Thus, a significant expansion of Chapter 16, "Risk Assessment and Epidemiology," has been made. Important and new terminology has been placed in **boldfaced text**. The glossary has been greatly expanded to cover a variety of medical and scientific vocabulary. The goal was to produce a more readable and useful document. The basic objectives remain to provide a comprehensive set of materials for use in teaching the course, to provide clear and concise explanations of topics and concepts in the field, and to present the material in a logical fashion.

A significant portion of the text remains devoted to industrial hygiene (IH). While this is frequently redundant with information presented in trainings, such as the 40-hour HAZWOPER or a number of introductory courses, I feel that the concepts and materials compliment those of the industrial toxicology portion of the text. These sections provide a review of IH and may, as appropriate, be lightly passed over. There are numerous exercises associated with these chapters. They can be completed for review and the text consulted for clarification, explanation, and expansion of these basic concepts. Time saved in the presentation of the IH material may be utilized to expand on Chapter 15, "Medical Monitoring, Treatment, and Management," and Chapter 16, "Risk Assessment and Epidemiology."

Health Effects

of

Toxic Substances

Second Edition

1 Industrial Toxicology
History and Hazards

Overview

The purpose of this chapter is to provide an introduction and historical perspective to toxicology. There is a relationship among toxicology, industrial toxicology and environmental toxicology. We are principally concerned with industrial toxicology.

It is necessary to be able to distinguish among a hazard, a hazardous material, and a toxic substance. To aid in this distinction, we shall explore the concepts of concentration, exposure, and absorption. The dose depends upon the concentration of the substance, the exposure time, and the absorption efficiency.

Toxicology is the study of the injurious effects of chemical and physical substances on living organisms. It is a subdivision of **Pharmacology**, which is the study of molecular and chemical interactions (drugs) with an organism. It is both an art and a science. We strive to scientifically increase our knowledge of the field, but all too often, due to insufficient data, we must make judgements when called upon to predict the likelihood of adverse responses following exposure to one or more agents in situations where there is limited and incomplete information. Toxicology embraces the study of effects of substances on all living organisms. Insights acquired from this study are now used for better treatment of emergencies, safer products, and prevention of unwanted reactions.

Many of the substances used, manufactured, or formulated in the industrial environment can adversely impact the health of humans. **Industrial Toxicology** is the study of the impact of these substances in and on the industrial environment. In the National Institute of Occupational Safety and Health (NIOSH) *Registry of Toxic Effects of Chemical Substances,* a toxic substance is defined as follows:

A toxic substance is one that demonstrates the potential to induce cancer, tumors, or neoplastic effects in man or experimental animals; to induce a permanent transmissible change in the characteristic of an offspring from those of its human or experimental animal parents; to cause the production of physical defects in the developing human or experimental animal embryo; to produce death in animals exposed via the respiratory tract, skin, eye, mouth or other routes in experimental or domestic animals; to produce irritation or sensitization of the skin, eyes, or respiratory passages; to diminish mental alertness, reduce motivation, or alter behavior in humans; to adversely affect the health of a normal or disabled person of any age or of either sex by producing reversible or irreversible bodily injury or by endangering life or causing death from exposure via the respiratory tract, skin, eye, mouth or any other route in any quantity, concentration, or dose reported for any length of time.

Most of the agents of concern to our studies may be classified as chemical or physical. Chemical agents include billions of tons of some 64,000 naturally occurring and synthetic commercial compounds, as well as thousands of food additives, medicinal drugs, and household products. Additionally, many new substances are produced by plants, animals, fungi, bacteria, and man each year. Physical agents includes dust and fumes, radiation, noise, and pressure.

History

Toxicology is the study of the nature and actions of **poisons**. The term is derived from the Greek word **toxikos** meaning of or for a **toxon** or bow. It, however, actually referred to the poisons in which arrows were dipped. We shall use the word toxon, explicitly, to indicate a specific toxic molecule, particle, substance, chemical or material which interacts with a specific target, i.e., the arrow hitting the bull's eye (also referred to as a LOCK and KEY mechanism). We will use toxin to indicate generic toxic substances as previously defined.

Anthropological and archaeological research has provided knowledge of the poisons used by primitive man. Venoms, poisonous herbs, and toxic materials were well known to ancient man and were intimately associated with species preservation. Myth, legend, and history indicate the growth of toxicological knowledge base over an extended period of time. The African Bushmen used a mixture of various species of herbs as poisons for arrow tips. Periodically, snake and black widow spider venom was added to this mixture. Other African tribes used seeds for poisons. South American Indians still utilize a wide variety of toxic substances to aid in hunting.

Each succeeding age had its own toxins and methods of administration. As early as 1552 B.C., Ebers Papyrus' scrolls, perhaps the earliest western medical record, listed over 800 recipes for remedies to specific diseases. Mention was made of the toxicity of heavy metals like lead, antimony, and copper, and included information on various herbs, e.g., hemlock, opium, and digitalis.

Hippocrates (460 B.C.), the Greek scholar, advanced the knowledge of poisons considerably. He identified arsenic, antimony, mercury, copper, and lead as toxic. Hippocrates was among the first to formulate basic, albeit primitive, principles of toxicology. Toxicology was starting its change from an *empirical* to an *applied* science.

Aristotle, Theophrastis (350–250 B.C.), described several plant, animal, and mineral poisons for use as suicide agents, for murders and for executions. (His associate, Socrates, died from drinking a poison hemlock solution.)

King Mithridates VI of Pontus (Roman Empire) experimented on criminals to expand his knowledge of toxicology. He claimed to have developed "an antidote for every venomous reptile and every poisonous substance." He is reputed to have regularly ingested a mixture of 36 ingredients as protection against assassination. Now a mithridatic refers to an antidotal or protective mixture.

It is alleged that Nero ordered the death of Britannicus, the son of Claudius. The first attempt failed, but the symptoms suggest that it was arsenic poisoning. Britannicus became suspicious and hired a taster. Later, Britannicus and the taster were fed a very hot soup. Unfortunately, Britannicus made a fatal mistake of cooling his portion down with water containing arsenic.

By 980 A.D., the Arabs had translated Greek works into Arabic, and had developed their own system of medicine. They were advanced chemists and developed novel analytical methods for processing medicines. Numerous agents were identified as toxic by Arab physicians. Our knowledge of poisons from the ninth to the fifteenth centuries (the Middle Ages) is derived from manuscripts which have come down to us from these periods.

A number of toxic agents were identified in the Middle Ages. The art of poisoning was becoming perfected, particularly in Italy. City council records of Venice and Florence contain ample testimony of the political use of poisons, including victims' names, prices, contracts, and deed of payment.

Catherine de Medici, a relative of the infamous Borgia poisoning family, is one of the most notable poisoners and, perhaps, the earliest experimental toxicologist. She noted the rapidity of toxic response, **onset of action**, the effectiveness of the chemical studied, **potency**, the degree of response of the parts of the body, **specificity and site of action**, and complaints of the victims, **clinical signs and symptoms**.

Paracelsus, Phillippus Aureolus Theophrastus Bombastus von Hohenheim, (1493–1541) was a Swiss physician and is credited as being the Father of Toxicology. He was a free spirit who wrote in German, rather than in Latin, and was more popular with gypsies and tradespeople than with his medical peers. Paracelsus was a man far ahead of his time, a bridge between alchemy and science. He was an iconoclast with a contempt for the medical doctrines and methods of the day. Thus, he earned the disdain of the medical establishment, which led to numerous moves of his medical practice. He died at the age of 48 from wounds suffered in a tavern brawl. He is most famous for his contention: "All substances are poisons; there is none which is not a poison. No substance is a poison in itself, it is the dose that makes a substance a poison. The dose differentiates a poison and a remedy. The dose makes the poison." He distinguished between acute and chronic toxic effects of metals, and described in detail the symptoms of chronic mercurialism. In 1567, 26 years after his death, Paracelsus's monograph on occupational diseases of miners and smelters was published.

Bernadina Ramazzini (1633-1714) is often considered to be the Father of Occupational Medicine. This Italian's specialty was **epidemiology**. Epidemiology is the study of the **incidence** (the number of new cases), **prevalence** (the number of existing cases), and movement of diseases in populations. A socially oriented physician, he performed **prospective** (looking for causality) studies by following his subjects into the mines, factories, and cesspools to see and experience the conditions under which they labored. His contribution to the medical profession was his counsel that, in addition to the usual questions, health professionals should inquire into the occupations of their patients.

As more toxic agents were discovered, they were utilized in heinous ways. Occupational exposures to toxic agents were also rampant. The nineteenth century saw the development of tests for identification of poisons. Claude Bernard used toxicants to determine the physiological basis of organ systems. In March of 1836 he developed an **analytical** test for arsenic. This effectively eliminated the use of arsenic as an intentional poisoning agent.

Orfila (1787–1853), a Spanish physician, was one of the first persons to attempt a systematic correlation between the chemical and biological information on poisons. He singled out toxicology and defined it as the study of poisons. He became involved in **forensic toxicology** by devising methods to detect poisons. He pioneered the use of **autopsies** to detect poisonings. In the 1880s, Orfila's classic text on toxicology, *Traite' de Toxicologie,* was published. The text was a major breakthrough in toxicology and this work provided a stimulus to the field.

Some industrial hazards have been known for centuries. In the first century A.D., the symptomatology for lead intoxication was identified. The Romans used only slave labor in the Spanish mercury mines at Almoden. A sentence to work at the mines was considered a death sentence. Mercury intoxication was identified in the hat industry in France in the seventh century, and became so widespread that the expression, "mad as a hatter", became a colloquialism.

Sir Percival Pott, in the late 1700s, discovered a cause and effect relationship of occupational cancer, in this case, scrotal cancer in chimney sweeps. As Ramazzini had suggested, he evaluated the profession and the type of work being performed. He observed that the sweeps would cover their hands with the coal tars from the chimneys. Subsequently, when they had an itch to scratch, they would coat the surface of the skin with coal tar. Over a period of time this led to the induction of scrotal cancer.

Exposure to toxic substances is a ever present hazard of modern technology. In addition to newly developed chemicals, many materials; first synthesized in the late nineteenth century, have found widespread industrial use. The hydrides of boron have been known since 1879, but the first report of their toxicity appeared in 1951, as a series of case histories of young chemical engineers who had been exposed to boron hydrides in the course of their work. Catastrophic events have resulted from worker exposures to benzene, lead, vinyl chloride, dibromo-chloropropane, kepone, polychlorinated biphenyls (PCB), and other industrial chemicals. Medical exposure (thalidomide), warfare (Agent Orange), accident (methyl isocyanate in Bhopal, India, 1984), and Love Canal are just a few examples of toxicological tragedies in our recent past.

Today, the Agency for Toxic Substances and Disease Registry (ATSDR) publishes a list called "Top 20 Hazardous Substances." (This is an annual evaluation activity required of CERCLA and is from the priority list of 275 substances.) Topping the list are the same heavy metals used and categorized as toxic throughout recorded history. In order, these twenty hazardous substances are as follows:

1. Arsenic
2. Lead
3. Mercury (Metallic)
4. Vinyl Chloride
5. Benzene
6. Polychlorinated Biphenyls (PCBs)
7. Cadmium
8. Benzo(a)pyrene
9. Benzo(b)fluoranthene
10. Polycyclic Aromatic Hydrocarbons
11. Cloroform
12. Aroclor 1254
13. DDT
14. Aroclor 1260
15. Tricholoroethylene
16. Chromium (+6)
17. Dibenz(a,h)anthracene
18. Dieldrin
19. Hexachlorobutadiene
20. Chlordane

In 1941, the American Conference of Governmental Industrial Hygienists (ACGIH) formed a committee to review available data on toxic compounds. They were interested in establishing exposure limits for employees working in the presence of airborne toxic agents. The committee publishes an annual list of compounds and recommended exposure limits entitled *Threshold Limit Values (TLV) and Biological Exposure Indices (BEI).* The primary purpose of the TLVs is to protect healthy male workers in chronic exposure situations and, therefore, not applicable to the unhealthy, women, and children. ATSDR reported in 1998 that women experience more adverse health outcomes from hazardous substances than men. They reported significantly more anemia and other blood disorders, skin rashes, and strokes compared to the national norms. There was greater frequency of diabetes and kidney and liver problems, and urinary tract disorders typically occur more often than in men exposed to the same substances.

In the early 1970s the United States Occupational Safety and Health Administration (OSHA) was established. OSHA is responsible for the adoption and enforcement of standards for safe and healthful working conditions for men and women employed in any business engaged in commerce in the United States. OSHA essentially adopted the then current TLVs, made them official federal standards, and referred to them as Permissible Exposure Limits (PEL). PELs are formally listed in Title 29 CFR, Part 1910, Subpart Z, General Industry Standards for Toxic and Hazardous Substances. These were last revised in 1989.

The American Industrial Hygiene Association (AIHA) has established a committee to develop Workplace Environmental Exposure Levels (WEELs) for some toxic agents. The agents covered are those which have no current exposure guidelines established by other organizations. The committee attempted to establish occupational exposure limits for materials not addressed by the ACGIH or OSHA, but of interest to various segments of industry.

Hazardous Substances

Hazardous substances, **hazardous materials**, and **hazardous wastes** are **Terms of Art** which are strictly defined by statute (legislative bodies) and regulation (agencies). **Hazardous materials** are defined by the U. S. Department of Transportation's labeling, packaging, and shipping regulations (49 CFR 171.8) as "a substance or material, including a hazardous substance, which has been determined by the Secretary of Transportation to be capable of posing an unreasonable risk to health, safety, and property when transported in commerce."

Hazardous substances are defined in the Comprehensive Environmental Response Compensation and Liability Act (CERCLA) as any chemical regulated by the Clean Air Act, the Clean Water Act, the Toxic Substances Control Act, or the Resource Conservation and Recovery Act (RCRA) (40 CFR 302.4). **Hazardous wastes** (40 CFR, 261.2) are solid wastes which are "any discarded material (garbage, refuse, sludge, or other discarded material, that is a solid, liquid, or confined gas which is abandoned, inherently waste-like or recycled) that is not excluded by P 261.4(a) or that is not excluded by a variance granted under P 260.30 and P 260.31." 40 CFR 261.4(b)(1) identifies "Household waste, including household waste that has been collected, transported, stored, treated, disposed, recovered (e.g., refuse derived fuel) or reused..." is excluded from the definition of a hazardous waste. Hazardous wastes may be either **listed** or **characteristic**. Listed wastes may be found in lists under 40 CFR 261.30 through 261.33. Characteristic wastes are defined; these definitions are found in 40 CFR 261.21 through 261.24.

Now, generally, we should consider substances that are toxic, corrosive, explosive, reactive, radioactive, and flammable as hazardous substances. However, we need to create further distinctions based upon what constitutes a hazard. A **hazard** is a much broader and more complex concept than just that of the presence of a hazardous substance. In general, a hazard must include the conditions of a substance's use. The hazard presented by a substance has two components. The first is the inherent ability of a substance to do harm by virtue of its nature. The second is the ease by which the substance can impact safe and healthful working conditions. This involves the establishment of contact between the substance and the subject. Together, these components address the probability that a substance will cause harm. For example, gasoline is a hazardous substance and presents a clear and present danger of both explosion and flammability; however, it is widely used as automotive fuel and, as such, rarely shunned because it is a hazard.

A variety a substances may be hazardous, while not being termed a hazard, because of their nature. In general, hazardous materials, substances, and wastes are not a hazard, as long as they are properly contained and managed. Hazardous classifications include explosivity, reactivity, flammability, combustibility, radio-activity, corrosivity, irritants, infectants, sensitizers, and photosensitizers. Exposure or contact with a hazardous agent is a hazard, because it presents a clear and present danger to health and safety.

Types of Toxicology

Toxicology is frequently subdivided into a number of specialties named for the principal mechanisms, systems, or areas being studied. Some of these divisions are described here.

Molecular toxicology involves the use of molecular biology to investigate the effects of substances at the level of Deoxyribose Nucleic Acid (DNA) and includes techniques to examine the effects of compounds on gene regulation and DNA mutagenesis. These methodologies are used to investigate the roles of proteins and enzymes in drug metabolism.

Biochemical toxicology uses the methods of biochemistry, cell biology, and cell physiology to determine the adverse affects of toxic substances on the normal functioning of cellular components. These methodologies are used to investigate how substances inhibit specific enzyme activity and to determine which enzymes are responsible for the metabolism of specific compounds.

Immunotoxicology examines the effect of substances on the immune system. Work involves examining the ability of substances to compromise immune function, elicit hypersensitive reactions, and induce autoimmune responses.

Reproductive toxicology examines the effects of drugs and environmental chemicals on reproductive fitness and how they affect reproductive capacity, and it attempts to identify agents that may adversely affect reproductive performance.

Genetic toxicology studies the interactions of chemical and physical agents with the process of heredity and investigates the types of mutations that an agent causes, along with its chromosomal effects.

Developmental toxicology is concerned with the investigation of chemically induced teratogenic effects (birth defects), examining detrimental effects produced following exposure of developing organisms to harmful substances.

Clinical toxicology is the practice of toxicology within a clinical setting.

Forensic toxicology is the practice of toxicology in an effort to determine cause of death, especially to link it to foul play.

Regulatory toxicology is the work of incorporating toxicological principals and data into the creation, drafting, and implementation of the myriad of laws and regulations dealing with chemical and physical agents.

Environmental toxicology examines the effects of substances present in the environment due to release following their use, creation as by-products, or decomposition of products.

Ecotoxicology examines the impact of substances present in or added to the environment or the various ecosystems within the environment.

Because of the focus of this text, we will concentrate on **occupational** or **industrial toxicology**.

Role of Industrial Toxicology

Industrial toxicology is a specific segment of **environmental toxicology** concerned with the harmful potential of the raw materials, intermediates, and finished products encountered by workers. Environmental toxicology is primarily concerned with the harmful effects of chemicals encountered by man in the total ecological system. Human exposure to atmospheric pollutants is generally incidental, whereas exposure to chemicals in industry is directly influenced by working conditions and hygienic practices of the specific industry. With the current diversification of industrial operations and chemicals, few occupations are entirely free from exposure to chemicals. When the toxic potential of the multitude of chemicals used in industry is considered, the significance of the problem of hazards in the workplace is of extreme importance to each of us.

Industrial toxicologists are involved with investigative research of chemicals utilized and produced by industry. Evaluations and characterizations of hazardous substances are being conducted in the attempt to define Permissible Exposure Limits (PELs). Both laboratory animals and humans have been utilized as test subjects for previous research. Because of social concerns and objections to the use of these model systems, alternative methodologies are under development. One difficulty of the current approach is that of extrapolation of test results to a human model system.

Most human exposure data is historical. Industrial **retrospective** studies are continually being conducted to identify substances to which humans have been exposed. However, the validity of such retrospective studies is questionable, because of the lack of explicitly defined human exposure levels. Nevertheless, reevaluation of retrospective data has been enlightening and has resulted in a decrease in permissible exposure levels for numerous substances.

Retrospective studies were, and are, essential in the identification and recognition of hazardous industrial agents. For example, retrospective studies conducted on the vinyl chloride incident at Goodrich Rubber in Akron, Ohio, determined the correlations between angiosarcoma (liver cancer) and the vinyl chloride monomer. If not for this study, this link probably would not have been established. The toxicologist takes into account the specific toxic properties of substances, as they reveal themselves in experiments and during accidental exposures. Experimental data generally are derived from animal experiments. Significant differences exist between the exposure of workers to toxins in the workplace and to animals in the laboratory. The selective and observational biases involved in the design and implementation of animal studies and interspecies response variations complicate data extrapolation. Available data are used to calculate the likelihood of harm to humans who may encounter the substance in a specified operation and concentration.

We have defined a hazard as a likelihood of damage or harm from a substance. Hazard evaluation and **risk assessment** (see Chapter 16) of substances utilize experimental data, susceptibility of workers, physical characteristics of substances, warning properties of substances, and conditions of exposure. The study of the susceptibility of workers who face potential exposure to substances is **industrial toxicology**. Different worker groups are particularly susceptible to certain toxic actions. For example, female workers of childbearing age should not be exposed to teratogenic or mutagenic agents. The expected frequency and duration of worker exposure is of significance. For a material, toxicity depends on rate, concentration, and time of exposure. This will be addressed to a greater degree in Chapter 4. The physical characteristic of a substance as it is encountered on the job is also of importance. Chapters 9 and 10 are a survey of common toxic substances.

The warning properties of substances, or lack thereof, need to be understood by industrial toxicologists. Many substances have their own alarm. When workers are exposed to hydrogen sulfide at levels of 4 or 5 parts per million (ppm), for instance, it provides a characteristic "rotten egg" odor. However, exposure at approximately 20 parts per million (ppm) can cause olfactory fatigue, degradation of the 'smell' receptors with a subsequent "loss" of smell and the "disappearance" of the characteristic warning order. Alternatively, there is methyl bromide, a widely used, highly toxic fumigant. Because it has no inherent warning property, humans are not able to sense it in the air, and central nervous system (CNS)(brain) depression, irreversible damage, and death have accompanied exposure to high concentrations. Such considerations are of extreme importance when performing risk assessments.

The Toxic Triangle

The interaction between the poison, the organism, and the environment is termed the Toxic Triangle (Figure 1-1).

We have been discussing the severity of any hazard. How hazardous a particular substance, or situation, may be depends upon the three elements represented at the apexes of the triangle. When the hazardous substance is toxic, we refer to its interactions within a toxic triangle. (The same analysis is, generally, valid for any hazardous material.) Previously, a distinction was made between a poison (common usage), a toxic (NIOSH), and a hazardous substance (CERCLA). To be classified as a hazard, it is necessary that the substance, or situation, be of potential danger to the health and safety of the organism. For example, some insecticides have mechanisms of actions that only affect insects, but not humans, or other higher animals. Therefore, they do not represent a hazard to you; nor would they be hazardous to your cat or dog. However, they are deadly to insects.

A hazard includes the conditions of a substance's use, the environment. Contact must be established between the substance and the subject, the poison and the organism, via a delivery system. Together, environmental components address the probability that a substance can cause harm. Of critical consideration are the specific work conditions, the environment, under which worker exposure may occur. The actual conditions of exposure include heat, humidity, vibration, noise, pressure, and extreme exertion. Since these can all affect absorption they are as important as the dose and concentration in predicting the responses of workers. As you will learn, Personal Protective Equipment (PPE) helps to provide a safe working environment. Being encapsulated in a "moon" suit usually provides a safe environment to a worker within a larger hazardous environment.

Another facet of exposure which must be addressed is the condition under which the data was collected. Much of the exposure data used to predict effects and specify hazard levels has been collected under specific conditions. A controlled condition (e.g., the laboratory) includes environmental factors in predicting the toxic properties of a substance. An uncontrolled condition (e.g., an accident) lacks specifications of a number a variables. The existence of multiple variables complicates the calculations required to accurately predict the magnitude of the hazard. For example, recommendations of exposure in the workplace are usually made for healthy adult males. How do these same standards apply to, for example, a sick, pregnant teenager? Additionally, workers may be exposed to several substances at once, which can cause unexpected reactions and **synergis-**

Figure 1-1

The Toxic Triangle

tic effects. (Synergism occurs when the magnitude of the effects is greater than that from simply the sum of the doses involved.)

Types of Toxic Effects

In addition to lethal effects of killing the cell or organism (e.g., Lethal Dose for half the population, LD_{50}) there are three more subtle types of effects that are characteristic of some agents. These are covered in greater detail in Chapter 8. Specifically these effects are as follows:

1. **Teratogenic**
   Teratology is Latin and deals with the study of monsters, i.e., congenital malformations (birth defects). It was not until 1941, when German measles and viral infections during pregnancy were linked to birth defects that this discipline was created. Methyl mercury and thalidomide are agents recently linked to birth defects.

2. **Mutagenic**
   Mutagens are agents that cause changes (mutations) in a cell's genetic code, or DNA. Typical events include chromosomal breaks, rearrangement, loss or gain, or changes within a gene. These changes can be lethal, sublethal (not effecting a cell until it attempts to divide), or carcinogenic. Benzene, ionizing radiation, and ethylene oxide (used in hospitals as a sterilant) have all been demonstrated as having mutagenic capabilities.

3. **Carcinogenic**
   Cancer refers to the uncontrolled proliferation of a population of cells in the body. Agents may directly cause, initiate, or promote cancer, or may indirectly promote cancer by preventing the body from effectively responding, causing immunosuppression.

Health Hazards

Chapters 11, 12, 13, and 14 deal with health hazards, monitoring of harmful agents, exposure limits, protection, and control. Initially, we will limit our discussion to toxic materials. Toxic materials, be they solids, liquids, gases, or vapors, can affect living creatures via four primary routes of entry.

1. **Inhalation**
   The process by which entry to the body is via the respiratory tract and lungs as a result of breathing.

2. **Ingestion**
   The process of consuming contaminated food or water, or otherwise permitting the oral intake of irritants or toxins.

3. **Dermal**
   The process by which hazardous materials cause injury to bodily tissues via direct contact with, or cause poisoning via absorption through the skin, mucus membranes, or other external tissues. Also included in this category is the passage of toxic materials into the body via puncture wounds or other breaks in the skin.

4. **Eyes**
   Because of the importance and role of vision in the industrial setting, the eye is frequently elevated to a unique status when evaluating health hazards. Effects include both absorption and loss of function.

Agent or Chemical	LD$_{50}$ (mg/kg)
Sucrose (Table sugar)	29,700
Ethyl alcohol (Grain alcohol)	14,000
Sodium chloride (Table salt)	3,000
Vitamin A	2,000
Vanillin	1,580
Aspirin	1,000
Chloroform	800
2,4-D	375
Ammonia	350
Carbaryl	250
Caffeine	192
Phenobarbital (Barbiturate)	162
DDT	113
Sodium nitrite	85
Arsenic	48
Sodium cyanide	6.4
Strychnine	2.5
Nicotine	1.0
Dioxin (TCDD)	0.001
Botulinum toxin	0.00001

Table 1-1

Approximate oral LD$_{50}$ values in rats

As first expounded by Paracelcus, while the correct amount of a substance may be required for health, too much of almost anything may be toxic. The amount of a substance required to kill you is termed the **Lethal Dose (LD)**. An LD$_{50}$ (mg/kg) is the amount of a substance, dose, per kilogram (2.2 pounds) of body weight necessary to kill 50% of a particular population. Table 1.1 provides some values for substances necessary for health (sodium chloride, vitamin A); useful or desirable, (sucrose, alcohol, aspirin, caffeine, nicotine); and toxic (cyanide and strychnine). These values are for rats; a similar range of toxicity may be expected for humans.

Concentration is a term used to let us know how "pure" a material is. A 100% concentration refers to a material or substance that is totally pure. If this pure material is adulterated by having a foreign substance added to it, the solution becomes diluted. Therefore, not as much of the substance will be available. Both the concentration and the nature of the adulterating materials are important factors when evaluating exposure risks.

Inhalation exposures occur from breathing. A volume of air inhaled may be expressed in liters (l) or Cubic Meters (m^3). The amount of material in the air is generally expressed in grams (gm) or milligrams (mg). (1000 mg = 1 gm, 1000 gm = 1 kilogram = 2.2 lb.) Concentration, expressed in mg/m^3, can be converted to parts per million (ppm). The constant 24.45 is required to convert between measuring systems.

$$\text{TLV in ppm} = \frac{(\text{TLV in mg/m}^3)\,(24.45)}{(\text{gram molecular weight of substance})}$$

The exposure depends upon the amount of time that the contaminated air was inhaled, seconds (sec.), minutes (min.), hours (hr.), work days (8 hr.), or days (24 hr.). Inhalation exposures may result from breathing any of the following:

- gases vented from containers,
- vapors generated from evaporating liquids (on land or in water),
- liquid aerosols generated during venting of pressurized liquids,
- mists or fogs generated from spilled acids,
- gases, mists, or fogs generated by chemical reactions,
- fumes from welding,
- dusts that become airborne due to an explosion or due to wind forces,
- combustion products or a burning hazardous material.

Ingestion exposures occur orally from consuming liquids or solids. Liquids are measured in liters (l), about a quart, and milliliters (ml), one/one-thousandths of a liter. Milliliters per liter, ml/l, or milligrams per liter, mg/l, are units of liquid concentration. Ingestion may result from:

- poor hygiene practices after handling contaminated materials,
- consumption of contaminated food or water,
- material trapped in mucous in the respiratory tract and swallowed.

Dermal contact results from exposure to hazardous materials in the environment. Contact may result from exposures to hazardous gases, liquids, or solids.

This may occur anywhere in the environment, either on land, in the water, or in the air. Effects may be local and/or cause systemic poisoning. Local effects frequently involve irritation or burns of the skin and mucus membranes such as those in the eyes or nose.

Systemic effects usually involve poisoning via absorption through the integument (skin). The hazards of poisoning due to the absorption of toxins through the integument is generally not well appreciated. Various gases, liquids, and solid materials have the capacity to rapidly pass through the integument upon contact. Although some materials may give warning of contact by causing some sort of burning sensation or irritation, others provide the victim with little or no warning. Absorption by bathing, swimming, or showering in contaminated water can be significantly higher than from just drinking the water.

Study Exercises

1. Define Industrial Toxicology and explain its role in the workplace.

2. Determine the relative time in history of the following people and discuss the value of their contributions to the field of toxicology.
   - Catherine de Medici
   - Paracelsus
   - Hippocrates
   - Mithridates
   - Ramazzini
   - Ebers
   - Nero
   - Bernard
   - Aristotle
   - Orfila

3. Distinguish between a hazardous material and a toxic substance.

4. Explain what is meant by the toxic triangle.

5. What elements are required to make a substance or situation a hazard? Discuss in reference to the toxic triangle.

6. Discuss the relationship among dose, concentration, and exposure.

7. Outline the differences between the four major routes of entry for poisons.

8. Discuss the commonly used units to express total dose and how each of these would be applicable towards the three major routes of entry.

9. Describe the commonly used units to express total dose for ingestion, dermal absorption, and inhalation.

10. Explain the differences between the exposure to toxic substances in actual work settings and how experimental animals are exposed in toxicological studies.

11. Define and differentiate among the following:
    - Dose
    - Chemical Concentration
    - Exposure Time

12. Find and copy a recent article related to an accident involving hazardous materials. Identify the hazardous material and explain why the material was hazardous. Explain how the material was released into the environment and who was put at risk by this event. Discuss different routes of exposure and how they could be minimized for this incident.

2 Pharmacokinetics
Exposure and Entry Routes

Overview

This chapter deals with the exposure phase of Pharmacokinetics. The concept of the body as a homeostatic unit, protected by a membrane barrier, is developed. The local effects of hazardous materials, versus systemic toxic effects, is explained. The most common routes of exposure and entry into the body are delineated. Each of the primary routes of exposure, the gastrointestinal system, the integumentary system, the eyes, and the respiratory system, are reviewed. The principles, risks, and examples of absorption for each of the primary routes of exposure are discussed.

Exposure and Absorption

Humans are warm blooded animals and, as such, are required to maintain a constant, **homeostatic**, environment. Basic to the establishment of this environment is the creation of a barrier between ourselves, the internal, and the outside, the external, environment. The maintenance of our internal environment requires the exchange of energy and material across this membrane barrier. Materials passing across this membrane into the body are **absorbed**. When a material first makes physical contact with our external membranous barrier, we have **exposure**. Absorption can only occur after exposure; exposure does not guarantee absorption. For example, swimming in a pool of water exposes you to water; however, little to none of this water is actually absorbed by your body.

Local and Systemic Effects

We have defined exposure as contact with the membranous barrier surrounding the organism. Subsequently, in general, interactions can be defined as local or systemic. For the former, damage tends to occur locally, upon contact, at the site of contact. Alternatively, or in addition to, the toxon is absorbed into, and disseminated throughout the organism.

For the latter, absorption is required before an effect can occur. This is called a **systemic effect**. A systemic toxon has a specific **target,** such as specific cells, a tissue (group of similar cells), an organ of a body system, the whole organ system, or the total organism, where the effect occurs. Nonlethal systemic poisonings are generally reversible interactions. The symptoms, interactions, frequently **remit,** disappear, after the toxons are cleared from the body. Lethal systemic poisonings, conversely, are frequently nonreversible interactions due to the nature of the mechanism of action or because the dose was sufficiently large to shut down a critical organ system, thereby causing the death of the organism.

Some toxicants act topically to cause irritation or **necrosis,** cell death and destruction. These may cause chemical lesions, and, thus, have a **local effect.** For example, both acids and bases can cause serious damage to the skin, although they are not absorbed through the skin. Poison oak is another example of a toxic substance that works primarily at a local level; however, some individuals who have become sensitized may also experience a systemic immune response. Some toxic substances have both local and systemic effects. For example, many gases, when inhaled, cause immediate **acute** irritation to the respiratory system with subsequent system failure. Both local and systemic exposure can be avoided or prevented by the proper use of Personal Protective Equipment (PPE).

Figure 2-1

Pharmacokinetics: absorption, distribution, storage, and excretion of toxicants. Metabolic reactions, changes to toxins, occur in a variety of locations, i.e., liver, kidney, and glands.

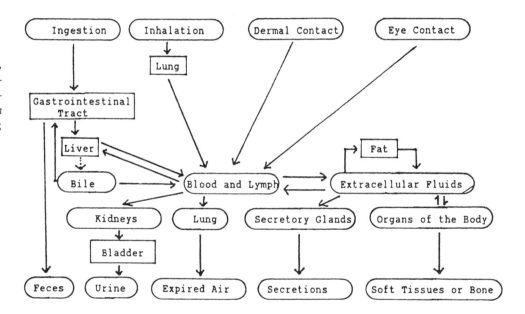

Pharmacokinetics

Pharmacokinetics is the study of passage of materials in, through, and out of our bodies (Figure 2-1). In general, the processes involved may be divided into five categories:

1. **Absorption**: the intake of materials across the membrane barrier.
2. **Distribution**: the transport of materials throughout the body.
3. **Localization**: the storage of materials in compartments throughout the body and the areas of interaction.
4. **Biotransformation**: the dilution, dispersal, or processing of materials.
5. **Excretion**: the disposal or ultimate fate of the materials.

Membranes

Our body is surrounded and protected by a membrane barrier. Since all substances that reach systemic circulation have had to cross a membrane barrier, it is important to understand a little about the membranes themselves and what enables a toxicant to cross them. Membranes (Figure 2-2) are spontaneously created phospholipid bilayers. Phospholipids are molecules which have a **hydrophilic** (water loving), or **lipophobic** (lipid hating), head and **lipophilic** (lipid loving), or **hydrophobic** (water hating), fatty acid tails. The fatty acid tails align themselves towards the interior of the membrane to form a moderately fluid lipid layer. The polar head groups are found on the surface of the membrane, facing the water environment. You may have experienced, observed, the result when salad oil is added to water or vinegar. Although you shake the two together and they may appear to mix, upon standing, they separate. This is due to the basic natures of lipophilic and hydrophylic substances. Like combines with like. Oil and water do not mix; unless you add soap or some other emulsificant. Similarly, with the membrane, the basic nature of the structure tends to prevent water from passing through the lipid portion of the membrane, and the water environment tends to prevent the passage of lipids to the membrane. Therefore, we have an impenetrable barrier. But, we do require the passage of materials across the membrane. For this reason proteins are embedded throughout the lipid bilayer; they have a variety of specific functions, several of which are important in the transport of nutrients and wastes across the membrane. (See Figure 2-2)

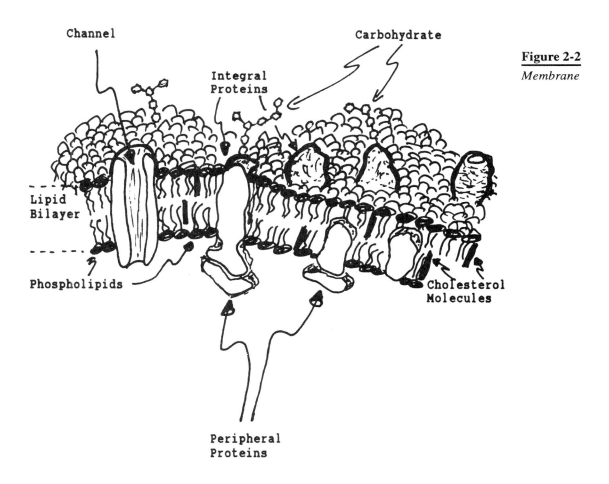

Figure 2-2

Membrane

Substances can be absorbed and passed across the membrane by four principal methods:

1. Diffusion or passive transport
2. Facilitated diffusion
3. Active transport
4. Special processes

Diffusion across or through a membrane is driven by a difference of concentration across the membrane; a high concentration (or large number of molecules) on one side, and a low concentration on the other. Molecules naturally move from areas of high concentration to low concentration. (Consider the movement, expansion, of an odor or gas released into the atmosphere.) Diffusion across a membrane is governed by the size and the nature, whether they are lipophilic or hydrophilic (basic or acidic), of the molecules.

Facilitated diffusion requires the use of a special carrier protein. However, it is not powered by **Adenosine TriPhosphate (ATP)**, the energy "coin" of the cell. Because the number of carrier molecules is limited, facilitated diffusion has a maximum rate at which it can occur. Facilitated diffusion can also be selectively inhibited by affecting the carrier molecules, a condition termed **inhibition**. Furthermore, the process stops at **equilibrium**, when the concentration is equal on either side of the membrane. Some sugars and amino acids are normally transported by facilitated diffusion.

Active transport requires the use of special carrier proteins and consumes ATP. It can be affected, as the facilitated transport process can, by inhibition of the carrier molecules. Additionally, because these molecules work and consume energy (ATP), they can establish a concentration gradient. For example, sodium and potassium gradients are created by a special pump, the sodium-potassium ATPase pump. These active transport processes are used both to maintain homeostasis and for the excretion and elimination of toxicants from the body.

Specialized transport processes are used to transfer molecules or materials across the membrane. The major contributor is a process known as phagocytosis. In this process, a special type of cell , a **phagocyte**, engulfs another cell, bacteria, virus, particles or debris. The "bubble" of material so formed is subsequently taken into the cell for processing or elimination. A variety of populations of phagocytic cells are found throughout the body.

Gastrointestinal (Digestive) System

Ingestion of contaminants is the least significant route of entry in the industrial environment, given good hygiene practices. However, the **GastroIntestinal (GI)** tract is a very important and effective route for the absorption of toxicants (Figure 2-3). Hazardous materials may be consumed when incorporated into food or drink, transferred from hand to lips or mouth by wiping motions, or may be incorporated into mucus of the surface lining in the respiratory system, expectorated, and swallowed.

The digestive tract is a tube 28 to 30 feet in length starting at the mouth and extending to the anus. Starting at the **cephalic**, head, end, it consists of the mouth and its associated structures, lips, tongue, teeth, and salivary glands; the esophagus; the stomach; the small intestine and its associated structures, the pancreas, liver, and gall bladder; and the large intestine. The opposite, **caudal**, tail end, terminates in the rectum and the anus.

The body, topologically, is a donut, with the digestive tract being the donut hole. Materials and substances placed within the hole are not incorporated into the body, only surrounded by the body. However, the digestive tract was created to be highly efficient at absorption and is quite good at its function. The digestive tract is lined with specialized **epithelial** cells which are only one or so layers thick. Many specialized strategies are employed to increase absorption. For example, for the small intestine, if the intestine's surface were flattened, it would cover a surface roughly the size of a tennis court; this large surface area greatly aides in absorption.

Absorption from the digestive tract is strongly site specific. The pH of the digestive tract varies from 1 to 3 (acidic) in the stomach to 5 to 8 (slightly basic) in the small intestine, depending upon the type and amount of food consumed. The degree of ionization of water soluble molecules is dependent upon the pH.

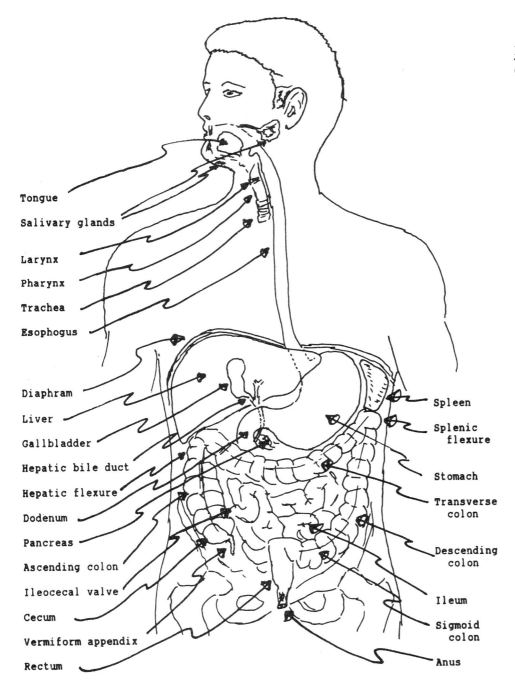

Figure 2-3

Gastrointestinal system

Tongue
Salivary glands

Larynx
Pharynx
Trachea
Esophogus

Diaphram
Liver
Gallbladder
Hepatic bile duct
Hepatic flexure
Dodenum
Pancreas
Ascending colon
Ileocecal valve
Cecum
Vermiform appendix
Rectum

Spleen
Splenic flexure
Stomach
Transverse colon
Descending colon
Ileum
Sigmoid colon
Anus

The ease of movement of a molecule through the membranes frequently depends upon its degree of ionization and, thus, the site pH. Aspirin, an organic acid, is well absorbed from the stomach. Compounds which are unstable in the acid pH of the stomach are denatured and can be rendered harmless.

Facilitated and active transport are used extensively in the digestive tract. Essential nutrients and electrolytes, sugars, amino acids, sodium and calcium, are so absorbed. Toxicants that mimic the molecular size, charge distribution, and configuration of essential nutrients can be transported by the carrier mechanisms present in the membranes of the epithelial cells. 5-Fluorouracil has been shown to be absorbed by a pyrimidine transport mechanism (normally used for nucleic acid transport). Many metals, like lead, appear to use this mechanism. Not only does lead gain entrance to the body in this manner, it is also stored in bone as is calcium.

The rate at which foodstuffs pass through the digestive tract can greatly affect the absorptive process. If the rate is slowed, the time available for absorption increases and, hence, the amount of material absorbed increases. Regurgitation or purgatives, which increase gastric mobility, quickly removes a substance from the tract and prevents or decreases absorption. Other factors which influence absorption are age, physical health (such as the presence of ulcers), nutritional status, and the chemical and physical composition, solubility, and interactions of materials in the digestive system. Lipids are normally solubilized by bile (from the liver and gall bladder) and absorbed from the small intestine. Fats can coat the stomach wall and delay absorption. For example, alcohol is normally absorbed from the stomach; drinking a glass of milk coats the stomach wall and greatly reduces the rate of absorption of alcohol. Activated charcoal is a vary good absorbant and in the gut will compete for materials and remove them from access and, therefore, exposure.

The large intestine's principal function is the absorption of water. Intestinal flora that inhabit the digestive system are important in the final processing of ingested materials. Intestinal flora have been shown to form carcinogenic nitrosamines from secondary amines. Antibiotics which affect the intestinal flora, frequently wiping them out, can cause diarrhea, a debilitating and potentially lethal condition. It is important to recolonize the tract with friendly organisms, such as the active cultures found in yogurt. If "good" organisms are not seeded, unfriendly, hostile, or dangerous organisms can colonize the tract and cause significant pain, grief, gas, and undo symptoms. Microorganisms, such as amoeba and giardia found in contaminated drinking water, can colonize the tract and cause severe intestinal distress.

Integument System

The skin covers the external surfaces of the body and presents the second most common risk of absorption of toxic materials. It is a large organ and is about 15 percent of the body by weight. The skin consists of an outer keratinized epithelial layer and a deeper dermal layer. It is a barrier well designed to resist external influences, bar the entrance of foreign organisms, materials, liquids, water, and prevents excessive water loss. It also serves to regulate body temperature and acts as a minor pathway for excretion (See Figure 2-4).

Because the skin is waterproof, ionic water soluble, hydrophilic materials are not readily absorbed. There may be slight absorption through sweat glands, hair follicles, or mucous membranes (around the eye). Breaks in the surface, cuts, scratches, abrasions, and burns greatly increase the risk of absorption. Punctures of the skin can readily introduce toxins systemically. Insect bites, snake bites, and high pressure liquid streams, high pressure cleaners and airless paint spray-

Figure 2-4

Integument system

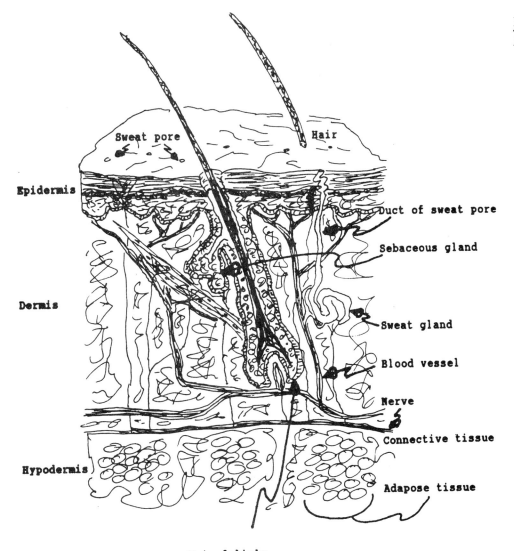

Hair folicle

ers can all deliver toxins. Sweat, perspiration, and gummy, sticky, or oily materials can increase exposure and the risk of absorption. By moistening the skin, lotions, hand and body creams can also increase the risk of absorption. Good personal hygiene (washing) can effectively control the risk of the absorption of hydrophilic materials.

Lipophilic substances, solvents, and oil can penetrate the skin via the passive diffusive processes. Hexane, benzene, carbon tetrachloride, nerve gases, and numerous insecticides have caused deaths in industrial and field workers after absorption through the skin. Solvents, such as some alcohols and hydrocarbons, can remove or deplete the skin of naturally occurring lipids, subsequently facilitating transport through the skin. Oils, poison oak, and coal tars can cause severe irritation and, in the case of the latter, induce cancer.

Eye

The eye (Figure 2-5) is a major sensory receptor and should be protected by the use of proper and appropriate PPE. The eye resides in a bony socket covered with moist epithelial tissue. Although a major function of these cells is secretion of lubricant to facilitate movement by the eye, they are also absorptive. The outer covering of the eye is also epithelial. The transparent cornea at the front of the eye is comprised of cells similar to those of the epidermis. Damage to these cells can impair the visual image. Located behind the cornea is the lens, which is used to focus light images on the retina at the back of the eye. Both the cornea and lens can be damaged by physical agents such as ultraviolet light, (e.g., sunburn of the cornea) due to acute exposure, or cataracts of the lens from chronic exposure. The light sensors, in the retina , known as rods and cones, and the optic nerve function as part of the nervous system and will be covered in more detail in Chapter 7.

Figure 2-5

The eye

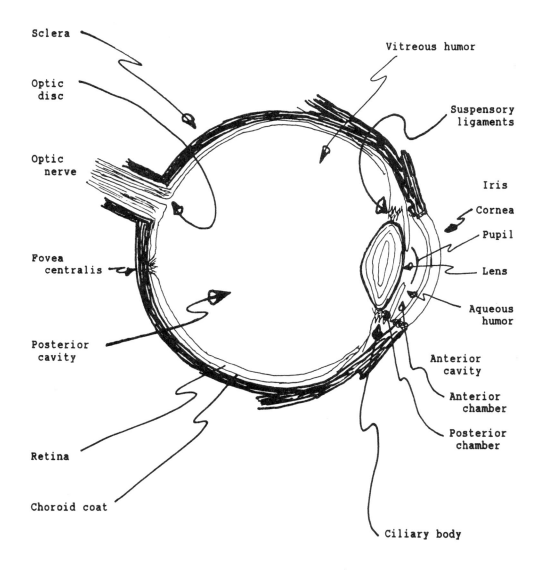

Respiratory System

The respiratory system (Figure 2-6) presents the highest risk of exposure to hazardous substances in the industrial environment. Gases and vapors, such as carbon monoxide, and volatile hydrocarbons are rapidly absorbed by the lung. Liquid and particulate aerosols, acid mists, fumes, coal or silica dusts, are deposited on the surface of the system or absorbed. Once inhaled, exposure is very great and absorption is very high.

The external respiratory system consists of a system of progressively smaller branching tubes which lead to the distal alveolar region. The tubes conduct gasses into the alveolar region and serve to clean, moisten, and warm the air. In

Figure 2-6

Respiratory system

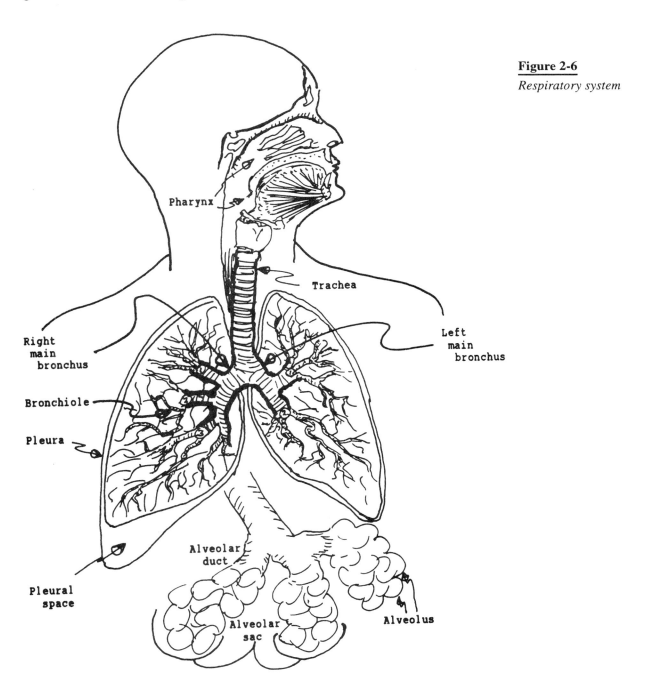

the alveolar region, rapid exchange of gasses between the exterior and the interior circulatory system occurs.

The first part of the system, the nose and pharynx, is lined with mucous membranes. In conjunction with the internal nose hairs, course particles 5 microns or greater are filtered from the air. In general this mucus-material mixture can be blown or sneezed out or wiped away. However, absorption may occur from this region, as with the snorting of cocaine.

The second portion of the system consists of the tracheal and bronchial tubes. These tubes are lined with mucous membranes and the surface is covered with microscopic hairlike structures called cilia. Particles 2 to 5 microns in diameter are captured in this region. The cilia beat upwards and move the mucal mixture to the back of the throat where it is swallowed, digested in the digestive tract, or **expectorated,** or spit out, orally.

The third portion of the respiratory system is the **alveoli.** Particles less than 2 microns in diameter reach this region; generally, those of lesser size are exhaled. However, some are deposited in the alveolar region by diffusion and Brownian motion. In this event, it appears that the smaller the size of a particle, the greater the chance it has being deposited in the alveoli. Particles of 0.02 microns are almost three times as likely to be deposited in the lung as particles of 0.09 microns. The small inorganic particles that settle into the smaller regions of the lung are frequently lethal, both locally and to the phagocytic cells which attempt to remove them. An interesting example is fly ash, a common by-product of the burning of coal. The burning coal acts to smelt and purify the contaminating elements in coal, which coalesce into small particles enriched as a single element. Nominally, such small amounts of a substance would not be toxic. However, when a phagocyte engulfs the particle, the amount per cell is very large, a lethal dose, and the cell dies, releasing the particle back into the lung. Not only do you lose the population of phagocytes, but you are left with the toxic substance that originated the damage in the lung. Frequently the net result of damage to the lungs is the formation of encapsulating scar tissue and the development of fibrous lesions of the lung. With these changes is a concomitant loss of alveoli and lung capacity.

The alveoli are designed to facilitate gas exchange. The rate of absorption of gases and vapors is largely a function of the blood-gas partition coefficient. Chemicals with high blood-gas partition coefficients are highly soluble in blood and, therefore, have a high rate of uptake into the bloodstream. Chemicals with a low blood-gas partition coefficient have a low solubility in blood and are poorly absorbed into the bloodstream. (Note: because the composition of blood is different than water, solubility in water should not be equated to solubility in blood; tables of blood-gas partition coefficients are available.) The degree to which gaseous materials are absorbed depends upon several factors:

1. Concentration of the substance in the air
2. Duration of exposure
3. Solubility of the substance in the blood and tissue
4. Reactivity of the substance
5. Respiratory minute volume (respiratory rate per minute x volume)
6. Alveolar surface ventilated

The deposition of particles (aerosols) on the lung surface occurs primarily from physical forces; these are inertia, sedimentation, and diffusion. Typically, the inhaled aerosols are variable in size and, therefore, a mass median diameter is ascribed to the aerosol to characterize both its physical size (diameter) and its density (mass). Particles with sufficient density will collide with the surface of the lung at points of branching and curvature. As the direction of air velocity changes, the inertial force of particles prevents them from following the airflow. The larger the mass, the more difficult it is to change direction and, thus, deposition of particles via impaction occurs. Particles that are of sufficiently small size to escape deposition via inertia may deposit on the surface of the lungs via sedimentation. Gravity is the primary determinant of sedimentation and the process occurs once the velocity of airflow becomes low. Extremely small particles are strongly influenced by the movement of individual gas molecules. This phenomena is termed **Brownian Motion** and is responsible for diffusion of these particles into the bronchioles and alveoli of the lung.

Breathing methodology can also affect the deposition of materials in the lung. For our purposes, we will examine four different capacities of the lung. These are **tidal volume**, **expiratory reserve**, **inspiratory reserve**, and **residual volume.** Normal breathing consists of repetitive small inhalations and exhalations, tidal volume, and is about 0.5 liters. Inspiratory reserve, the extra air that can be inhaled with a deep breath after you have inhaled normally for tidal volume, is 3.0 to 3.3 liters. Expiratory reserve is the extra air that you can exhale after you have exhaled your tidal volume, and it is about 1.2 liters. The residual volume is most of what is left after you exhale all that you can—about 1.0 liters. This amount can be forcibly, externally displaced, but it cannot be expelled by your normal muscle actions. [**First-aid note:** The mechanical force involved in the Heimlich maneuver uses the expiratory reserve and the residual volumes to remove obstructions from the conducting tubes.] An interesting aside is that if you inhale a substance, it diffuses throughout all of the space here described. If you exhale all that you can, there is still a significant amount of substance left— 40%, with normal tidal inhalation and expiratory reserve, 80% with just tidal breathing (for each breath).

Non-normal breathing methodologies may increase exposure and absorption. Hyperventilation increases the exposure by inhaling larger amounts of air and increasing the minute respiratory volume. Increased minute respiratory volume increases the probability of deposition via impaction. Holding your breath, and then taking large breaths, can increase the exposure and absorption compared to normal tidal breathing. First, the deeper breath takes substances further into the lung and, second, holding your breath allows for greater absorption.

Asphyxiants produce a toxic effect by interfering with the supply of oxygen to the tissues. Decreasing the partial pressure of oxygen in the alveoli decreases the amount of oxygen entering the blood. Substances which dilute the oxygen in the air so as to reduce its partial pressure and decrease the amount, or partial pressure, of oxygen in the blood are simple asphyxiants. For example, carbon dioxide, nitrogen, nitrous oxide, and methane can act in this manner. Other agents, such as carbon monoxide, interfere with the ability of the blood to transport oxygen. Carbon monoxide and oxygen compete for the same binding sites on hemoglobin. Unfortunately, carbon monoxide is several hundred times better at it than is oxygen. Therefore, a much lower concentration of carbon monoxide than oxygen in the air can significantly decrease oxygen delivery to the tissues. Fortunately, the binding is reversible, and fresh air or concentrated oxygen can flush the carbon monoxide out of the body. However, life support may be required in the interim.

Chemical agents that irritate the linings of the respiratory system may cause asphyxiation in several ways. Inflammation or smooth muscle irritation constriction may block the conducting tubes, limiting the minute respiratory volume, asthmatic symptoms. Hydrogen chloride, sulfur trioxide, and ammonia gases can cause these effects. A secondary effect of inflammation is fluid leakage from the tissue due to damage and the activation of nonspecific immune response mechanisms. Leakage accumulating in the alveolar region of the lung physically prevents the exchange of gases (especially oxygen, which has a low water solubility) and causes drowning. Chlorine, bromine, ozone, phosgene, and nitrogen dioxide can, additionally, cause these secondary effects.

Normally, the mucal-cilial escalator mechanism and the phagocytes scavenging in the alveoli, can clear 90 percent of the inhaled particles in an hour. Coughing speeds the progress of clearance. Practices which damage the cilia, principally tobacco smoke inhalation, decrease a workers ability to respond. Both smoking and inhalation of toxic heavy metal particles (e.g., fly ash) adversely impact a phagocyte's performance and may even be fatal for them.

Study Exercises

1. Explain the difference between exposure to a hazard and the absorption of a toxic substance.

2. Explain the difference between a local effect and a systemic effect.

3. Identify the principal factors that affect the rate of systemic absorption.

4. Identify the most common hazard exposure routes.

5. Rank the exposure routes based on their frequency, based on their potential for systemic absorption, and based upon their potential toxicity.

6. Identify the principal parts of the Digestive System and explain their functions.

7. Identify the types of effects that can occur after ingesting a toxic substance.

8. Identify the principal layers and parts of the Integumentary System (skin) and explain their functions.

9. Identify the principal types of local effects on the skin.

10. List four factors that determine the severity of a chemical burn.

11. Identify the principal parts of the eye and explain their functions.

12. Identify the principal types of local effects on the eye.

13. Identify the principal parts of the Respiratory System and explain their functions.

14. Identify the principal types of local effects on the respiratory tract.

15. Explain the difference between simple asphyxiation and chemical asphyxiation, and provide an example of each.

16. Define the following terms: fibrosis, edema, ulceration, defatting agent, necrosis, narcosis.

17. Identify the principal types of airborne hazardous substances.

18. Explain why particulates between 0.5 and 5.0 microns in diameter are of primary concern in respiration.

19. Identify a different hazardous or toxic material for which each of the major pathways of exposure is at risk. Explain factors that increase or reduce the absorption rates on exposure.

3

Pharmacokinetics
Distribution, Localization, Biotransformation, and Elimination of Toxics

Overview

This chapter continues our discussion of the movement of toxic substances through the body. After a substance is absorbed, it is transported systemically by the circulatory system. The physical and chemical properties of the substance affect its partitioning into the bloodstream, its accumulation in different cells, tissues, and organs of the body, its transformation, and its excretion. The physical and mental health status of the body greatly impact these processes.

Pharmacokinetics II

Only a fraction of the **total body burden (TBB)**, the amount of a chemical compound absorbed into the body, reaches the site of action, the target tissue. Specifically, as previously discussed, a toxon must reach the molecular receptors at a particular site in order to be efficacious. The **pharmaceutical dose** is defined as that amount of a substance which is available for absorption in the exposure phase. The pharmacokinetic, or toxokinetic phase, deals with the partition of a substance throughout the body. As such, it includes all of the processes involved in the relationship between the pharmaceutical availability and the concentration of the substance reached in the various body fluid compartments and in the target tissue. Figure 3-1 shows the partitioning of the total body burden.

The absorbed dose, TBB—that amount of a material or a substance actually absorbed by the organism—becomes biologically available. The TBB is the effective dose that reaches general circulation. However, more important than the

Figure 3-1

Pharmacokinetics II

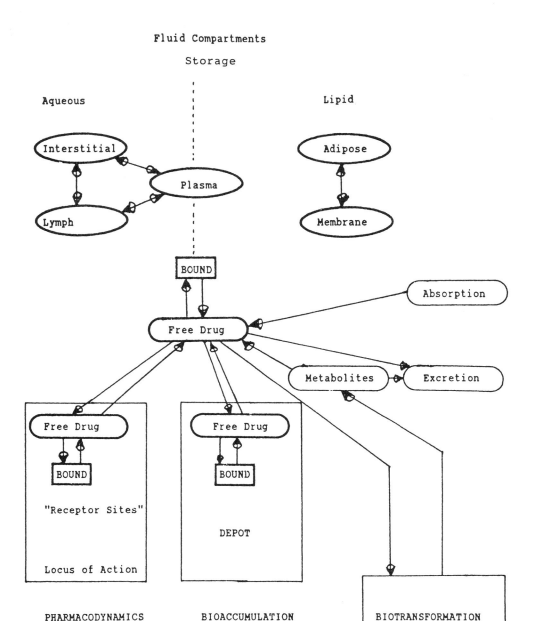

amount, are the time-concentration relationships, i.e., how much is available where and when. In particular, we are interested in the specific concentration available at the site of action at any particular time. Consider alcohol. Both the amount consumed and the rate of consumption are important in determining when you feel good, when you get drunk, and when you may die, dead drunk. If you feel like you've had too much, you can stop drinking alcohol. However, even after intake is terminated, alcohol continues to be absorbed from the digestive system, and blood levels continue to increase. Alcohol tends to be eliminated at a constant rate and is not greatly affected by other activities. Therefore, drinking coffee produces a wide awake drunk and exercising creates a moving drunk. Such is the pharmacokinetics of alcohol in your body.

Distribution

The distribution process involves the movement of substances throughout the body. It includes absorption of, transport of, and binding to tissue components, distribution through the organs, and excretion. Absorption, distribution, and excretion, as a rule, depend upon transport through membranes. The different compartments of body fluid, into which substances are distributed, are separated, or compartmentalized, by lipid membranes. We previously discussed membranes in conjunction with separating our internal environment from the external environment. Likewise, compartments are separated by a membrane. These membranes are similar in structure, being composed of lipid bilayers with the inclusions of proteins, carbohydrates, and biomolecules created by combinations of these molecules. A number of these biomolecules form pores, holes in the membranes, known as **fenestrations**. Understanding the underlying membrane transport mechanisms is key to understanding distribution throughout the body and between compartments.. Let us consider two types of transport through the membrane: **passive**, which requires no external "work," and **active,** which requires the expenditure of energy, or work.

Passive transport is governed by the membrane structure. Small hydrophilic molecules, such as urea, and inorganic ions, such as sodium and chlorine, can often diffuse through the small protein pores embedded in the membrane. Larger organic molecules, possessing a degree of lipophilicity, can diffuse through the lipid phase in the membrane. Highly hydrophilic organic substances pass the membranes poorly. Passive processes are a result of diffusion and are driven by concentration gradients. Materials tend to move from regions of high concentration to regions of low concentration. The relationship tends to be linear, i.e., the greater the concentration gradient, the faster the transport.

$$V = c(S_o - S_i)$$

Where V is the velocity of transport, c is the appropriate diffusion constant, S_o is the concentration of the material on the outside, and S_i is the concentration of the material on the inside. (Figure 3-2)

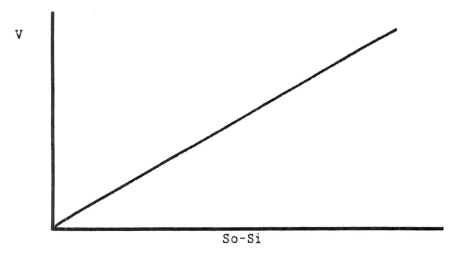

Figure 3-2

Free diffusion. A graphic representation showing the linear relation between the velocity (V), rate of diffusion, and the concentration gradient (S_o-S_i). There is a proportional increase in the rate of diffusion with an increase in the concentration gradient.

Active transport requires a carrier across the membrane. (Figure 3-3) Active transport requires energy. An interaction takes place between the molecules of the substance and the carrier molecules. The number of molecules that can be transported at any one time depends upon the number of the carrier molecules, or binding sites, available. The speed with which the carrier operates (i.e., the turnover per binding site) is also of consideration. The capacity of the system, rate of transport, thus, depends upon both the number of active sites and the turnover speed. Therefore, the kinetics of active transport are significantly different from those of passive transport.

Figure 3-3

Transport by carrier molecules. The rate of transport is proportional to the amount of pharmacon-carrier complex. As the concentration increases, the rate of transport attains a maximum value (Vmax) as a result of the saturation of the carrier system. Km is the dissociation constant for the pharmacon-carrier complex.

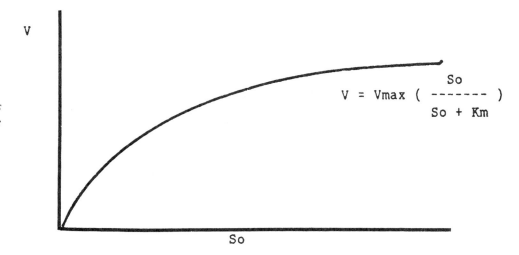

$$V = Vmax \left(\frac{So}{So + Km} \right)$$

An important consequence of capacity is **saturation**. At a given concentration, the system becomes saturated and is unable to transport any more of a substance. The rate of transport does not remain proportional to increasing concentrations; it reaches a maximum velocity. In passive transport, when a combination of substances is transported, the diffusing molecules do not interfere with one another and different substances are transported independently. In active transport, substances that use the same active transport mechanism will interfere with one another; they will compete with one another for the available carrier molecular sites. A consequence of this **competition** is that the transport of one substance will be inhibited by the presence of other molecules that use the same carrier.

Active transport systems are reasonably selective in regards to the substances they transport. The carrier site is responsible for the selectivity. These systems generally transport substances that dissolve easily in water, **hydrophilic** substances (water loving). Systems exist for acids, bases, and sugars. **Lipophilic**, lipid loving, substances are not actively transported; rather, they diffuse through the lipid phase of the membrane. Lipophilic substances that are bound to proteins cannot diffuse through the membrane. Large protein molecules cannot cross membranes; therefore, substances that bind to proteins are limited in their transport. Usually lipophilic organic substances, especially organic acids and bases that have large lipophilic groups, bind to proteins. These factors play an important role in determining the distribution of substances in an organism.

The blood of the circulatory system serves as the distribution, communication, and collection agent between cells and the external environment. It is comprised of the **formed elements** and the **plasma**.

The formed elements are **Red Blood Corpuscles (RBC), White Blood Cells (WBC, Leukocytes),** and **Platelets (Thrombocytes).** RBCs are red because they contain a chromic molecule, hemoglobin, which is designed for the purpose of carrying oxygen to the tissues. Insufficient RBCs or hemoglobin results in a decreased oxygen carrying capacity, a condition known as **anemia**. WBC serve as custodians to keep the body clean and to fight invaders. They function as the immune system. Abnormalities in WBCs or their numbers are conditions termed **leukemia**. Platelets, also known as thrombocytes (although they are not actually cells), serve to control blood loss and are central to blood clotting.

Approximately half of the blood is plasma, 40 to 60%. About 90% of plasma is water, in which a variety of organic and inorganic substances are dissolved. However, although the composition of plasma is mostly water, the behavior of plasma is quite unlike that of water. Dissolved substances include salts, anions, and cations; glucose and other sugars; and a variety of proteins: albumin, globulins, prothrombin, transferrin, and lipoproteins. The proteins are of particular importance because they serve as binding sites to many materials, a phenomenon which stores, transports, and distributes these materials. The liquid compositions of the blood plasma, interstitial fluid, and the lymph are quite similar; these are both extracellular, **interstitial** fluids. The fluid within the cell, **intracellular** fluid, has a significantly different composition.

The capillaries are the smallest vessels in the circulatory system. Every functional cell within the human body is located within 50 microns of a capillary. One function of the capillary wall is to retain the formed elements while allowing passage of the plasma. For this reason, their structure as a barrier is moderated by holes, or fenestrations, between cells. All substances that are not macromolecular and occur in a dissolved form in the blood readily cross the capillary wall. This is true independent of the hydrophilicity and lipophilicity of the substance. The brain has a special lipid-rich membrane created by a population of **glial** cells, **astrocytes**, part of the **neuroglia** of the nervous system, surrounding it. This barrier prevents direct passage of many blood components between the circulatory system and the **central nervous system (CNS)**. Therefore, some materials which are generally transported by the circulatory system to other cells and tissues of the body, e.g., protein-bound materials which pass through the fenistrations, may not pass into the brain. This phenomenon is referred to as the **blood-brain barrier**. (Lipid-soluble materials, such as ethanol, readily diffuse through the membrane.)

Localization

Localization refers to the storage of materials in compartments throughout the body, including the target sites or areas of interaction. Bioaccumulation refers to the increase in the concentration of a substance or materials in specific organs, tissues, or sites at levels above which they would normally be expected to be present. Storage and bioaccumulation are not synonymous; storage refers to the nominally benign deposition, bioconcentration, of a chemical or substance at some anatomical site while bioaccumulation refers to the process whereby there is an increase of the amount of a toxon at an anatomical site, depot. For example, lipids are stored in fat or adipose tissue, and calcium is stored in bone, while lead is bioaccumulated into our bones.

Central to understanding localization is an understanding of compartments and depots. We shall refer to compartments in delineating localization and to depots for bioaccumulation. These may or may not be synonymous. Next, it is important to distinguish what makes a compartment or depot. In general, these may be groups of molecules, fluids, cellular structures, cells, tissues, organs or some combination of these elements. What is important is that these compartments, or depots, for our considerations, have similar structures or functions such that they may be treated collectively. Finally, it is important to understand enough about the structure and function of a depot to predict which molecules, substances, materials, or toxons accumulate there.

Now, bioaccumulation is, inherently, neither good nor bad. Whether it is beneficial or detrimental depends on the context in which it is viewed. The storage of a toxon at some anatomical site can be benign in that it does no harm while residing in a storage site. For example, lead, a heavy metal, is stored in bone where it does no harm; lead in nerve tissue causes serious problems. The removal of lead from blood by bone, decreases the blood lead levels. However, bone is a dynamic structure which is constantly remodeling. As prolonged lead absorption saturates the bone storage depots, the remodeling increases blood lead levels which transports lead to the nervous system where toxic effects occur. Compare the lead scenario to the fate of **strontium**, a radioactive element with chemical properties similar to calcium and lead. As with lead, strontium accumulates in the bone. However, bone is home to the **hemopoietic**, or blood-forming, tissues. The walls of the home are made of bone. Now, while the radiation products of strontium are very weak—they can be stopped by a simple piece of paper—there is not even this separation between the strontium in the bone and these blood-forming houses, or cells. Therefore, the decaying strontium irradiates the hemopoietic tissue and induces cancer, i.e., leukemia.

Pharmacokinetics describes an active dynamic process, i.e., all of the processes are occurring simultaneously. After a period of time, an equilibrium is reached. Consider your intake of food and water. Normally, this is balanced by your energy needs and excretory processes, such that you maintain a relatively constant weight and size. Over a period of time, your inputs balance your outputs, creating a dynamic equilibrium. So true is the process for any substance, material, or toxon; the inability to establish a dynamic equilibrium can lead to disease and death.

There are a number of compartments of aqueous solutions in the body. The major compartments are the blood (blood plasma and lymph), interstitial fluid (between the cells), and intracellular fluid (within the cells). Lesser compartments are the fluid in the central nervous system, the cerebrospinal fluid, the aqueous humor of the eye, and the synovial fluid of the joints (articulations). These fluids tend to be similar to blood plasma, except that they have been filtered by an additional membrane barrier and contain additives which allow them to perform specific functions. The pH and toxicity of the various fluids show minor variation, due to the necessity of homeostasis. However, there are exceptions, (e.g., uric acid tends to accumulate in synovial fluid and, after a critical concentration has been reached, crystallizes). Crystals of uric acid in the joints, can be extremely painful, a clinical condition termed **Gouty Arthritis**.

There are two major compartments of lipids in the body. These are the interior of the cell membranes (lipid bilayers) and fat (adipose) cells. The fats of the membranes are diglyceraldehydes with polar head groups. These molecules are one of a class of molecules known as **surfactants.** Surfactants are schizoid molecules which are lipophilic at one end and hydrophilic at the other. Because of this structure they tend to accumulate at the interface between oil and water. In sufficiently high concentrations, the lipophilic portions will attract one another to form droplets called **micelles**. The hydrophilic outer surface of the micelle allows it to be solubilized. (This is how soaps and detergents function.) Adipose tissue is the storage site for fats and lipids, triglyceraldehydes. Some fats, HDLs and LDLs, and cholesterol also circulate in the blood.

A special aspect of the pharmacokinetic phase is the binding of drugs and toxic substances to plasma proteins, and other tissue components. The albumin of the blood plays a major role. The bound form of the substance is generally at equilibrium with a low concentration of the free (unbound) form. The free concentration in the interstitial fluid, and at the target sites, remains relatively low because of this plasma binding. Protein binding results in a low concentration of the substance in the tissues and a relatively high concentration of the substance available for elimination.

The binding to plasma proteins is a relatively indifferent process. Compounds that are completely unrelated chemically have common binding sites. There appears to exist groups of binding sites for weak acids and for weak bases. This means that all weak acids compete for the same binding sites. Such competition for binding sites can cause the release of normally protein bound substances. For instance, the displacement of bilirubin by sulfonamides causes high tissue concentrations of bilirubin.

Biotransformation

Biotransformation reactions are metabolic processes that change the structure and characteristics of a chemical compound. The toxic effects of many substances are highly dependent upon the metabolic fate of the chemical compound in the body. There are many ways in which the body can biotransform a toxic chemical. A set , or series, of processes that biotransform a chemical compound into an end product is termed a **pathway**. These pathways are all mediated by **enzymes**, specialized proteins, in the cells which function as factories to convert a variety of chemicals compounds into **metabolites,** the end product of the pathway. Biotransformation has been observed in the liver, kidney, lung, intestine, skin, testis, placenta, and adrenal glands. The liver, however, is the principal organ of biotransformation and contains the greatest variety of pathways and the largest capacity.

Generally, biotransformation leads to detoxification. Unfortunately, this is not always the case. Some intermediate chemical compounds formed in the process may actually be of far greater toxicity that the original molecule. Alcohol, both methanol and ethanol, is initially converted by the body into aldehydes, formaldehyde and acetaldehyde respectively. These intermediate products have been associated with blindness (methanol) and cirrhosis of the liver (ethanol). Such events are termed as **bioactivation**.

Biotransformation reactions are general divided into Phase I and Phase II type reactions. Phase I reactions are so named because they are, generally, the first biotransformation step in what is often a multi-step process. Ideally, this process leads to the eventual excretion and elimination of the biotransformed products. They are **catabolic** reactions; catabolic reactions are breakdown reactions, such as **oxidation**, **reduction**, and **hydrolysis**.

Phase II reactions are those enzymatic processes that utilize the products of phase I reactions to impart further structural changes. The structural changes are frequently associated with increasing water solubility, therefore, increasing excretion and elimination. Phase II reactions are synthetic, or **conjugation**, reactions. Synthetic reactions are those in which an additional molecule (e.g., sugar or amino acid) is covalently bound to the parent molecule. The conjugate product so formed is usually more water soluble, or lipophobic. **Metabolites** are molecular substances which have been biotransformed. Biotransformation mechanisms are generally directed towards producing metabolites which are more water soluble than the parent molecules and, thus, easier to excrete and eliminate.

A variety of factors can affect biotransformation. Of particular importance in industrial toxicology are the condition of the liver, nutritional status of the host, and competing processes. The liver is the primary site of biotransformation. Conditions affecting the liver, such as cirrhoses of the liver, decrease its efficiency and, thus, its ability to biotransform. Drugs such as acetaminophen, a popular **nonsteroid anti-inflammatory drug (NSAID)** for head and body aches, interferes with the metabolism of ethyl, grain, alcohol and has been associated with liver failure and death. Nutritional deficiencies in vitamins, coenzymes, and proteins can reduce the ability of the body to synthesize the key enzymes required for biotransformation. Competing processes can be varied; alcohol, nicotine, and other drugs are toxicants and can directly or indirectly affect the biotransformation processes. In some instances there are genetic differences (i.e., those that may completely lack a particular enzyme) that are essential to a specific metabolic pathway. Such an individual may exhibit an idiosyncratic reaction to a chemical. Accurate prediction of adverse or toxic effects is made more difficult by these individual variabilities.

Excretion

The liver, kidney, and lungs are the major excretory organs. The ability of the body to rid itself of toxic chemicals is largely dependent upon the physical and chemical properties and characteristics of the substance. Chemicals which have very low blood-gas partition coefficients, as with poor blood solubility or high vapor pressures, may be effectively eliminated via exhalation. Chemicals that are highly water soluble are generally eliminated by excretion into urine or bile. Those lacking these characteristics tend to accumulate in the body unless they are biotransformed.

The kidney is the most important excretory organ in terms of the number of compounds excreted. However, the liver and the lung are of greater importance for certain kinds of compounds. The lung is active in the excretion of volatile compounds and gases. The liver, because it is the major biotransforming organ, may even excrete metabolites before they have a chance to reach the systemic circulation. This is called a first-pass effect. Conversely, metabolites excreted in the **bile,** an excretion fluid of the liver, which is transported via the bile duct to the small intestine of the digestive tract, may be reabsorbed from the digestive tract and, thus, maybe indefinitely recycled.

Kidney

The kidney has a unique approach to excretion. It indifferently filters the plasma with all of its dissolved and bound substance from the blood. Subsequently, it does not try to determine what is bad, toxic, but, rather, concentrates on what the body needs and reclaims these substances, a process referred to as **re-uptake**. To this end most of the plasma portion of the blood is allowed, by ultrafiltration, to pass into the functional unit of the kidney, called a **nephron**. Subsequently, the cells lining the nephron reabsorb specific molecules the body needs. Blood sugars and proteins, sodium and chlorine, potassium and hydrogen, and water are actively sequestered back into the body. The remainder of the molecules, except those that passively diffuse back across the membrane, exit via the urine.

Passive diffusion from the urine to body fluids occurs according to the usual principles previously discussed. Therefore, lipid-soluble compounds are subject to reabsorption after having been filtered by the kidney. The degree of reabsorption of **electrolytes**, materials which dissolve in water, are strongly influenced by the pH of the urine. Electrolytes, all ionizable acids and bases, have a pK_a value, related to the dissociation constant K.

The degree of ionization in a body fluid depends on the pH of the medium and the pK_a of the acid or base. (The pK_a of an acid or base is an intrinsic property of a molecule. Therefore, for any chemical compound the pK_a is a constant.) When the pH is equal to the pK_a, half of the acid or base is present in the ionized form and half in the non-ionized form. At pH less than the pK_a, acids are less completely ionized. At pH greater than the pk_a, bases are less completely ionized. Thus, weak organic bases are strongly ionized in an acid environment and remain so in the case of oil/water partition in the aqueous phase. In a basic environment, weak organic bases are undissociated and nonpolar. Thus, weak organic bases in a basic environment will accumulate in the organic phase; i.e., in lipid, oil, ether, or benzene. Just the opposite occurs for weak organic acids.

Fluid pH is of paramount importance when evaluating the potential for passage through lipid biological membranes. Fluid pH plays a critical role in the absorption from the gastrointestinal tract and the reabsorption from the renal tubules. Passage is poor for weak organic bases from an acid environment and poor for weak organic acids from a basic environment. The reason, as previously explained, is that these chemical compounds are highly ionized under these conditions. Some control can be exerted over the rates of excretion, **clearance rates**, of weak acids and bases by adjusting urine pH. For example, alkalinization of the urine (ph 7-8) by the administration of bicarbonate has been used to treat salicylic acid (aspirin) and barbituric acid poisoning.

Alkalinization of the urine decreases excretions of weak bases. This may also give rise to complications with the misuse of **Central Nervous System (CNS)** stimulating agents, such as amphetamine and methamphetamine. Central stimulating agents are frequently used as doping agents to increase performance in certain sports.

Cyclists often use these stimulating agents in combination with alkalinizing substances, such as sodium bicarbonate. This is because the formation of lactic acid during intensive muscular exertion is one of the prime factors causing symptoms of fatigue. By inducing an alkalosis with sodium bicarbonate, the acidosis caused by the lactic acid is counteracted. Research with horses at Ohio State University has shown a 4 percent improvement in the fastest speeds and a 13 percent improvement in duration after sodium bicarbonate ingestion. The alkalosis, however, also produces an alkaline urine. Under these circumstances, the excretion of substances such as amphetamine in the urine is greatly inhibited.

An alkaline urine results in the reabsorption of amphetamine in the tubules, where it occurs mainly in the un-ionized form. The delayed excretion results in accumulation of amphetamine, which may reach toxic concentrations, even at typically nontoxic doses. **Uremia** is the situation where high blood toxon levels are reached due to decreased kidney function or urine excretion efficiency. A secondary consequence is that the decreased excretion of amphetamine makes it difficult to identify amphetamine in the urine during control tests.

A similar situation holds true for nicotine. Acidification of the urine (pH 5) by intake of high doses of vitamin C (10 gm/day) or by other means, enhances nicotine excretion and thus encourages cigarette consumption. Alternatively, alkalosis, induced by the intake of sodium bicarbonate, which causes alkaline urine, reduces the cigarette consumption to maintain a constant blood nicotine level. This reduced consumption is linked to the enhanced reabsorption of nicotine from the alkaline urine. There is a recycling of the nicotine and, thus, fewer cigarettes are needed to maintain the maintenance plasma nicotine level.

These relationships also apply to other toxic amines, such as aromatic amines like aniline. Workers regularly using systemic antacids containing bicarbonate are at elevated risk. This is because in such situations, the retarded excretion of aromatic amines could lead to toxic concentrations, TBB, although exposure is below permissible exposure levels, PELs.

Diuretics are agents that increase the rate of urine formation. Despite this simple definition, a large number of physiological and pharmacological factors are involved. Normally, the volume of urine is largely determined by the concentrations and types of solutions delivered to the renal tubule. Therefore, water and various electrolytes and non- electrolytes can act as diuretic agents when given in excess.

Mercury chloride and organic mercurial compounds depress tubular mechanisms responsible for the active resorption of sodium and, thus, lead to increased water excretion. Acetazolamide inhibits carbonic anhydrase, causing increases in the urine volume and pH (acidic to basic). The dominant actions of the thiazides, ethacrynic acid, xanthines (coffee, tea, and chocolate), pyrimidines, and triazines is to increase the renal excretion of sodium and chloride and an accompanying volume of water. Aldosterone antagonists inhibit the hormonal control mechanisms which determine water excretion rates. Beer contains a diuretic agent produced by the fermentation process. In general, by shifting the dynamic equilibrium of kidney function, diuretics affect renal **clearance rates**. Since toxicity dose levels are based upon "healthy" individuals, they may no longer apply when kidney clearance rates are impacted.

Liver

The liver, in addition to being a major metabolizing organ, is a major excretory organ. The liver has the capacity to eliminate toxicants absorbed from the gastrointestinal tract by metabolism or excretion before they reach systemic circulation on a first-pass basis. This is because food laden blood from the digestive tract is first ported to the liver before passing into general circulation. Additionally, the liver can metabolize compounds already present in the systemic circulation preventing them from circulating further. The products, metabolites, of hepatic cells may either pass into the blood or be incorporated into the bile.

There are at least three active systems for transport from hepatic cells to bile. There is one for organic acids, one for organic bases, and one for neutral organic compounds. There is probably a transport system for metals. Compounds with molecular weights greater than 300 are more frequently found in the bile than in the urine. However, which compounds are excreted into the bile and which are excreted into the urine is not known with certainty.

The efficiency of liver bile as an excretory route varies. Bile metabolites excreted into the digestive system may be resorbed by the body or absorbed and adsorbed by roughage. Bile efficiency may be diet dependent, thus explaining recommendations to increase the amounts of fruits and vegetables in the diet. Compounds excreted may also be metabolized by the intestinal flora, e.g., hydrolysis of glucuronides, to compounds more easily absorbed. Certainly, liver disease can reduce the excretory as well as the metabolic capacity of the liver. Conversely, there are drugs, like phenobarbital, that produce an increase in bile flow rate. Compound excretion rates, tied to bile flow rates, would increase under these circumstances.

Lungs

The third major excretory organ is the respiratory system. The lung is the key organ of the excretion of volatile liquids and gases. Pulmonary excretion is essentially the reverse of pulmonary absorption. Excretion is by passive diffusion and is directly proportional to blood concentration levels. Compounds with low solubility in the blood are perfusion-limited while those with high solubility are ventilation-limited. Increasing the ventilation, or minute respiratory rate, increases the excretion rate for ventilation-limited compounds such as chloroform. Since chloroform is a highly lipophilic compound, typical of such compounds, it may be present in expired air for an extended period after absorption.

Other Routes of Excretion

The integumentary system accounts for a variety of minor routes of excretion. Sweat, hair, nails, and milk excrements, while not necessarily significant quantitatively, can still be important qualitatively. Sweat is an alternative to kidney excretion, although not as elaborately moderated. Saunas have seen use for some time as detoxification from overindulgence. The hair appears to sequester heavy metals and, thus, can be used diagnostically to monitor absorbed dose, TBB over a period of time. Nails are only rarely of interest as a route of excretion because loss by this route is slight. However, because of the extended period of growth, they may be of diagnostic value. Milk may be a major route of excretion for some compounds. Milk tends to have a 3 to 5 percent fat content and, therefore, compounds that are lipophilic may be excreted in a mild to significant extent. Some highly lipophilic toxicants found in milk are the chlorinated hydrocarbons, such as polychlorinated biphenyl (PCB) and DDT. Certain heavy metals, such as lead and strontium, may be bioaccumulated and excreted in milk through the calcium transport process. Such milk is not fit for infant, human, consumption.

Study Exercises

1. Identify the principal factors that affect the distribution of a toxic substance in the body.

2. Explain what is meant by the blood-brain barrier. Provide an example of a material that would pass through the barrier. Explain why it would pass. Provide an example of a material that would not pass and explain why it would not.

3. Identify the principal sites of localization or accumulation.

4. Differentiate between a toxic substance's site of action and the inactive sites of accumulation.

5. Explain the function of making a toxic substance more water soluble in a detoxification reaction.

6. Define the following terms: target organ, reservoir site, compartment, depot, polyuria, uremia, anemia, clearance test, biliary secretion.

7. Differentiate between a Phase I (destructive) reaction and a Phase II (conjugation) detoxification reaction.

8. Explain what is meant by a bioactivation reaction.

9. List what factors could affect the rate of metabolism of a toxic substance.

10. Explain how the kidney acts to remove toxic substances from the bloodstream, and what factors affect its ability to do so.

11. Explain how cirrhosis of the liver can affect both the transport of a material to its target tissue and its metabolism. Why would this impact the material's toxicity?

4 Dose-Effect and Time-Effect Relationships

Overview

This chapter explores the response of living organisms to different amounts of toxins over varying periods of time. It provides an introduction to the terminology and methodology used in experimental toxicology, especially in regard to animals. Special attention is given to dose-effect and time-effect relationships. The factors that affect the toxicity of a substance, as well as the factors that affect the variations in the response to a given substance are covered. The threshold theory of toxicity is presented, and those effects (such as cancer) that may not exhibit thresholds are introduced. The three types of animal studies and the ways these studies differ from real-world, human exposures are delineated.

Dose Response

Earlier we paraphrased Paracelsus' contention that no chemical or substance is entirely safe; it is the dose that is important. It is important to quantify both the amount of substance available at the target site and the length of time it has been there. As you should realize by now, only a small fraction of the amount of the substance that a worker is exposed to is absorbed, and only a small fraction of the absorbed dose, total body burden, actually reaches the target site; the rest may be stored, bound or otherwise bioaccumulated. Also, from our discussions of kinetics, you should realize that the body is dynamic and, as fast as a substance is ingested, it is distributed, transformed, and excreted. Therefore, generally, the effective dose varies greatly with time.

Let us explore some time based relationships. First, let us assume that there are some limitations in absorption and transport and that some excretion is occurring. When a substance is absorbed, its concentration builds up in the body. (Figure 4-1)

The shape of the curve is classical and resembles the one described for V_{max} cited in the previous discussion on carrier saturation.

Figure 4-1

For a single dose, the concentration in an organism will reach some maximum level, Sm, and then decrease over time. After a period of time, some maximum concentration, Sm, is reached.

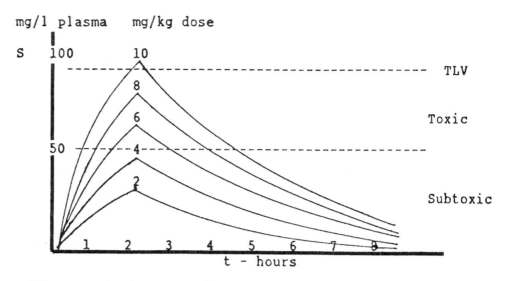

When a worker is removed from exposure, absorption stops. The retention time of a substance in the body is characterized by its **half-life**. The half-life, $t_{1/2}$, is the time it takes one-half of the substance to leave the body, by any exit route. (Figure 4-2)

Figure 4-2

Decrease in concentration in an organism over time where $t_{1/2}$ is the time required for one-half of the substance to be removed from the body

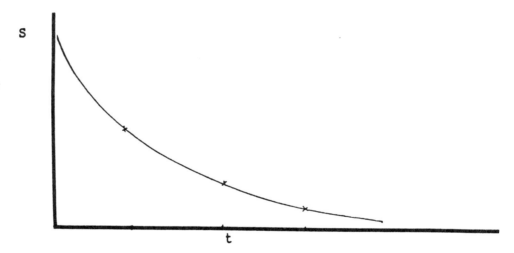

The shape of this curve is classical and resembles the inverse of the curve previously discussed. In this case, it is very difficult to reach a "zero" concentration, since the best we can do is to remove one-half of the material over any period of time. An important question is "How long will it take to reduce the concentration below a specified level?"

Let us specify a **threshold level**, a level or value that we do not want to exceed, a threshold. Let us assume constant exposure and absorption at some fixed rate during the day and excretion based upon the half life. Let us examine three cases: too much, that is where the total body burden (TBB) increases and threshold is exceeded; the maximum, where absorption and excretion are balanced and the total body burden stabilizes at threshold; and acceptable quantities, where we stay below threshold at acceptable levels. We will assume an eight-hour workday with exposure, 16 hours off, without exposure, and a five-day workweek.

In the first case, absorption is occurring faster than excretion and the available dose keeps building up over time. Over a period of time the threshold level is

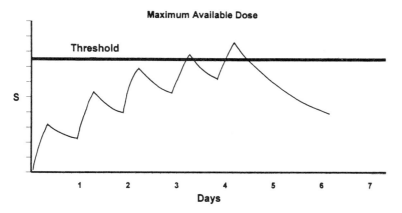

Figure 4-3

Situation where the dose rate exceeds the clearance rate and the concentration in the organism builds to beyond threshold levels

reached. (Figure 4-3)

The second situation is where, over an extended period of time, the absorption and clearance rates are equal. In this situation, the substance builds to a maximum level. Providing all parameters remain constant, the absorbed dose is subthreshold. Unfortunately, this is generally the best a person can do. The downside

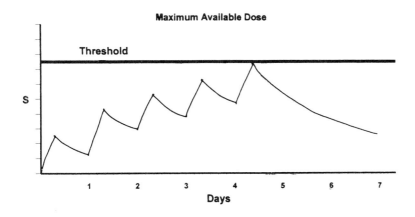

Figure 4-4

The absorption and clearance rates are balanced and the maximum available dose is sub-threshold.

is that there is no margin for error; any added stress or activity can alter the balance and quickly increase the available dose. (Figure 4-4)

The third situation is where the absorption rate is less than the clearance rate. Here the substance is cleared rather quickly and the available dose (TBB) is kept relatively low, i.e. below threshold. (Figure 4-5)

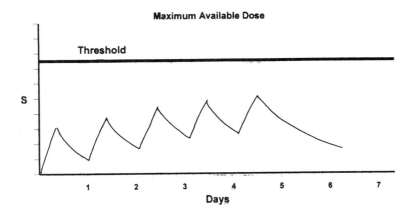

Figure 4-5

Situation where the absorption rate is less than the clearance rate and the available dose remains low

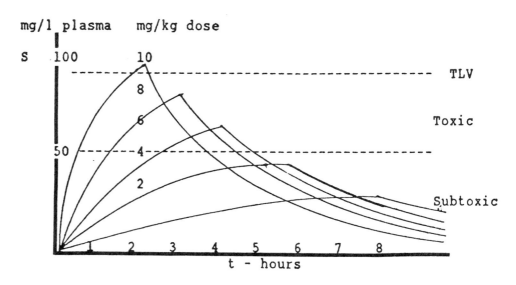

Figure 4-6

Effects of varying the rate of absorption

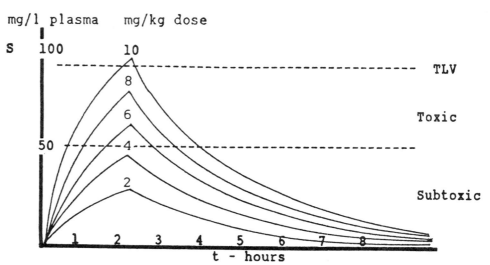

Figure 4-7

Effects of varying the rate of elimination

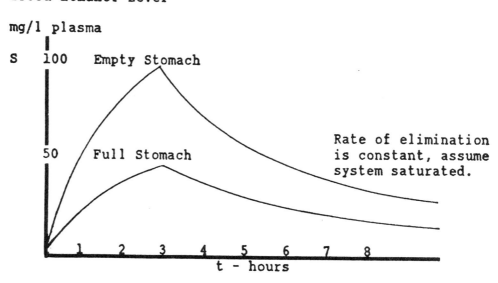

Figure 4-8

Retarded absorption using Ethanol as an example, assuming a full and an empty stomach

Two principal factors affect the balance of these dose-time relationships. The first is the dose rate. The more of a substance that is absorbed, the higher the available dose. The second is the clearance rate or the $t_{1/2}$ of the substance. Increasing the clearance rate, shortening the $t_{1/2}$, means that the build-up levels are less. Conversely, decreasing the clearance rates, the $t_{1/2}$, implies an increase in the available dose. (See Figures 4-6, 4-7, and 4-8)

End Points

A common end point is death. This end point is commonly used in toxicity testing and in laboratory studies. However, for workers, this end point is rather extreme. Additionally, it tells us little about the sublethal effects, e.g., irritation, rashes, coughs, aches, and pains. Therefore, a more useful and frequently used end point is the **effective dose**. This is the dose where a particular effect is seen or felt, such as with tear gas or a similar irritant. We are not interested in the dose at which death will occur so much as in the dose at which the eyes are affected. For a worker, it is important to keep the level below that at which irritation will occur.

Observable Effects Level, NOEL and LOEL

Another important set of levels are the No Observable Effects Level, NOEL. (No Observable Adverse Effect Level, NOAEL) and Lowest Observable Adverse Effects Level, LOAEL. Ideally, in an industrial setting, the NOEL will never be exceeded. There are two methods for obtaining this goal. The first is engineering controls. This methodology prevents the exposure levels from reaching effective levels and will be dealt with in more detail in Chapter 12. The other method is by the use of personal protective equipment. PPE is designed to allow workers to operate in unsafe environments safely.

Dose-Response

Death of the species being tested is the response most commonly used in preliminary toxicological testing. When a substance is undergoing initial evaluation, the specific toxic ranges must be characterized. Generally, a dose is administered, and either increased or decreased until the critical range is found. This range is from when all animals survive to when all animals die.

Experience has shown that biological variations in response to a substance by individual members of a species are generally small. Biological variability be-

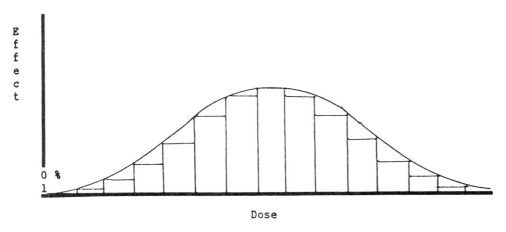

Figure 4-9

A bar graph illustrating the relationship between increasing dose and the percent of the population affected. The dotted line represents a curve fitted to the data.

tween species can be quite large. The principal goal of experimentation is to quantify the response. This is the purpose of the dose response studies. A quite separate problem is to relate these study results to exposure in the workplace.

One important concept is that of **threshold**. Threshold assumes that below a certain dose, no effect occurs. This is generally true for **reversible** reactions. Once the toxic substance has been removed, the effects are mitigated. However, this is not true for **irreversible** reactions, such as when a strong acid or base eats away at the skin or tissue. Sometimes exposure is cumulative, such as when the body is exposed to radiation or carcinogens. One difficulty in evaluating carcinogenicity is that there may be no threshold and the concepts designed to evaluate risk must be altered. Chapter 16 will evaluate risk factors which are applicable in some of these instances. (See Figure 4-9)

If we have a group of animals and administer a poison, some of them will die. If we increase the dose, more will die. For a large enough dose, they will all die. We can use a bar graph to plot this relationship. By inspection we can determine what dose is required to kill half of the animals. This is then the lethal dose for 50 percent of the animals or LD_{50}. Similarly, we can determine a lethal concentration of a breathable gas, LC_{50}, for half of the animals. If we use a different end point, such as when we can first see or determine an effect, we can establish an effective dose for 50 percent of the animals or ED_{50}.

Another manner of presentation of the data is to consider the cumulative effects. That is, plotting the relationship between increasing the dose and the percent of the population affected. This generally results in a "S" shaped curve. This observed relationship is due to the interaction between the pharmacon, toxon, and the molecular site of action, the target receptor. Technically, this is the Law of Mass Action and is represented mathematically as:

$$A + R = RA$$

where	R is the receptor site
	A is the toxin, substance, or chemical
	RA = % of receptor sites occupied
If we let	r = number of sites
then	RA/r = maximum of 1
and	E = function of (RA)
where	E is the observable effect we discussed previously.

What this expression means is that the observable effect we are measuring is dependent upon the dose that is administered. To better visualize the relationship

Figure 4-10

Relationship between increasing dose and the percent of the population affected. Notice that there exists a region below threshold where there is a lack of data. Extrapolation of the curve into this region is problematical.

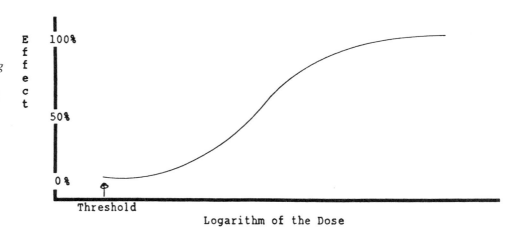

we manipulate the scale of the graph. Using a Log-of-the-Dose scale provides for a symmetrical "S"-shaped response curve. (Figure 4-10)

Potency ranks various toxins by their LD_{50}. A more potent toxin is effective at lower doses. For three toxins, A, B, and C, the most potent is the most deadly at the lowest dose, while the least potent requires the largest dose. (Figure 4-11)

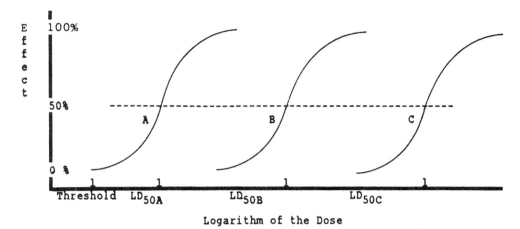

Figure 4-11

By plotting the cumulative dose-response curves (log dose), one can identify those doses of a toxicant or toxicants that affect a given percentage of the exposed population. Comparing the values of LD_{50}, LD_{50B}, and LD_{50C} ranks the toxicants according to relative potency for the response monitored.

The dose-response relationship curves provide information about the toxicity of a toxicant. However, there are certain caveats when using these data. Note that the slope of the curves is also important in analyzing the effects of toxicants. For example, while both toxicants A and B have the same LD_{50}, A has a steeper slope. Thus, while A is more toxic at higher doses, at lower doses B has more impact on a greater number of workers and would be more dangerous due to this lower threshold. (Figure 4-12)

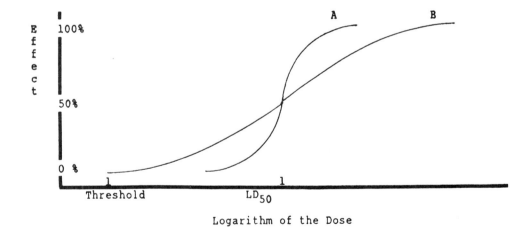

Figure 4-12

The shape of the dose-response curve is important. If one found the LD_{50} values for toxicants A and B from a table, one would erroneously assume that A is (always) more toxic than B. The figure demonstrates that this is not true at low doses.

Once someone is exposed to a toxicant, the width of the dose-response interval is as important as threshold. The difficulty with toxicant A and a steep dose-response curve is that this is much less variability among individuals. Therefore, once a toxic level is reached, the margin for error of toxicant A is much less than for substance B. Second, acute toxicity as generated by lab tests may not accurately reflect chronic toxicity. For example, both toluene and benzene cause depression of the CNS; for this acute effect, toluene is the more potent of the two. However, benzene is of greater concern under conditions of chronic, long-term exposure because it is carcinogenic, while toluene is not.

Margin of Safety

Four terms commonly used in pharmacology and toxicology are the margin of safety, the therapeutic margin, the therapeutic index, and the safety index. The **margin of safety** is defined as the lethal dose, 5%, minus the effective dose, 95%. The **therapeutic margin** is the lethal dose, 50%, minus the effective dose, 50%. The **therapeutic index** is the lethal dose, 50%, divided by the effective dose, 50%. And the **safety index** is the lethal dose, 5%, divided by the effective dose, 95%.

The therapeutic margin and the safety margin provide information about the spread of the effective dose and lethal dose curves. Ideally, if there is a choice, it is best to use substances which have wider margins. (Figure 4-13)

Figure 4-13

Graphic representation showing the relationship between the therapeutic margin and safety margin.

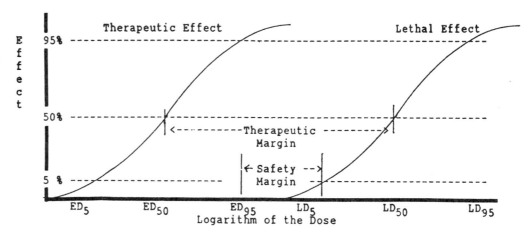

The therapeutic index and safety index are numbers. The larger the number, the safer the substance.

For a variety of reasons, toxicity testing is performed on nonhuman model systems. An important aspect of experimental toxicology is the extrapolation of animal model system data to humans.

There are a number of problems involved in extrapolating model system data to humans. First, because of the expense of conducting animal studies, the number of animals is minimized. This factor increases the likelihood of statistical errors which can skew the results. Second, because the normal life-span of laboratory model animal systems commonly used is significantly shorter than that of humans, high doses over relatively short periods of time are given. These factors may significantly skew the kinetics and, therefore, the results.

Animal testing is based on two principles. The first is that the results are applicable to humans on the basis of dose per unit value. The second is that overexposure of animals to toxic agents is necessary to discover the possible hazards to humans. Toxicity tests are not designed to demonstrate that a chemical is safe, but rather to characterize what toxic effects a chemical can produce.

Generally, the following information is needed to evaluate animal models and to extrapolate the data to human model systems:

1. **Animal species**

   The selection of an animal model system is very important for evaluating the specific toxicities under consideration. Dr. Bruce Ames and the University of California developed the "Ames" test. It operates under the hypothesis that chemicals which are mutagenic in bacteria are carcinogenic to humans. This is a very great leap from bacteria to man, yet there is a certain elegance and utility to the concept. The rat is frequently utilized as a model for toxicologic response; however, the thermoregulatory response of the laboratory rat following acute exposure to toxic agents is interesting. The rat responds to acute intoxication by lowering its core temperature, **hypothermia**, using both physiological and behavioral mechanisms. Moderate hypothermia results in a decreased toxic response and increased survival. A similar response has not been reported in humans. Pig skin is very similar to human skin. Thus, it is a good model for evaluating the impact of agents on human skin.

2. **Route of administration**
   If agents are administered to a model system intravenously (IV), how can you correlate this to a worker whose only exposure risk is topical dermal? On the other hand, if the dermal exposure risks are well delineated, how does this relate to exposure through injection under the skin by an airless paint sprayer?

3. **Time period**
   It would seem reasonable to have exposure, dose-response and lifetime studies of similar duration for different models. However, how do you, or can you, correlate effects in an animal with a several-year life-span with one that lives for 50 or a 100 years?

4. **Vehicle used to dissolve or suspend material**
   Absorption can vary significantly depending upon the medium or form of the material. Alcohol on an empty stomach is quickly absorbed. A glass of milk before or mixed with the alcohol greatly slows the rate of absorption and the peak blood alcohol levels.

5. **Dose per unit rate**
   Doses are frequently quoted in, e.g., grams of a substance per kilograms of body weight. Does it make sense to use this approach with animals (like mice) that are small, weigh only a few hundred grams, and have a high basal metabolic rate?

Study Exercises

1. Describe the dose-response relationship.

2. Explain the variations in response within the population.

3. Describe how the relative toxicity of substances is most commonly expressed.

4. List 5 factors that affect the toxicity of a substance.

5. Explain the threshold concept.

6. Define the following terms: toxicity, threshold dose, NOEL, LD_{50}.

7. Identify those toxic responses that may not exhibit thresholds.

8. How is cancer experimentally identified in animal studies?

9. Explain the differences in the responses commonly observed following an acute exposure to a toxic substance and chronic exposure to that same toxic substance.

10. Explain why higher doses of a toxic substance must be used in animal experiments in order to statistically determine the responses that would occur in a larger population.

11. List five hazardous or toxic materials for which you are at risk at home or in the workplace. (See Chapter 9) Using the NIOSH "Pocket Guide to Chemical Hazards" and Material Safety Data Sheets (MSDSs), create a table that shows the LD_{50}/LC_{50} for the material, the amount of material in your product, and the volume you would have to be exposed to during a given time period to reach the LD_{50} or LC_{50}.

5

Classification, Type, and Limits of Exposure

Overview

This chapter provides information on characterizing exposure. We will consider the principal factors which affect absorption due to exposure and differentiate the various types of exposure. This chapter provides an introduction to how the principles of toxicology are applied to situations in the real world. Exposure limits have been adopted by a number of different groups. These limits are designed to protect us against harm from both short-term and long-term exposures. We will define exposure limits and discuss the units for expressing these exposure limits.

Absorption

Five principal factors affect absorption due to exposure:

1. Conditions of exposure

2. Duration of exposure

3. Route of exposure

4. Concentration (pharmaceutical dose)

5. Form of toxon (physical and chemical properties)

An exposure may be accidental or intentional. Accidental exposures are more difficult to characterize because the concentration and duration are frequently unknown. Major spills or discharges of toxic materials may pose a significant threat to public health. This is because the resulting contaminant concentrations in the local area may be so high that only a second or two of exposure will cause adverse effects. A difficulty with some accidental exposures is the lack of decontamination measures. In situations where there is a risk of exposure (e.g., in a chemical laboratory) remediation technology like eye wash fountains and showers are installed. In areas where risk is considered negligible, such mitigation processes are unavailable and the exposure duration may be significantly lengthened. Intentional exposures are designed to be maintained below some TLV or PEL. Either PPE or engineering controls are used to maintain exposures below the permitted levels. (See Chapter 13)

Duration of exposure affects the pharmacokinetic process within the worker. The longer the exposure, the greater the absorption and, therefore, the longer it will require for detoxification and excretion. While the toxic substance remains in the system, there is potential for damage, the pharmacodynamics discussed in Chapter 6.

The route of absorption is important because it helps to determine both the speed of absorption and the duration of the exposure. Injections are very quick to reach circulation and distribution, providing virtually instantaneous access to circulating blood. Inhalation of a vapor into the lungs and absorption from the mucus membranes of the respiratory tract provide very quick access to the blood. Absorption from the digestive system tends to be slower and is distributed over a longer period of time. Absorption through the skin is the slowest process; substances may, inadvertently, be left in contact with the integumentary system for extended periods of time.

Functional Groups

The properties of organic compounds are highly influenced by their functional groups. Hydrocarbons are molecules made out of hydrogen and carbon, such as oil. Adding a hydroxyl group turns an alkane like ethane into an alcohol, grain alcohol, which is found in beer, wine, and whiskey. Carbonyl groups such as aldehydes, ketones, and carboxyl are found in sugars. A carboxyl group is an

Figure 5-1

Classes of organic compounds and functional groups

Class	Typical Functional Groups		Occur In:
Hydro-carbons	Methyl:	$-\overset{\overset{\text{H}}{\vert}}{\underset{\underset{\text{H}}{}}{\text{C}}}-\text{H}$	Fats oils waxes
	Ethyl:	$-\overset{\overset{\text{H}}{\vert}}{\underset{\underset{\text{H}}{}}{\text{C}}}-\overset{\overset{\text{H}}{\vert}}{\underset{\underset{\text{H}}{}}{\text{C}}}-\text{H}$	
Alcohols	Hydroxyl:	$-\text{OH}$	Sugars
Carbonyls	Aldehyde:	$-\overset{\overset{\text{O}}{\Vert}}{\text{C}}-\text{H}$	Sugars
	Keytone:	$-\text{C}-\overset{\overset{\text{O}}{\Vert}}{\text{C}}-\text{C}$	Sugars
	Carboxyl	$-\overset{\overset{\text{O}}{\Vert}}{\text{C}}-\text{OH}$ or $-\overset{\overset{\text{O}}{\Vert}}{\text{C}}-\text{O}^- + \text{H}^+$	Sugars
Amines	Amino:	$-\overset{\text{H}}{\underset{}{\text{N}}}-\text{H}$ or $-\overset{\overset{\text{H}}{\vert}}{\underset{\underset{\text{H}}{}}{\text{N}^+}}-\text{H} + \text{OH}^-$	Amino Acids Proteins
Phosphate compounds	Phosphates	or $-\overset{\overset{\text{O}}{\Vert}}{\underset{\underset{\text{O}}{}}{\text{P}}}=\text{O}$	ATP

organic acid, bestowing acidic properties on a molecule. Amines, which are basic, bestow alkaline properties, and are linked with carboxyl groups to create amino acids. Phosphates are used for energy storage, ATP, and for hooking molecules together such as with DNA and RNA.

The concentration can be important for several reasons. First, a more concentrated substance will tend to be absorbed more rapidly (or until saturation has been reached in a carrier moderated system). Second, a shorter exposure period is required for a more concentrated substance than for a lesser concentrated substance. Third, if there is a threshold, then it may require a more concentrated exposure to exceed the threshold. Carbon dioxide and carbon monoxide are both found in low concentrations in the atmosphere. However, at elevated levels, both of these gases may be lethal. Concentrated acids and bases can cause irreversible skin damage. Mild bases (e.g., soaps) are commonly used to cleanse the skin.

The form of a toxin has much to due with both its ability to be absorbed, the route of administration, and the period over which it is absorbed. Absorptive considerations were previously discussed in Chapter 2. (Figure 5-1) Classes of organic compounds and functional groups provide a synopsis of some of the more common classes of organic compounds. Generally, the behavior of these classes of substances is similar. Therefore, an understanding of the behavior of these classes can be used predictively to estimate the behavior of specific substances. Generally, the lipophilicity and hydrophilicity of a substance can affect its ability to be absorbed (i.e., what causes a substance to dissolve), in what (e.g, oil or water), and its rate of passage across barriers (e.g., the skin). Physical factors also affect the rate at which a substance will dissolve: medium (surface area, powder or large lumps); temperature (high temperature promotes dissolution and mobility); and humidity (helps promote dissolution by hydrophilic substances). Molecular size also affects a substance's physical properties. Substances with low molecular weight are gasses, while high molecular weight substances are solids. Substances which have intermediate molecular weights are liquids.

Exercise Draw up a list of substances that you know are liquids (e.g., water) and determine their molecular weights. Compare those weights to those of gases (e.g., oxygen) and solids.

Exposure Classes

Exposures are generally classified by the period over which they occur. This may be as short as a few seconds or as long as many years. Industrial toxicology commonly uses these five exposure classes:

1. acute

2. subacute

3. subchronic

4. chronic

5. long term

An acute exposure is a single event with rapid absorption. Although this type of exposure is usually short, it may produce irreversible effects. For example, inhalation of carbon monoxide, which rapidly attaches to the hemoglobin of the red blood corpuscles, can cause loss of consciousness and death (irreversible). An injection of a substance or the prick of a needle or nail (puncture wound) is an acute exposure that occurs in a very short time period, often a matter of seconds or less. Working for a day in an environment full of toxic vapors is an acute exposure. The effects may be reversible (survival), or irreversible (complications or even death).

Subacute exposures closely resemble acute exposure, except they result from 13 to 40 doses or exposures. The doses are frequently repeated and extended exposures over a period of several hours or days.

The subchronic exposure results from 30 to 90 exposures while chronic exposures are generally produced by greater than 90 exposures. These two classes are perhaps the most damaging. The exposure is generally with some frequency, daily or weekly, and over an extended period of time. If the frequency is sufficient, the exposure may become effectively continuous over the time period of concern. This is because of the nature of the time-dose effect relationships of an individual toxon; these were previously discussed in Chapter 4.

Generally, in an industrial setting, exposures of greater than 16 hours a day are rare. Therefore, there is usually a daily and, frequently, a weekend recovery period. However, it should be understood that for any single exposure in a chronic series, one may observe some acute effects. Additionally, repeated, long-duration, low-level exposures can saturate or exceed the body's capacity for detoxification and excretion. Alternatively, the worker's capacity for detoxification and excretion may become impaired due to other contributing factors, leading to levels exceeding threshold.

Long-term exposures are not typically found in the industrial environment; they may be found in the home or used for laboratory experiments. They are greater than one year in length. An example might be a small child kept in a lead-dust polluted house for a number of years.

In general, for chronic and long term exposures, if the rate of administration exceeds the rate of elimination, the toxic substance will accumulate in the body. In situations where the rate of detoxification and/or elimination is extremely low, the chronic dosage may result in an accumulation of virtually the total dose over the duration of the exposure. The total dose would then be equivalent, at the end of the exposure period, to the total dose given at a single time.

Asphyxiant

An unique situation arises around the body's need for oxygen. In this situation, it is necessary to consider exposure to oxygen deprivation. Technically, this is known as asphyxiation and, as discussed previously, may be either simple asphyxiation and chemical asphyxiation. Normally, the air is 18 to 20% oxygen. When oxygen levels drop below these levels, as is commonly found during smoggy days in some large cities, respiratory volume increases, muscular coordination diminishes, attention wanes, and clear-thinking requires more effort. In the 12 to 15% range, shortness of breath, headache, dizziness, quickening of the pulse, quick fatigue, and loss of muscular coordination of skilled movements occurs. In the 10 to 12% range, nausea and vomiting, inability to exert, and

Table 5-1 *Stages of asphyxiation*		
	First Stage	21 to 14% oxygen by volume: increased pulse rate and respiratory rate, some disturbed muscular coordination. The 14% oxygen content corresponds to an altitude of about 10,000 feet.
	Second Stage	14 to 10% oxygen by volume: faulty judgment, rapid fatigue, and insensitivity to pain. The 10% oxygen content corresponds to an altitude of about 15,000 feet.
	Third Stage	10 to 6% oxygen by volume: nausea and vomiting, collapse, and permanent brain damage.
	Fourth Stage	Less than 6% oxygen by volume: convulsions, respiratory arrest, and death.

paralysis of motion occurs. At 6 to 8% oxygen levels, there is collapse and unconsciousness and at less than 6%, there is death in 6 to 8%. (Table 5-1)

Additionally, there may be exacerbation of the depletion effects of oxygen due to the presence of other substances. Smog is typically created by the presence of a variety of pollutants. These pollutants serve both to decrease the oxygen levels in the atmosphere and to expose the worker to potentially toxic substances. Performance is decreased by both the loss of oxygen and the exposure to toxics.

Exposure Limits

For too long the environment was considered a bottomless sink for all of man's wastes. In 1970, the federal government confronted this situation with attempts to clean the air and water with the enactment of the Clean Air Act and the Clean Water Act. In 1976, it made further strides with the Resource Conservation and Recovery Act (RCRA). These laws were enacted to preserve the environment, protect the general population from pollution, and maintain public health.

In 1970, the federal government also initiated laws to protect workers in the workplace by regulating health and safety issues. The Occupational Safety and Health Act (OSHA) of 1970 went into effect on April 28, 1971. OSHA was the first uniform federal safety and health regulatory action. Before this time a variety of state and local regulations existed, however, they were rarely enforced.

The Act resulted in the formation of several new government organizations. The Occupational Safety and Health Agency (OSHA) was formed as part of the Department of Labor to implement and enforce workplace standards. The National Institute for Occupational Safety and Health (NIOSH) was created as the research agency within the Department of Health and Human Services. NIOSH is responsible for research, training, and education in the areas of safety and health.

Typically, federal regulations may be enforced by state agencies, provided that state regulations meet or exceed federal regulations. CAL-OSHA is the California State agency which takes the lead in these areas. California encourages workers to become actively involved in their safety and health in the workplace. The Division of Occupational Safety and Health (DOSH) also provides consultation services to employers and employees though a consultation service. Both state and federal regulations must be adhered to in states with both state and federal regulations.

Determination of Exposure Limits for Workers

In 1941, the American Conference of Governmental Industrial Hygienists (ACGIH) formed a committee to review available data on toxic compounds. They were interested in establishing exposure limits for employees working in the presence of airborne toxic agents. The committee publishes an annual list of compounds and recommended exposure limits entitled *Threshold Limit Values (TLV)* and *Biological Exposure Indices (BEI)*. The primary purpose of the TLVs is to protect workers in chronic exposure situations.

Exposure limits established and published by the ACGIH are of several different types:

Threshold Limit Value - Time Weighted Average (TLV-TWA)

> The time-weighted average concentration for a normal 8-hour workday and a 40-hour workweek, to which nearly all workers may be repeatedly exposed, day after day, without adverse effect.

Threshold Limit Value - Short Term Exposure Limit (TLV-STEL)

A time-weighted average concentration to which workers should not be exposed to for longer than 15 minutes at a time and which should not be repeated more often than four times per day, with at least 60 minutes between successive exposures. This limit supplements the TLV-TWA where there are recognized acute effects from a substance whose toxic effects have been reported from high short-term exposures in either humans or animals. These values ensure that there will be none of these effects:

• Irritation
• Chronic/irreversible tissue damage
• Narcosis

Threshold Limit Value-Ceiling (TLV-C)

The concentration in air that should not be exceeded during any part of the working exposure. Ceiling limits may supplement other limits or stand alone.

Skin

Notation after listed substance indicates the potential for absorption through the skin, eyes, or other membranes and the possibility for excessive absorption, which may invalidate TLVs.

Excursion Limits

Airborne concentrations derived from recommendations where data is insufficient to establish a TLV-STEL. Short-term exposures should exceed three times the TLV-TWA for no more than a total of 30 minutes during a workday, and under no circumstances should they exceed five times the TLV-TWA, provided that the TLV-TWA is not exceeded for the 8-hour workday.

In the early 1970s, the Occupational Safety and Health Administration was established. OSHA is responsible for the adoption and enforcement of standards for safe and healthful working conditions for employees engaged in any business in the United States. OSHA essentially adopted the then current TLVs, made them official federal standards, and referred to them as **Permissible Exposure Limits (PEL)**. PELs are formally listed in Title 29 CFR, Part 1910, Subpart Z, General Industry Standards for Toxic and Hazardous Substances. These were last revised in 1989.

The American Industrial Hygiene Association (AIHA) has established a committee to develop **Workplace Environmental Exposure Levels (WEELs)** for toxic agents which have no current exposure guidelines established by other organizations. Essentially, the committee is attempting to establish occupational exposure limits for materials not addressed by the ACGIH or OSHA, but of interest to various segments of industry.

Table 5-2

Toxicity rating scale

Category	Signal Word on label	LD$_{50}$ oral ppm	LD$_{50}$ dermal ppm	Probable Lethal Dose
I Highly toxic	Danger - Poison (skull and crossbones)	< 50	< 200	A few drops to a teaspoon
II Moderately toxic	Warning	50–500	200–2000	A teaspoon to one ounce
III Slightly toxic	Caution	> 500	> 2000	Over an ounce

Dose	Abbreviation	Metric	Abbreviation	Approximate amount
Parts per million	ppm	milligrams per kilogram	mg/kg	one teaspoon per 1,000 gallons
Parts per billion	ppb	micrograms per kilogram	ug/kg	one teaspoon per 1,000,000 gallons

Table 5-3

Measures of exposure

There are two WEEL limits for most materials. The first is an 8 hour TWA value. The second is a short-term TWA for exposures to either 1 or 15 minute duration.

NIOSH defines **Immediately Dangerous to Life or Health (IDLH)** levels as the maximum airborne contaminant concentrations from which one could escape within 30 minutes without any escape-impairing symptoms or any irreversible health effects. These generally far exceed TLVs and PELs.

Emergency Planning

The Committee on Toxicology of the National Research Council (NRC), an operating arm of the National Academy of Sciences (NAS), has published a list of **Emergency Exposure Guidance Limits (EEGLs)** and **Short-term Public Emergency Guidance Levels (SPEGLs)** as guidance in advance planning for the management of emergencies. SPEGLs are concentrations whose occurrence is expected to be rare in the lifetime of any one individual. These values reflect an acceptance of the statistical likelihood of a nonincapacitating reversible effect in an exposed population while avoiding significant decrements in performance. EEGLs differ from SPEGLs in that they are intended to apply to defined occupational groups such as military or space personnel.

Several major chemical companies formed a task force in 1986 to develop Emergency Response Planning Guidelines (ERPG) values for selected toxic materials. Guidelines have been completed for ammonia, chlorine, chloroacetyl chloride, chloropicrin, crotinaldehyde, diketene, formaldehyde, hydrogen fluoride, perfluorisobutylene, and phosphorous pentoxide. Several one-hour maximum exposure levels are being used.

Under the **Superfund Amendments and Reauthorization Act (SARA)** of 1986, the US EPA established a list of hundreds of **Extremely Hazardous Substances (EHA)** subject to emergency planning, community right-to-know, hazardous emissions reporting, and emergency notification requirements. The EPA specified **Levels of Concern (LOC)** for these substances. The LOC is one-tenth the established IDLHs or estimated IDLH.

The National Fire Protection Association (NFPA) publishes the *Fire Protection Guide on Hazardous Materials*. Substances are ranked on a scale of 0 to 4, where 0 is for materials which, on exposure under fire conditions, would offer no health hazard beyond that of ordinary combustible material, and 4 is for materials where a few whiffs of the gas or vapor could cause death. Additionally, with a 4, the gas, vapor, or liquid could be fatal upon penetration of the fire fighters' normal full-protective clothing, which is designed for resistance to heat.

Toxicity Testing

Central to setting exposure limits is the determination and classification of a substance's toxicity or hazard potential. Substances are tested to establish the type of toxicity and the dose necessary to produce a measurable toxic reaction.

Since different species of animals respond differently to chemicals, a substance is generally tested in mice, rats, rabbits, and dogs. The results of toxicity test in these animals are used to predict the safety of substances on humans.

Toxicity tests are based on two premises. The first is that information about toxicity in animals can be used to predict toxicity in humans. Experience indicates that data obtained from a number of animal species can be useful in predicting human toxicity. Single animal-study projections tend to be inaccurate. Second, by exposing animals to large doses for short periods of time, long term low-dose human exposures risks can be assessed.

The types of toxicity tests are as follows:

Acute toxicity studies
Single oral or injection doses providing an LD_{50} value in 14 days.

Subchronic toxicity studies
Studies over a 90- to 150-day period starting with the NOEL dose of the acute study to determine a **Maximum Tolerated Dose (MTD)**. The MTD is the highest dose of a substance that does not alter the animals' life span and it should not have any severely detrimental effects on the animals' health.

Chronic toxicity studies
Studies testing for mutagenesis, carcinogenesis (oncogenesis), teratogenesis, and reproductive toxicity.

Mutagenesis is the alteration of genetic structure or mutations. Mutagenic effects are passed on to future generations. In order for mutagenic effects to be passed on, they must occur in reproductive cells. Generally, mutagenicity testing cannot be done directly on humans, but relies on bacterial, cell culture, and animal testing.

Carcinogenesis means an uncontrolled growth or proliferation, a loss of growth inhibition, or loss of the ability to die gracefully, **apoptosis**, of cells. The terms *tumor, cancer,* and *neoplasm* all refer to this condition. Over an 18- to 24-month period, animals are dosed daily up to the MTD. Substances are administered as per expected human exposure. A control group is maintained and examined for the spontaneous development of tumors. Pathological microscopic examinations are performed on the animals. Four criteria provide evidence of carcinogenicity:

1. Tumors occurring more often in treated animals than in untreated ones

2. Tumors occurring sooner in treated animals than in untreated ones

3. Treated animals developing different types of tumors than treated ones

4. Tumors occurring in greater numbers in individual treated animals than in individual untreated ones

Teratogenesis is the production of birth defects. A teratogen is anything that is capable of inducting changes in the structure or function of offspring (child) when the embryo or fetus is exposed before birth. Teratogens affect the normal development of the embryo or fetus but not the reproductive genetic characteristics of the offspring. For testing, first male rats are exposed to the substance for 60 days and female rats are exposed for 14 days. The animals are then mated and the females are dosed throughout pregnancy and until the offspring are weaned. This test is for nonspecific reproductive toxicity. Next, two species of pregnant animals are dosed during the most sensitive stage of pregnancy, **organogenesis**, when the organ systems are under development. Fetuses are delivered by Cesarean section and examined for abnormalities. Finally, pregnant animals are dosed during the last third of the pregnancy and during weaning. These tests assess potential delivery and early growth associated effects. Teratogens typically have a TLV; this value, generally, has no relationship to carcinogenesis or mutagenesis.

Reproductive toxicity is assessed by exposing male and female rats to a toxic substance. The rats are then mated and the number of offspring counted. Harmful effects on fertility are evaluated by determining whether a substance will decrease the number of offspring produced. Subsequent testing is performed to determine if the effects are primarily male-oriented, female-oriented, or equally oriented.

Behavioral studies are also used. If animals appear to be behaving strangely, or substances induce slight depression or drowsiness, studies are conducted to determine if these effects could impact a worker's performance around various hazardous materials, including dangerous equipment.

Study Exercises

1. Describe the five principal factors which affect absorption.

2. Describe the five exposure classes.

3. Explain how the exposure to an asphyxiant is unique.

4. Describe how the workplace exposure limits were developed. Compare this with the development of exposure limits for air and water.

5. Describe the following terms associated with workplace exposure limits.
   - OSHA
   - CAL-OSHA
   - PELs
   - NIOSH
   - EELs
   - ACGIH
   - TLVs
   - AIHA
   - WEELs

6. Most workplace limits were designated for what exposure route? Why?

7. Convert between ppm and mg/m^3.

8. Describe the differences between the following types of exposure limits:
   - TWA
   - STEL
   - Ceiling

9. Explain how it is possible for a worker to be harmed from a toxic exposure, even when the TWA limit has not been exceeded.

6

Action of Toxic Substances
Pharmacodynamics

Overview

The purpose of this chapter is to introduce the mechanisms of actions of toxons. Specifically, we shall examine mechanisms whereby a toxic substance, i.e., a TOXON, acts at a specific site of action, a receptor site, or at receptor sites throughout the body.

Pharmacodynamics

The pharmacodynamic phase comprises the interaction between the molecules of a toxic substance, toxons, and the specific sites of action, the receptor sites. In general, a set of events occur subsequent to an interaction at a specific site of action. First, there are processes which lead to the initiation or induction of a stimulus in the target organ. Consequentially, there are a set of events, evoked by this initial event, usually chemical processes, which result in the final effect. The degree of biological action, the magnitude of the result, is related to the concentration of the toxon in the target site.

It is important to understand three points about this process:

- First, that the results of the stimulus are independent of the specific properties of the toxon that set off the chain of events.

- Second, that a variety of different toxons with different properties can initiate a chain of events which lead to similar results.

- Third, the specific site of action, where the toxon acted, and the effector organ, where the result or effect is observed, need not be identical. For example, strychnine acts on the cells of the central nervous system (CNS), but the resulting convulsions are observed in the skeletal muscles. However, the effects of strychnine are due to a chain of events and, thus, also measurable in changes of the action potentials at the **motor end plate**s, electromyograms, and electroencephalograms.

Types of Actions

Interference with the Action of Enzyme Systems

Enzymes are organic catalysts that perform the essential tasks of the biochemical process used by cells. Enzymes are complex, three-dimensional protein molecules with specific sites that are used by the substrates to be converted. Interference or blockage of the sites, as well as changes in the three dimensional structure, affect enzyme function and interfere with the biological functions. Interference may be either reversible (temporary), or irreversible (permanent).

Irreversible Enzyme Inhibition

Irreversible enzyme inhibitions involve **covalent bonding** of the toxon to the enzyme, frequently at the substrate active site. The classic example of this type of interaction are the organophosphates originally developed as a neurotoxic chemical warfare agents and, subsequently, used as insecticides.

These toxons form an irreversible covalent bond at the active site of the enzyme known as acetylcholinesterase. Acetylcholine is a **neurotransmitter** (passes information between nerve cells, nerves and muscles, and nerves and effector cells). Acetylcholine is released by the neurons at the surface of the muscle, diffuses across a small gap, binds to receptors on the surface of the muscle cell, and causes the muscle to undergo contraction. Subsequently, the acetylcholine is released and either taken back by the neuron for reuse or broken down by the enzyme acetylcholinesterase. Since acetylcholinesterase is the enzyme which breaks apart acetylcholine and, therefore, stops its action, this will terminate the transmission and, thus, the contraction. Without the action of acetylcholinesterase, neuro-stimulation continues. For example, the skeletal muscles continue to contract, the body goes rigid, breathing becomes impossible and death by asphyxiation quickly follows.

These compounds are broad spectrum and toxic to most animals, e.g., man and other mammals, birds, reptiles, fish, insects, and worms. Lethal effects may be mitigated by the administration of atropine between the exposure and absorption stages. This is a very short period. The military utilizes a thigh slap pack. This is affixed to the pants over the thigh muscle, upon being gassed, the pack is slapped, which forces an injection of the antidote into the thigh muscle.

A variety of heavy metals, e.g., mercury, arsenic, and lead, are less selective enzyme inhibitors. Typically, there is a covalent bonding between these metals and the SH groups essential to the functioning of a large variety of enzymes. Chemicals like dimercaprol which possess SH groups, act as a chelating agent and can accomplish a degree of reactivation of the enzyme by uptake and withdrawal of the heavy metal ions.

Reversible Enzyme Inhibition

Reversible enzyme inhibition is generally caused by competition between the normal substrate and an antimetabolite. These compounds are closely related chemically to the normal substrates of the enzyme such that they fool the enzyme into using them. Unfortunately, the enzymes are unable to metabolize them and, therefore, waste their time. This wasted time prevents the enzyme from performing its normal duties expeditiously. For example, methotrexate, a folic acid antagonist, is used as a **cytostatic** agent, to inhibit cell growth, in the treatment of cancer. These antimetabolites can inhibit the synthesis of amino acids required to produce proteins and nucleic acid required for the production of DNA and RNA.

Uncoupling of Biochemical Reactions

The body oxidizes a variety of compounds in order to release energy. Normally this energy is stored in the form of high-energy phosphate bonds, i.e., Adenosine TriPhosphate (ATP). ATP is the "energy coin" of the cell and provides the energy almost universally used by biochemical processes. Uncoupling agents interfere with the ATP production cycle and liberate the energy as heat, thereby producing a fever. Dinitrophenol, used in the past to treat obesity, and dinitroorthocresol, used as a weed killer, act in this manner.

Several herbicides, paraquat and diquat, interfere with the transfer of hydrogen to Nicotine Adenine Dinucleotide Phosphate (NADP), a catalyst in oxidation-reduction reactions. Both glycolysis and the Krebs cycle produce NADPH; therefore, animal toxicity is based upon the organism's ability to uncouple redox catalysts.

Lethal Synthesis

Lethal synthesis is the process whereby a substance is accepted for conversion competitively with a normal substrate with the production of a lethal by-product. Ethanol, or grain alcohol, is degraded by a series of enzymes, one of which produces acid aldehyde. Methanol, or wood alcohol, is degraded by this same set of enzymes. However, its intermediary is formaldehyde, a preservative which favors the optic nerve, thereby inducing blindness. The administration of ethanol mitigates the toxic effects of methanol by competitive inhibition.

Fluoroacetic acid is very similar to acetic acid, except that one of the hydrogen atoms is replaced by a fluorine atom. Fluoroacetic acid is accepted in the place of acetic acids in the Krebs citric acid cycle. Fluorocitric acid is formed and acts as an inhibitor of the enzyme aconitase, thereby blocking the cycle. This set of events blocks the formation of NADPH, inhibits the normal function of the mitochondria, and ultimately prevents the production of ATP.

The cell normally carves up fatty acids, alkyl compounds, by twos, a process known as beta oxidation. Organic fluoroalkyl compounds of even numbers produce fluoroacetic acid, while those of odd numbers produce fluoropropionic acid. While both of these end products are toxic, even numbered compounds can be a hundred times more toxic than odd numbered compounds.

Removal of Metallic Cofactors

Various metal ions such as iron, copper, zinc, manganese, and cobalt are cofactors of coenzymes of enzyme systems. These metals are integral to the function of the enzyme; without them the enzyme cannot function. Chelating agents form complexes with metal ions to prevent them from functions as cofactors. The water soluble chelating agent dimercaprol was discussed earlier. Lipophilic chelating agents, such as the hydroxyisoquiniline derivatives and the dithiocarbamates, exist and are potentially toxic.

The dithiocarbamates are used as activators in vulcanizing and as antioxidants in the rubber industry. Workers exposed to these substances were found to develop nausea, violent headaches, and comas after drinking moderate amounts of alcohol. This is because the dithiocarbamates bind copper ions and inactivate copper-dependent enzymes such as acetaldehyde dehydrogenase, a normal step in the degradation of ethanol. A dithiocarbamate derivative, disulfiram, has been used for antibuse, a chemical treatment for the prevention of alcohol abuse.

A number of fungicides use dithiocarbamate derivatives. They accelerate the transport of copper, cobalt, manganese, and iron into the cell through the lipophilic cell membranes. This results in the disturbance of the cellular biochemical processes and, thus, metal poisoning on enzyme systems.

Inhibition of Oxygen Transfer

In the mitochondria of the cell, on the walls of the cristae, are the chains of the cytochrome oxidase system. The cytochrome oxidase system uses iron molecules to catalyze the reaction of hydrogen and oxygen to make water. This is the last step in the breakdown of glucose, glycolysis, and the production of ATP. Hydrocyanic acid, HCN, and hydrogen sulfide, H_2S, bind to the iron in the enzyme and prevent its function. This plugs up the system and prevents ATP production by asphyxiation, even though sufficient oxygen is present in the tissues.

Blockade of Hemoglobin Oxygen Transport

Carbon Monoxide Poisoning

Carbon monoxide and oxygen have the same binding site on the hemoglobin molecule. Although a carbon monoxide bind is reversible, it has a much greater affinity to bind than oxygen. That is, it binds much tighter. In fact, about 210 times as great as oxygen. That means that it takes only $\frac{1}{210}$ as much CO to bind to the same number of hemoglobin molecules as oxygen. Therefore, 0.1% atmospheric pressure of CO will bind the same number of hemoglobin molecules as 21% of oxygen in the atmosphere.

The TLV for CO is 50 ppm, which equates to about 8% saturation. At about 5 to 10% saturation, carboxyhemoglobin is formed by heavy smoking. After 10% saturation, there is a tightness across the forehead, headache, and dilatation of cutaneous blood vessels. After 20%, the headache becomes throbbing, and after 30%, severe, and weakness, dizziness, dimness of vision, nausea and vomiting, and collapse occur. After 40% saturation, there is an increase of respiration and pulse. At 50%, coma and convulsions occur, and after 60%, depressed heart action. Greater concentrations lead to slowed respiration, respiratory failure, and death.

Formation of Methemoglobin

Many compounds—such as aromatic amines, aniline, sulfonamides, acetanilide, phenazone, p-aminosalicylic acid, nitrofurantoin, quinine, primaquine, azo compounds, nitro compounds, and nitrites—can oxidize hemoglobin to methemoglobin, which has no oxygen binding ability. Although the formation of methemoglobin is natural and is generally reduced in the erythrocytes back to hemoglobin, the regenerative process is limited. When methemoglobin builds up, the transport of oxygen decreases markedly.

Beta thalassemia, a genetic deficiency found in many Mediterranean populations, reduces the capacity to regenerate hemoglobin. Neonates also have a limited ability to regenerate methemoglobin. When babies are fed on spinach grown in soil rich in nitrogen-containing fertilizers, methemoglobinemia may result. Reduced ability to convert methemoglobin results in higher toxicity and added risk to affected individuals.

Formation of Sulfhemoglobin

Sulfhemoglobin and methemoglobin exhibit certain similarities. Sulfhemoglobin is formed by the presence of sulfur-containing compounds, such as the sulfonamides. Sulfhemoglobin is also formed when blood is exposed to hydrogen peroxide and sulfur-containing compounds. Sulfhemoglobin has no oxygen carrying capacity.

Hemolytic Processes

Hemolysis is the destruction of the formed elements of the blood. Surfactants and hydrazine derivatives can cause a breakdown in the membranes of the erythrocytes. Such membrane damage results in the release of hemoglobin. Free hemoglobin loses its ability to transport oxygen. Additionally, the natural sequence of events that reclaims hemoglobin from damaged RBCs produces bilirubin and jaundice. Frequently, levels of urobilin in the urine are raised due to the increased blood bilirubin levels, caused by the hemolytic processes or liver damage.

Interference with the General Cellular Function

Anesthetic Actions

Lipophilic substances, such as the clinically used ether, cyclopropane, and halothane, accumulate in cell membranes and have anesthetic actions due to depression of cellular activity. This action is likely due to the inhibition of the transport of oxygen and glucose. The cells of the CNS are particularly sensitive to depressed oxygen and glucose levels, which cause narcotic effects.

Not only commonly used general anesthetics, but all nonpolar organic solvents—such as gasoline, halogenated organic derivatives, and organic solvent-based glues and products—act as anesthetics. Generally, the vital centers that control respiration and cardiac actions are only depressed by relatively high concentrations of these solvents. While the line between narcosis and death is relatively large for clinically used substances, i.e., ether and halothane, it is much smaller for organic solvents commonly found in the workplace. Thus, uncontrolled exposures of these substances carry substantial lethal risks.

Interference with Neurotransmission

Nerves use chemicals, called neurohumoral transmitters, to transmit information to other nerves, muscles, and effector cells. The endocrine system also uses chemicals, hormones, to transmit information between cells and groups of cells. Curare (the first toxon, used as arrow poison), atropine, nicotine, norephedrine, and estrogen are naturally occurring botanical substances which affect neurotransmission and cellular function.

Most of the neurons of the peripheral nervous system, both voluntary and autonomic branches, use either acetylcholine or norepinephrine, noradrenalin, as transmitters. Those systems that use acetylcholine are referred to as cholinergic; the individual receptors are either muscarinic or nicotinic receptors. Those systems that use norepinephrine are referred to as adenergic (this misnomer was created by early physiologist which mistakenly identified noradrenaline as adrenaline); the individual receptors are either epinepherinic or norepinepherinic. Chemicals that mimic one of these neurohumoral transmitter's actions are known as **mimetic**, while those that prevent transmission are inhibitors or antagonists.

Curare blocks the cholinergic receptors in the neuromuscular junction and acts to relax skeletal muscle. Atropine and nicotine also inhibit cholinergic receptors. Puffer fish toxin, tetrodotoxin, and shellfish (mussel) poisoning affect cholinergic transmissions. Botulinus toxin interferes with the release of acetylcholine in the neuromuscular junctions causing paralysis.

As previously mentioned, acetylcholinesterase inhibitors act as enzyme inhibitors. Monoamine oxidase inhibitors perform a similar function in the adenergic nerves. Other chemicals inhibit the production, release, reuptake (as in cocaine), or metabolism of neurohumoral transmitters. Any and all of these effect the function of the nervous system. Chemicals affecting the same mechanisms, similar effects per dose, are termed additive. Those substances that affect different mechanisms or systems may, together, exhibit synergistic effects, where the combined effects are much greater than either does separately. (**Potentiation** has been used to indicate multiplicative effects; see potent.) Those chemicals that cause opposing effects or responses are termed **antagonistic**.

A wide variety of chemicals can affect various portions of the nervous or endocrine system. Mescaline, LSD, THC, arecoline (as in betel nuts) and philosybin are classified as hallucinogens and affect neurotransmission in the CNS. Heroin, codeine, and morphine are narcotics which affect hormonal effectors in the CNS. Estrogenic mimetics affect estrogen receptors located throughout the body. Pubescent boys fed chicken stimulated to grow with estrogen, may develop feminine characteristics, a condition known as gynecomastia.

Interference with Nucleic Acids

Information is stored and transported throughout the cell by nucleic acids. Within the nucleus, DNA stores all of the information necessary for the creation, maintenance, upkeep, and functioning of a cell. The DNA must be duplicated before a cell can divide. This information may also be transcripted onto **messenger RNA** (mRNA) for transport into the cell's cytoplasm. Messenger RNA is translated by the ribosomes to produce proteins.

DNA Synthesis

Interference with the duplication process of DNA inhibits cell division. Acridine derivative and antibiotics of the actinomycin group bind to the DNA chains. Polyvalent biological alkylating substance react with corresponding groups in the DNA chains or form chemical bridges or dimers between the double DNA strands. Numerous enzymes control and moderate the duplication process and may be inhibited. Such events prevent the successful duplication or separation of DNA strands.

RNA Synthesis

Transcription of RNA from DNA is required to utilize the information stored as DNA. Interference with the enzymes that control and moderate this process will prevent or corrupt the transcription process. For example, rifampicin inhibits RNA polymerase. Many viruses subvert these processes for their own use, producing RNA entirely unrelated to the cell's needs.

Protein Synthesis

The information carried by RNA is translated by ribosome to construct amino acid chains, or proteins. There are a variety of mechanisms that can be subverted by toxic chemicals. Puromycin, an amino acid mimetic, blocks chain growth. Streptomycin, kanamycin, and gentamicin interfere with mRNA translation. Tetracycline, chloramphenicol, and lincomycin interfere with the ribosomal action.

Lack of an amino acid to incorporate into a chain prevents protein production, a condition known as Kwashiorkor Disease.

Conversely, many hormones affect RNA and protein synthesis to control cell, tissue, and body growth. Additions to, or lack of, these chemicals lead to regulatory imbalances of RNA and protein synthesis and, thus, ill effects.

Cytostatic Actions

Substances inhibiting the normal process of cell division are known as **cytostatic** agents. Cytostatic agents tend to act systemically; rapidly proliferating cells and tissues are most affected. The tissues and cells most affected are the integument (skin and hair); bone (hemopoietic tissues); and the digestive tract (intestinal mucosa). Alkylating agents (ethyleneamine groups) interact with amine and hydroxyl groups on cross-linking strands such as DNA, preventing cell division. Triethylene melamine (TEM) is used in the textile industry to cross-link synthetic macromolecular filament.

Immunosuppressive Actions

Our body's defense against invasion and disease is orchestrated by the immune system. The weapons used include the complement system, specific immune response (antigen-antibody complexes), inflammation, fever, and clotting. Agents which suppress the immune defense mechanism are called **immunosuppressants**. Hormonal substances such as glucocorticoids, are used as anti-inflammatants, but cause decreased resistance to disease. Heparin, an anticoagulant, as is warfarin, is used as a rodenticide. A number of substances suppress the production of white blood cells, leukocytes, causing leukemia.

Hypersensitivity Reactions

The immune system is tasked with determining self from nonself. Because of the constantly changing nature of the assaults upon the body, the immune system has evolved with the capacity to learn. At each new nonlethal assault, the immune system learns effective responses. Subsequent assaults are met by the defense system with rapid weapons deployment. This sequence of activation and deployment events is known as **sensitization**.

Hypersensitivity is just the overactivation of this sensitization process. Smallpox and other vaccinations are common examples of the function of the sensitization process. Usually, the sensitization is specific in that it activates the specific immune response system; however, frequently, it is has a general impact in that it is based upon some nonself attribute of the insulting agent. This, then, may lead to cross-sensitization, where sensitization to one agent also leads to sensitization for similar agents.

Sensitization is not always wanted or considered beneficial. The general sensitization reaction may be an allergic or anaphylactic response. Allergies are unwanted by most and, frequently, desensitization is sought. An allergic reaction to a bee sting can be lethal—initially an annoyance, but increasing in intensity with each episode, until a deadly anaphylactic response is elicited. Eczema is another unwanted generalized allergic type reaction. **Farmer's lung** is related to the inhalation of an organic dust. Often due to the inhalation of spores of thermophilic actinomycetes, which flourish in the warm environment of damp hay, symptoms occur 5 or 6 hours after exposure and are characterized by fever, malaise, and chills with aches and pains. Chronic exposure leads to weight loss, shortness of breath, and fibrosis of the lung. Toluene diisocyanate, used to manufacture polyurethanes, also produces allergic symptoms. In fact, practically any substance, under appropriate circumstances, can induce an allergic response.

A special type of sensitization is due to light. This is generally induced when the skin is exposed to light after previous exposure to some chemical substance. This reaction may be photoallergenic, photosensitizing, or phototoxic. Certain chemicals in medications and products can make the skin more sensitive to the ultraviolet (UV) light from the sun. Once photosensitized, exposure to light may induce a **photoreaction**, where the skin becomes red, itchy, bumpy, swollen, or blistered. Photoreactions are not the same as a sunburn. If you have a sunburn, the sun is just affecting the skin, or dermis. However, with a photoreaction, the sun acts synergistically with a chemical in or on your skin to cause the reaction. Photoreactions typically last longer than a sunburn, three days or longer, and can be more painful.

Photoallergies are induced by photochemical alterations to products soliciting a specific immune response. Subsequent exposure to these substances in sunlight can induce allergic skin reactions such as eczema. The drugs griseofulvin and promethazine, deodorants and disinfectants such as tetrachlorsalicylanilide, hexachlorophene and bithionol, sunblocks, aminobenzoic acid and digaloyltrioleate, blancophores, and wash whiteners are capable of these effects.

Photosensitization occurs upon the first exposure to both a substance and light. The skin reacts by **dilation** of **subcutaneous** capillaries with a resultant **erythema** characteristic of sunburn. Washing whiteners and furocoumarins (psoralenes) are photosensitizers. Subsequent to exposure, there is an increased pigmentation of the skin, typical of a suntan; the tan is frequently blotchy and discolored. The sensitization can remain long after exposure to the causative agent has terminated.

Phototoxic reactions occur when a substance absorbed locally by the skin is biotransformed into a toxon by a photochemical reaction. The nature of the response depends upon the nature of the toxon.

Most photosensitizing medications carry a warning. Over-the-counter skin and hair products that may make your skin more sensitive to sunlight include the following:

- perfumes and perfumed soaps

- after-shave lotions

- dandruff shampoos that contain coal-tar ingredients

- skin-bleaching creams

- some alpha hydroxy and beta hydroxy lotions

- Sunscreens that contain para-aminobenzoic acid and sunscreens
  with fragrances

Over-the-counter products may not carry a warning. To be safe, read the labels and avoid products that contain known photosensitizers, such as quinoline, musk ambrett, 6-methyl-coumarin, bergamot oil, and sandalwood oil. The herbal remedy St. John's wort may cause photosensitivity in some people; it is also synergistic with prescription antidepressant drugs. Nonsteroidal anti-inflammatory drugs, NSAIDs, in dosages used for the treatment of arthritis, may also cause sensitization.

Always take precautions when sun exposure may occur. Avoid peak, noon, sun hours, use a sunscreen, or better yet, a sun block such as zinc oxide or titanium dioxide, a wide brimmed hat, dark glasses, long-sleeved and legged, tightly woven fabric. If exposed, take a cool bath or shower and use over-the-counter pain relievers as for sun burn.

Direct Chemical Irritation of Tissues

Local effects due to direct contact with chemical irritants were discussed previously in Chapter 2. They are due to the chemical reaction with various tissues. Typically, they can impact on the skin, and on mucus membranes of the eye, nose, throat, trachea, bronchi, and alveoli. The skin, or dermis, is affected by mustard gases and blistering agents, resulting in **dermatitis.** Polychlorinated biphenyls (PCBs) is notorious for producing chloracne. Phenyls (salicylic acid), alkaline solutions (sodium and potassium hydroxide), strong acids (nitric and hydrofluoric acids), etch tissue. Chlorine and carbonyl chloride (phosgene) gasses can blister the skin and cause edema of the lung. Sulfur dioxide, nitrous oxides, and ozone irritate mucus membranes.

A number of gases cause irritation of the conjunctiva of the eye and **lacrimation,** or tearing. Those effective at extremely low doses—acrolein, bromacetone, bromacetophenone, chloracetophenone, 2-chlorbenzylidene malononitril (CS), and chlorpicrin—are called tear gases.

Tissue Toxicity

Cell degeneration accompanied by the formation of **vacuoles** (empty or fluid-filled spaces), accumulation of fat, and **necrosis** (cell damage and destruction) is due to direct damage to cell structure. Such symptoms are exhibited by the liver and kidney due to the sequestration of toxic substances in these organs. Such chemical lesions can be produced by chlorinated alkanes, chloroform and carbon tetrachloride, and halogenated aromatic compounds, like bromobenzene. Necrotic cell death is accompanied by immune system responses such as redness, swelling, inflammation and pain; apoptotic cell death is considered natural and does not stimulate an immune response.

Sequestration of Toxic Substances

As discussed previously in Chapters 2 and 3, toxic substances may be deposited and stored in certain tissues for extended periods of time. Common depots are adipose tissue for DDT, DDE, and PCPs, and bone tissues for strontium, thorium, lead, and silver. Dust and crystals embedded in the lung pose significant health risks. Silicon, silicosis, beryllium, and coal (anthracosis) dust induce fibrosis of the lung and lead to chronic **bronchitis** and **emphysema.** Asbestosis is a particularly dangerous form of pneumoconiosis since, in addition to the aforementioned conditions, it can induce malignant growths, like cancer, in the lung.

Electrolyte Imbalances

The body's electrolytic homeostatic balance can be upset in many ways. The body, in its attempt to maintain a constant aqueous and temperate environment for its cells, regulates water, salts, dissolved substances, and metabolic activity. These are major tasks for the kidney and integumentary systems. Hypotonic dehydration, or sunstroke and heat stroke, are due to the loss of electrolytes and may result in vomiting and diarrhea. Replacement of lost fluid by water leads to a hypotonic hydration and water intoxication. Abdominal and leg cramps, known as heat cramp and miner's cramp, may follow. Alternatively, diarrhea itself leads to water loss and dehydration, which is potentially lethal. (For more information, see Chapter 15.)

Chemical Interaction

A dosage with more than one chemical, may or may not alter the effect of the individual chemicals. If two or more chemicals or substances act to produce a greater effect, they are said to be synergistic. If two substances produce a lesser reaction or effect, then they are antagonistic. For example, caffeine, a stimulant, acts to counteract the effect of a sleeping pill.

Four major types of antagonism form the basis by which antidotes are determined. The first type is functional and produces a smaller or lesser effect by competition for the same effector site. The second type is chemical, in that the two substances have opposite effects and inhibit each other. The third type, dispositional, functions when the biological transport mechanisms are altered and less toxin reaches the target organ or site. The fourth type involves receptor antagonism when two or more agents bind to different sites on the target organ and produce lesser effects or antagonistic responses.

Multiple substances acting via the same sets of mechanisms tend to be additive such as when the sum of the effects of the doses are the sums of the dosages of the individual agents. In general, additive substances will produce an effect which is the sum of the effects of the two substances taken individually. The effect of taking one Anacin and one Tylenol is similar to the effect of taking two Anacin or two Tylenol. Dosages of chemicals such as epinephrine (adrenaline) and norepinephrine (noradrenalin) tend to be additive.

Potentiation is a term best avoided; *potent* and *potency* are defined later. Potentiation is sometimes used to indicate substances that produce an effect in excess of the effects observed when either of the substances is taken individually. Antibuse is a pill (drug) taken to discourage the drinking of alcohol. Taking antibuse has, in general, no effect on an individual. However, a person who has a single alcoholic drink while on antibuse will become very ill. This is a potentiation of alcohol intoxication.

Study Exercises

1. Explain how toxons can interfere with the action of enzyme systems
   A. Irreversible enzyme inhibition
   B. Reversible enzyme inhibition
   C. Uncoupling of biochemical reactions
   D. Lethal synthesis
   E. Removal of metallic cofactors
   F. Inhibition of oxygen transfer

2. Explain four different means of blockage of hemoglobin oxygen transport.

3. Explain how, and give examples, of toxons interfering with general cellular function.

4. Explain how toxons interfere with RNA and DNA processes and with the process of cell division.

5. Discuss the importance of the immune system.

6. Explain hypersensitization activities and reactions.

7. Describe and give examples of what is meant by direct chemical irritation of tissues.

8. Describe tissue toxicity.

9. Review the sequestration of toxic substances.

10. Define the following types of interactions between toxic substances:
    - additive
    - synergistic
    - potentiation
    - antagonistic
    - sensitizing

11. One of your workers is responsible for applying an epoxy coating to the surface of one of your product parts. Recently he went on a two-week vacation. After his return he started to complain of a number of symtoms ranging from skin rash to breathing difficulties. Is he faking it? What might be the basis for this sequence of events?

7   Target Organ Effects

Overview

This chapter introduces the major organs and organ systems affected by toxic substances after systemic absorption. The human body is divided into ten systems. We shall review the general organization and functions of five of these: the cardiovascular, nervous, liver, excretory, and immune systems. Additionally, we shall be examining, in some detail, the composition and functions of blood. In Chapter 6, we described the dysfunctions initiated by toxons in the pharmacodynamic phase. These dysfunctions impact the normal functioning of our body systems. In this chapter we provide a better understanding of the specific consequences of certain toxons on our body. To help illustrate these consequences, we include examples of a wide variety of toxons.

Nervous System

The nervous system consists of two general populations of cells. The first is the neurons. Neurons are highly specialized cells responsible for the transmission of information. They are so specialized that they require a second population of cells, the glial or nurse cells, known as **neuroglia**, to care for them and provide a variety of functions. An important distinction between neurons and glial cells is that neurons do not divide. People are born with their maximum number of neurons. As they age, that number decreases as the neurons die.

A typical neuron has a cell body and a number of cell processes. One process is the axon. Each neuron has only one axon. The attachment of the axon to the cell is an enlarged area known as the axon hillock. Information leaves the neuron via the axon. Each neuron may have none, or up to 50,000 processes, called dendrites. Information may arrive at the neuron via the dendrites, the cell body, or the axon hillock.

There are three general populations of neurons. Sensory neurons transduce information and pass it on the other neurons. Motor neurons conduct information to the effector cells, e.g., muscles, which then initiate some action. Internuncial or intermediary nerve cells process information provided by the sensory neurons and activate the motor neurons.

Information is transmitted along the membrane of the neuron by a mechanism known as depolarization. Normally, sodium-potassium ATPase pumps that are located in the membrane pump sodium out of the cell and potassium into the cell. This action causes an increase in potassium inside the cell relative to the outside of the cell. Because more sodium is pumped out than potassium pumped in, the inside of the cell is negative relative to the outside of the cell, typically -70 to -80 millivolts (mV). Within the membrane are sodium and potassium channels; the sodium channels pass sodium while the potassium channels pass potassium. These are called voltage-controlled channels because, when a certain voltage threshold (-50 mV) is reached, they open. At threshold, the sodium channel opens quickly, allowing sodium to rapidly flow into the cell. Subsequently, the potassium channel opens and allows potassium to exit the cell. This causes a rapid depolarization of the membrane to +30 to +40 mV. The magnitude of this depolarization is sufficient to cause other local voltage-controlled channels to reach threshold and open, and so on, until the impulse, or wave of depolarization, reaches the end of the axon, the synapse. This sequence of events is referred to as a neural impulse.

Figure 7-1

Sites of action and the interrelationship of miscellaneous agents at the neuromuscular junction and adjacent structures.

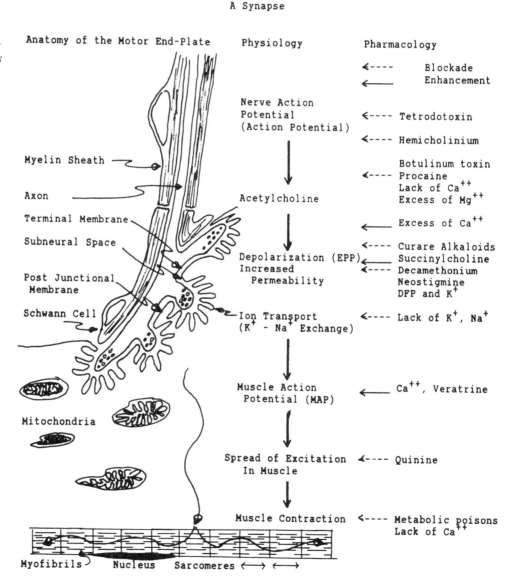

A synapse (Figure 7-1) is the point of contact between the neuron and an adjoining cell. No actual physical contact is made at a synapse; rather there is a gap of several hundred angstroms. At the synapse, the action potential allows the release of a chemical, the **neurohumoral transmitter**, which diffuses across the synaptic gap to elicit a response from the synapsed cell. The chemicals, or **neurotransmitters,** released from the **presynaptic** terminal into the gap, diffuse across the gap to the specific **postsynaptic** terminal receptors. Acetylcholine, previously mentioned, is a common neurotransmitter found in many synapses. At the synapse, neurotransmitters may be either broken down, such as in the action of acetylcholine esterase, or taken up for reuse, i.e., **reuptake**.

The nervous system is divided and categorized by location and function. The **Central Nervous System (CNS)** consists of the brain and spinal cord. Branches of nerves radiating out from the CNS to the periphery of the body are collectively called the **Peripheral Nervous System (PNS)**. Functionally, the portion of the nervous system over which we exert conscious control, is called the voluntary nervous system. The voluntary nervous system controls skeletal muscle action and is also known as the **Somatic Nervous System (SNS)**. A large number of body functions (e.g., heart beat and digestion) are not generally under conscious control; this system is referred to as the **Autonomic Nervous System (ANS)**. Since each nerve can only carry one type of information and affect one type of response, effective control of response requires paired systems, one to excite and one to inhibit. The ANS, therefore, has two branches, the **Sympathetic Nervous System (SNS)**, or **adrenergic**, and the **Parasympathetic Nervous System (PNS)**, or **cholinergic**, which elicit opposite organ responses.

There are five functional populations of glial cells. The astrocytes are responsible for creating the membranous, **blood-brain barrier**. The ependymal cells are responsible for filtering blood plasma to create the **cerebral spinal fluid (CSF)**, used in place of blood plasma by the CNS. The **microglia** perform functions analogous to the white blood cells, which cannot pass the blood-brain barrier in the CNS. The **oligodendrocytes** and **Schwann cells** create sheaths of myelin around nerve cell process to increase the speed of conduction.

Neurotoxins selectively impair the structure and function of the nervous system. They may affect the following functions:

- protein synthesis
- propagation of the nerve impulse
- neurotransmitter activity
- maintenance of the myelin sheath

Blocking Agents

Some toxons act to block the gates, potassium or sodium channels. They are called *blocking agents*. These blocking agents prevent the influx of sodium into the cell during the depolarization event. Tetrodotoxin (puffer fish), and saxitoxin (dinoflagellates) are examples.

On the other hand, botulinum toxin functions in a different manner; it actually prevents the release of acetylcholine at neuromuscular and PNS synapses. This is also a blocking agent.

Depolarizing Agents

Toxons may increase the membrane permeability to sodium, preventing polarization. Batrachotoxin (frog alkaloid) is an example of a toxon that functions by this mechanism.

Dichlorodiphenyl trichloroethane (DDT) and pyrethrins increase sodium permeability at the synapse. This effect exaggerates each impulse with repeated firings, eventually leading to convulsions.

Stimulants

Strychnine prevents the inhibitory neuronal activity of glycine, a competitive antagonist at postsynaptic sites, leading to CNS excitability.

Picrotoxin (Anamirta cocculus, fishberries) is a competitive antagonist to the inhibitory neurotransmitter Gamma Amino Butyric Acid (GABA) in the CNS.

Methylated xanthines (caffeine, theophylline, theobromine) prevent the breakdown of **cyclic AMP**. C-AMP controls the activity of the sodium-potassium ATPase pump and, thus, polarization levels.

Depressants

Volatile organic solvents affect membrane fluidity and, thus, fluxes of sodium, potassium, and calcium.

Alcohol decreases sodium and potassium conductance in the CNS, which serves to block impulse conduction.

Barbiturates depress neuronal metabolism and respiration to decrease neuronal activity in general and also to decrease synaptic neurohumoral transmitter release.

Receptor Antagonists

Anticholinergic compounds (atropine, belladonna, acopolamine) competitively bind with the cholinergic postsynaptic receptors to prevent response.

Antiadrenergic compounds (phenoxybenzamine, phentolamine, propranolol, and tolazoline) bind with the adrenergic postsynaptic receptors to prevent response.

Anticholinesterase Agents

Organophosphate insecticides and nerve gases (diazinon, malathion, and parathion) bind irreversibly to acetylcholinesterase, preventing the breakdown of acetylcholine.

Carbamate insecticides (aldicarb, carbaryl, and sevin) bind reversibly to acetylcholinesterase, causing temporary breakdown prevention.

Anticholinesterase inhibitors (adrophonium, neostigmine, and physostigmine) competitively inhibit acetylcholinesterases.

Neuromuscular Blocking Agents

Curare is a competitive postsynaptic acetylcholine antagonist at the neuromuscular junction.

Succinylcholine causes a persistent depolarization of the muscle cell membrane.

Anoxia

The neurons within the CNS have a high metabolic rate with very limited anaerobic metabolic reserve. Therefore, loss of oxygen by asphyxiation or by loss of ATP, such as by cyanide or hydrogen sulfide poisoning, is rapidly manifested by loss of function.

Demyelinating Agents

Agents (such as cyanates, hexachlorophene, lead, salicylanilides, and thallium) which disrupt or destroy the myelin sheath cause dullness, restlessness, muscle tremor, convulsions, loss of memory, epilepsy, idiocy, neuritis, palsy, muscle weakness, sensory disturbances, and hair loss. Multiple sclerosis (MS) is a disease of the myelin sheath.

Anti-PNS agents

A number of substances (arsenic, carbon disulfide, dinitrobenzene, ethylene glycol, hexane, methanol, methyl mercury, tetraethyl lead) can affect the PNS causing weakness of the lower extremities, abnormal limb sensation, visual and hearing disturbances, irritability, and loss of coordination.

Lesioning Agents

A number of substances (acetylpyridine, DDT, manganese, and mercury) may cause necrosis to the brain cells resulting in convulsions, personality disorders (mad as a hatter), and loss of fine motor coordination (Parkinson's, disease).

Evaluation of Injury to the Nervous System

Assessment of nervous system damage is difficult and requires a functional evaluation. Evaluatory steps include the following:

- Review of patient history
- Evaluation of mental status
- Evaluation of cranial nerve function
- Evaluation of motor system and reflex function
- Evaluation of sensory systems
- Observation of stance and gait
- Coordination testing
- Blood, urine, and **Cerebral Spinal Fluid (CSF)** testing
- Biopsy and histopathological examination

Blood

Blood is comprised of **formed elements** (40-60%) and blood **plasma** (60-40%). The formed elements include the **red blood corpuscles (RBCs)**, **erythrocytes**; the **white blood cells (WBCs)**, **leukocytes**; and the **platelets**, **thrombocytes**. Plasma is mostly water with a variety of dissolved proteins, sugars, and salts.

The formed elements are produced from stem cells found in the bone marrow, liver, and in lymphatic tissues. RBCs are primarily responsible for oxygen transport. Anemia is a decrease in the number of RBCs or the oxygen transport molecule, hemoglobin. WBCs protect against foreign materials, scavenge debris, and are an integral part of the immune system. Leukopenia is a decreased number of WBCs. Platelets provide for blood clotting to protect against hemorrhage. Thrombocytopenia refers to a decreased number of platelets. Poietins is a hormone that controls and regulates the production of the formed elements by their particular stem cell lines. (See Table 7-1 for list of chemicals affecting blood.)

Table 7-1

Chemicals affecting blood

Acetaminophen	Dextropropoxyphene	Pamaquin pantaquin
Acetanilide	Diazepam	Paracetamol
Acetazolamide	Diethylstilbestrol	Penicillin
Alkylating agents	Digitoxin	Phenacetin
Amidopyrine	Dimercaprol	Phenobarbital
Amitriplyline	Dinitrophenol	Phenylbutazone
Ampicillin	Disulfiram	Phenylhydrazine
Antozoline	Ethacrynic Acid	Potassium iodide
Arsenicals	Furazolidine	Potassium perchlorate
Arsine	Gold compounds	Primaquine
Arsphenamine	Hydroxyquinolone	Probenecid
Aspirin	Indomethacin	Procainamide
Barbiturates	Insecticides	Propylthiouracil
Benzene	Insulin	Quinacrine
Bismuth	Isoniazid	Quinidine
Busulfan	Lead	Quinine
Butyl cellusolve	Lindane	Rifampicin
Carbon tetrachloride	Mepacrine	Salicylates
Carbutamide	Mephenytoain	Stibophen
Cephaloridine	Meprobamate	Stilbestrol
Cephalosporins	Mercurial diuretics	Streptomycin
Chloramphenicol	Mercurials	Sulfa drugs
Chlordane	Methimazole	Tetracycline
Chloroquine	Methyl chloride	Thiourea
Chlorpromazine	Methylene blue	Tolbutamide
Colchicine	Naphthalene	Toluene dyisocyanate
Corticosteroids	Nitrobenzene	Toluidine blue
Cyclophosphamide	Nitrofurantoin	Trinitrotoluene
Dapsone	Nitrogen mustards	Vitamin K
DDT	Oxyphenbutazone	(water-soluble analogs of)
Desipramine		

Erythrocytes

Erythropoiesis, the process whereby RBCs are produced, is controlled by Renal Erythropoietic Factor (REF). REF is produced in response to hypoxic conditions in the presence of cobalt. Production of REF is halted in hyperoxic conditions. Iron and cobalt deficiencies inhibit RBC production. Chemicals such as surfactants destroy the membrane of the RBC. Deficiencies in folic acid or vitamin B_{12} and bone marrow dysfunction inhibit RBC production.

Numerous substances inhibit the transport of oxygen by the hemoglobin. CO competes with oxygen for the hemoglobin site. Nitrites, nitrates, aromatic amine, and chlorate compounds oxidize the iron in the form of hemoglobin forming methemoglobin. Heinze bodies are clumps of denatured hemoglobin, covalently bound to the interior of the RBC membrane. These are caused by arsine, chlorates, methylene blue, naphthalene, phenylhydrazine, and primaquine.

Thrombocytes

Platelets are small corpuscles which are the first line of defense against blood loss. They work in conjunction with blood proteins, like fibrinogen, to produce blood clots. A number of anticancer drugs affect platelet production levels. Warfarin prevents fibrin formation, and acetylsilic acid (aspirin) inhibits platelet aggregation.

Leukocytes

There are five WBC populations, each with its own special tasks. Many are phagocytic, i.e., eaters of cellular debris and invading organisms. Several are involved in the general immune response, inflammation, pain, and fever. Several are involved in a specific immune response, the antigen- antibody complex formation. Chemicals such as benzene, chloramphenicol, and butazone cause leukemia, a cancerous proliferation of WBCs.

Endangered Liver Areas

Centrilobular Hepatotoxins	Midzonal Hepatotoxins	Periportal Hepatotoxins
Acetaminophen	Anthrapyrimidine	Acrolein
Aflatoxin	Beryllium	Albitocin
Bromobenzene	Carbon tetrachloride	Allyl alcohol
Carbon tetrachloride	Furosemide	Arsenic
Chloroform	Ngaione	Iron
DDT	Paraquat	Manganese
Dinitrobenzene		Phosphorous
Trichloroethylene		

Table 7-2

Hepatotoxic agents

Endangered Liver Cells

Hepatocytotoxic	Cholestatic
Acetaminophen	Anabolic steriods
Aflatoxin	Arsphenamine
Allyl alcohol	Chlorpromazine
Bromobenzene	Diazepam
Carbon tetrachloride	Estradial
Dimethylnitrosamine	Mepazine
Phosphorous	Thioridazine
Urethane	

Table 7-3

Hepatotoxic agents

Dyscrasias

Over 1000 substances have been identified as being responsible for one or several types of blood disorder.

Evaluation of Hematoxicity

A variety of routine diagnostic tests are available to analyze blood. These include tests to measure the following:

- **Hematocrit**: percentage of RBC volume
- **Hemoglobin**: concentration of hemoglobin in RBCs
- **RBC count**: RBCs per microliter of blood
- **WBC count**: WBCs per microliter of blood
- **Platelet count**: platelets per microliter of blood
- **Blood plasma component measures**: glucose, urea, lead

Liver

The liver (Figure 7-2 and 7-3) is the largest gland in the body and has seven distinct sets of functions. (See Table 7-4) All of the blood supplied to the digestive tract passes via the portal system to the liver. All of this blood is run past the cells of the liver for processing. Processed blood is, subsequently, sent to the heart. Material removed from the blood, biotransformed, or produced by the liver cells, is either stored or secreted into the bile canaliculi, to become bile.

Many chemicals are known to induce liver damage. Centrilobular hepatotoxins include bromobenzene, carbon tetrachloride, chloroform, and DDT. Midzonal hepatotoxins include beryllium, carbon tetrachloride, and paraquat. Periportal

Figure 7-2

Schematic representation of a liver lobule from the external portal vein blood entrance to the internal central vein blood exit. Different toxons have been determined to have effects specific to zonal, periportal, midzonal, or centrilobular areas.

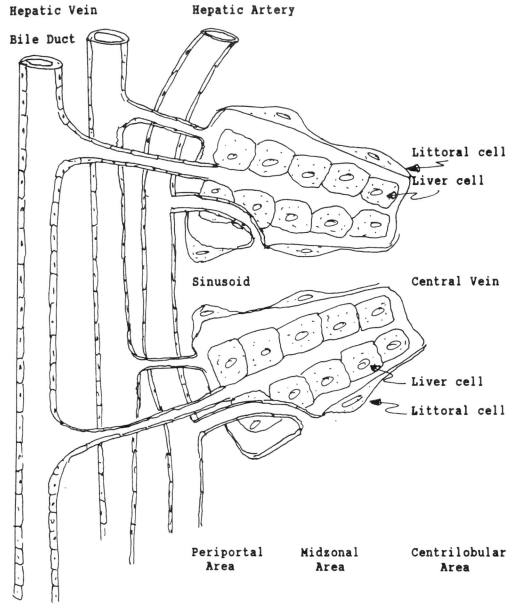

hepatotoxins include acrolein, arsenic, iron, and phosphorous. Necrotic agents include carbon tetrachloride, phosphorous, and urethane. Cholestatic agents include anabolic steroids, chlorpromazine, diazepam, and thioridazine. Carbon tetrachloride, dimethylnitrosamine, and thioacetamide damage the cell membrane and the endoplasmic reticulum. Dichloroethylene, hydrazine, and phosphorous damage the mitochondria. Aflatoxin, beryllium, galactosamine, and nitrosamines damage the nucleus.

Cirrhosis

Cirrhosis of the liver is a progressive disease characterized by diffuse damage to the cells with nodal regeneration and fibrosis. The resulting liver dysfunction is frequently manifested by jaundice and portal hypertension.

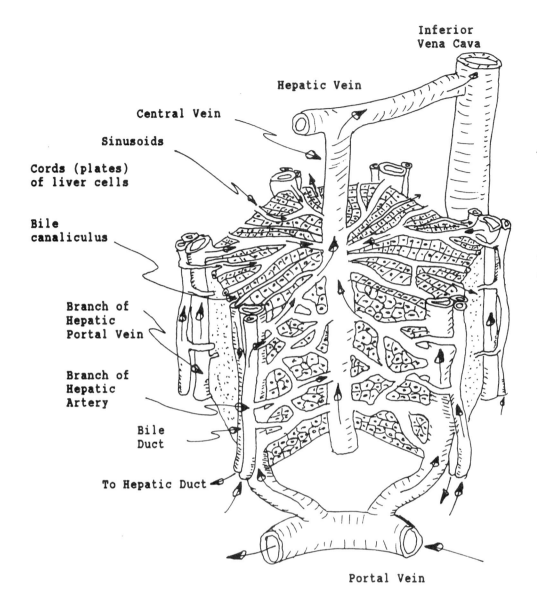

Inferior
Vena Cava

Hepatic Vein

Central Vein

Sinusoids

Cords (plates)
of liver cells

Bile
canaliculus

Branch of
Hepatic
Portal Vein

Branch of
Hepatic
Artery

Bile
Duct

To Hepatic Duct

Portal Vein

Figure 7-3

Microscopic anatomy of a liver lobule. Blood flows from the digestive system via the portal vein to the liver. Portal blood is mixed with oxygenated blood from the hepatic artery to flow through the sinusoids of the liver for processing. Blood exits the lobule via the central vein, which coalesces into the hepatic vein. The hepatic vein joins the inferior vena cava which delivers blood from the lower portion of the body to the heart.

Evaluation of Liver Injury

A number of tests are used to diagnose the state of the liver. Dye clearance tests, such as bromosulfophthalein and indocyanine green, monitor clearance times. Prothrombin clot time is an indication of the ability of the liver to produce blood clotting proteins. Serum albumin tests monitor the ability of the liver to produce proteins, such as albumin. Bilirubin levels, and thus clearance rate, is a measure of liver function. The presence of certain enzymes, in the blood, e.g., aminotransferases and serum alkaline phosphatase, indicate a leaking and defective liver cell's membrane, and therefore, potential injury.

Heart

The heart (Figures 7-4 and 7-5) is a four-chambered organ responsible for pumping blood throughout the body. Two of these chambers work together: the atrium stores incoming blood while the ventricle pumps the blood out. One set, on the

Table 7-4

The functions of the liver

Liver Functions

1. Excretory and degradative functions
   A. Biotransforms many endogenous and foreign organic molecules
   B. Excretes, via the bile, many endogenous and foreign organic molecules and trace metals
   C. Secretes bilirubin and other bile pigments into the bile
   D. Destroys old erythrocytes

2. Digestive functions
   A. Synthesizes and secretes bile acids, which are necessary for adequate digestion and absorption of fats
   B. Secretes into the bile a bicarbonate-rich solution to help neutralize acid in the duodenum

3. Organic metabolism
   A. Converts plasma glucose into glycogen and triacylglycerols
   B. Converts plasma amino acids to fatty acids, which can be incorporated into triacylglycerols
   C. Converts fatty acids into detones during fasting
   D. Synthesizes triacylglycerols and secretes them as lipoproteins
   E. Produces glucose from glycogen and other sources and releases it into the blood
   F. Produces urea by catobolism of amino acids, or proteins, and releases it into the blood

4. Cholesterol metabolism
   A. Synthesizes cholesterol and releases it ino the blood
   B. Converts plasma cholesterol into bile acids
   C. Secretes plasma cholesterol into the bile

5. Clotting functions
   A. Produces many of the plasma clotting factors, including prothrombin and fibrinogen
   B. Produces bile salts which promote absorption of Vitamin K from the intestinal tract

6. Endocrine functions
   A. Secretes insulin-like growth factor I (IGF-1) which promotes mitosis and bone growth
   B. Contributes to the activation of Vitamin D
   C. Forms triiodothyronine from thyroxine
   D. Secretes angiotensinogin, which is acted upon by renin to form angiotensin
   E. Metabolizes hormones

7. Plasma protein synthesis
   A. Albumin
   B. Acute phase proteins
   C. Binding proteins
   D. Lipoproteins

right side, pumps blood towards the lung, while the other set, on the left side, pumps blood out to the body. The heart utilizes a specialized type of cardiac muscle called the **myocardium**, which has the dual capacities of speed and endurance. Specialized cardiac muscle cells, **Purkinje fibers**, are responsible for conduction impulses throughout the heart muscle. A nucleus of these cells forms a node called the **pacemaker**, which sets the pace of the heartbeat. The impulse passes to the atrioventricular (AV) node located on the AV wall where it is

Figure 7-4

Excitation and conduction system of the heart

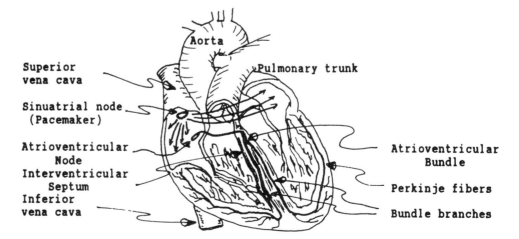

Superior vena cava
Sinuatrial node (Pacemaker)
Atrioventricular Node
Interventricular Septum
Inferior vena cava
Aorta
Pulmonary trunk
Atrioventricular Bundle
Perkinje fibers
Bundle branches

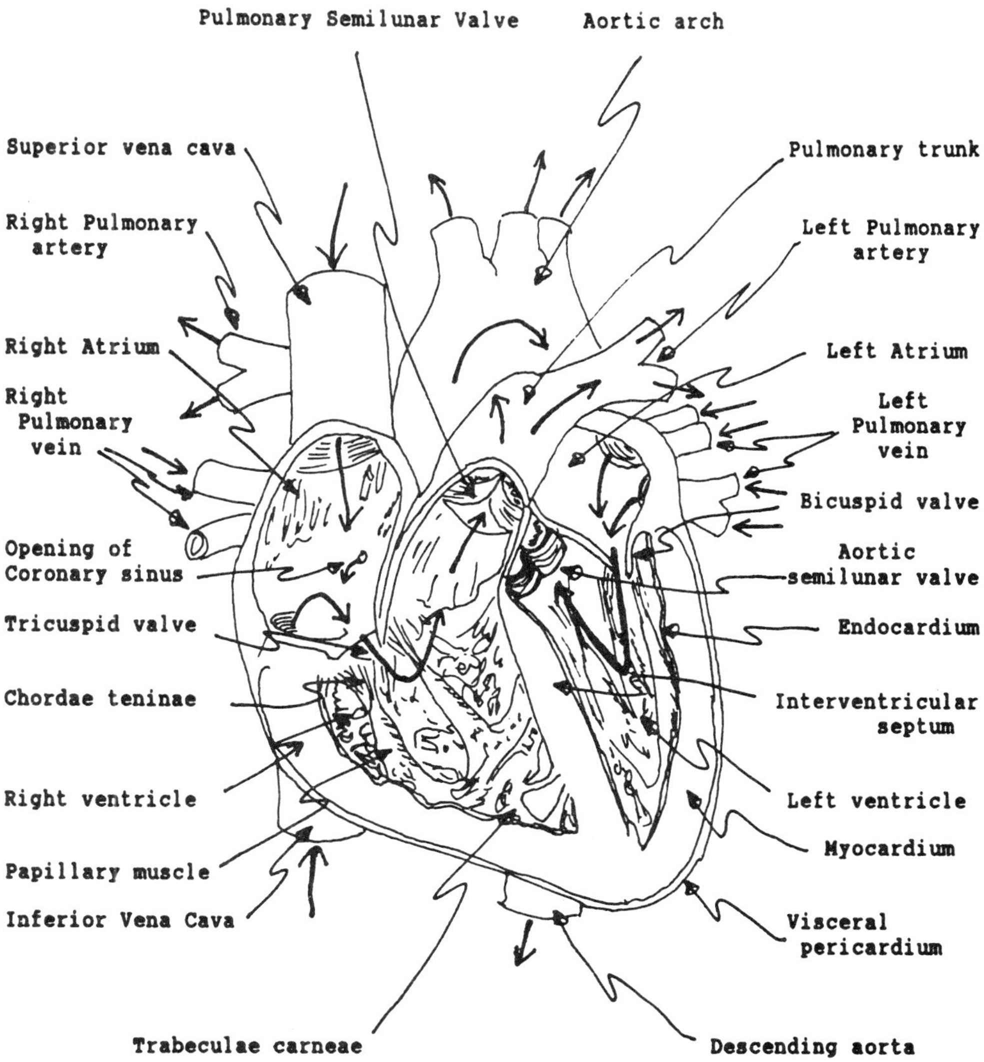

Figure 7-5
Frontal section of the heart

Pulmonary Semilunar Valve

Aortic arch

Superior vena cava

Right Pulmonary artery

Right Atrium

Right Pulmonary vein

Opening of Coronary sinus

Tricuspid valve

Chordae teninae

Right ventricle

Papillary muscle

Inferior Vena Cava

Trabeculae carneae

Pulmonary trunk

Left Pulmonary artery

Left Atrium

Left Pulmonary vein

Bicuspid valve

Aortic semilunar valve

Endocardium

Interventricular septum

Left ventricle

Myocardium

Visceral pericardium

Descending aorta

coupled to the ventricles. Bundles of fibers, called the **bundles of His,** lead from the AV node and cause the heart muscle to contract in an organized manner in order to pump blood. If the conduction throughout the heart is not exquisitely orchestrated, then the pumping action becomes inefficient and impeded.

The electrical activity of the pacemaker is controlled by Na^+, K^+, and Ca^{++} ions, their concentrations, and membrane pumps, gates, and permeability for these ions. Alterations of any of the factors can cause arrhythmias. Certain metals, such as barium, manganese, nickel, lead, cadmium, and cobalt, adversely impact the heart by disrupting electrical or metabolic activity. Lipid soluble organic solvents accumulate in cardiac cells to alter generation and conduction of action potentials. A number of gases, such as carbon monoxide and methane, impact tissue oxygen availability and, therefore, cell metabolism and operation.

The speed of the heartbeat is affected by hormones such as adrenaline, which increases heart rate, and SNS nerves, which depress heart rate. Adrenergic and cholinergic mimetics can thus affect heart rate. Potassium loss (vomiting, diarrhea, and diuresis) leads to cardiac arrhythmia (loss of heartbeat). Digitalis (foxglove plant) increases the force of contraction of the myocardium leading first to **tachycardia,** excessively rapid heartbeat, and subsequently to **atrial or ventricular fibrillation,** contractions without actual pumping of blood.

Quinidine, a water-soluble alkaloid (found in cinchona bark) depresses the metabolic activities of all cells, but is most pronounced in the heart where it may slow conduction, prolong refractory period, cause heart block, and cause ventricular fibrillation.

Thiocyanates may cause myocardial necrosis.

Nitrites and nitrates dilate coronary vessels and cause **hypotension** (low blood pressure).

Kidneys

The kidneys are paired organs located **retroperitoneally** on each side of the spine in the small of the back. They receive approximately 25% of the cardiac output. This amounts to about 1.25 liters per minute. Of the blood flow, approximately 125 ml is filtered through the glomerulus and less than 1 ml actually becomes urine. Due to this large blood flow, substances in the blood are found in high concentration in the kidneys. Since about 10% of resting oxygen consump-

Figure 7-6

A nephron, the functional unit of the kidney

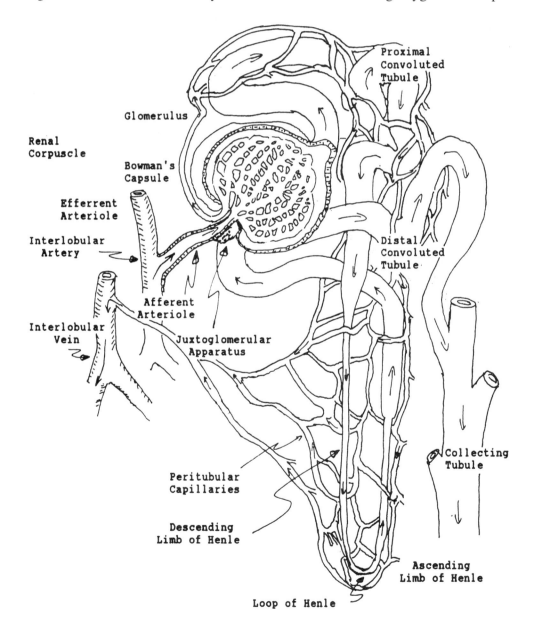

tion is required for proper kidney function, the kidney is sensitive to agents reducing oxygenation. Barbiturate intoxication causing hypotension results in kidney **ischemia** and ultimately **renal** failure and death.

The functional unit of the kidney is the **nephron** (Figure 7-6). Toxons affecting the nephron are called nephrotoxins. The nephron consists of a **glomerulus** apparatus, the **proximal convoluted tubule**s, the **Loop of Henle**, the **distal convoluted tubules**, and the **urine ducts**.

The kidney has a unique view of its job of purifying the blood and excreting wastes; it first filters plasma from formed elements and then resorbs a few specific substances. The process starts in the glomerulus apparatus by filtering about 20% of the plasma from the blood. Subsequently, in the proximal tubule, it resorbs the sugars and proteins. In the Loop of Henle, it resorbs water, and finally, in the distal convoluted tubules, it adjusts inorganic ion (sodium and potassium) concentrations and pH. As previously discussed, pH affects the resorption of substances, depending upon their pK_a, here in the distal convoluted tubules. The final water concentration of the urine is adjusted in the collection ducts. In general, whatever is left within the tubules goes on to become urine.

Cadmium

The kidney is the organ most sensitive to the toxic effects of cadmium, where it interferes with protein resorption by the proximal convoluted tubule. **Proteinuria**, protein in the urine, is a symptom of such damage. Cadmium is sequestered by the kidney and liver.

Mercury

Like cadmium, mercury affects the proximal tubule, initially causing **polyuria** (large amounts of urine), and proteinuria followed by **anuria** (no urine), edema, and death.

Lead

Lead likewise affects the proximal tubules, leading to **glucosuria** (glucose in the urine), **aminoaciduria** (amino acids in the urine), and **hyperphosphaturia** (excess phosphates in the urine). Short-term changes are reversible with chelation; long-term exposure causes irreversible interstitial fibrosis and tubular atrophy and dilation.

Halogenated Hydrocarbons

Carbon tetrachloride, chloroform, bromobenzene, tetrachloroethylene, and trichloroethylene all affect the proximal tubule function. Additionally, they may cause general necrosis.

Obstructive Agents

Some chemical agents can be concentrated in the tubular fluids to levels above their solubility and, thus, they may crystallize, obstructing the passage and causing physical damage. Methotrexate and sulfonamide drugs are such nephrotoxins. Ethylene glycol is metabolized to oxalic acid by the body, becoming an insoluble calcium oxalate salt in the tubules. Oxalate, found in the leaves of plants (e.g., rhubarb), may cause nephrotoxicity.

Acute Events

Arsine gas causes massive **hemolysis** (cell destruction) of RBCs with an associated **hemoglobinuria** (hemoglobin in the urine) and renal failure. Heroin may cause direct lysis of muscle cells with a massive release of **myoglobin**. Streptococcal death and lysis by antibacterial agents may lead to a bolus of streptococcal toxin being released into general circulation; unfortunately, this is nephrotoxic and may lead to acute **nephritis** (inflammation and dysfunction of the nephron).

Immune System

The immune system consists of the WBCs and several organs such as the spleen and the thymus. The agents impacting WBCs, previously discussed, depress the immune system and its function.

Immunosuppressants are used for some cancer therapies and to prevent organ transplant rejection. Mercaptopurine and thioguanine are used in cancer therapy. The vinca alkaloids (periwinkle plant) depresses WBC formation. Azathioprine is used as an adjunct for renal transplants. The adrenocorticosteroids are Immunosuppressive.

Over two thousand chemicals are used as food additives. A number of these, such as aspartame, the benzoates, butylated hydroxyanisole (BHA) and butylated hydroxytoluene (BHT), food dye and colorings, monosodium glutamate (MSG), nitrates, nitrates, methyl, ethyl, propyl, butyl parabens and sodium benzoates, and sulfites have been linked to adverse reactions in sensitive individuals. These chemicals may act synergistically with other chemical substances to change effective doses or to make a worker more sensitive to other chemicals. Therefore, what you eat for lunch may affect how you react to exposure on the afternoon shift. For example, eating Chinese, Japanese or southeast Asian cooking, with a high dose of MSG, may sensitize a person for a headache which may be triggered by a whiff of solvent. The American Academy of Allergy, Asthma and Immunology maintains that exposure to food additives and environmental chemicals is the leading cause of sensitization for allergies and asthma.

Study Exercises

1. Describe the following aspects of the nervous system:
   - its general organization
   - the blood-brain barrier
   - the three functional categories of neurons
   - the symptoms associated with narcosis
   - the three types of neural damage caused by neurotoxins and the clinical manifestations

2. Describe the blood, including the following:
   - the general composition of whole blood
   - the principal types of formed elements and their functions
   - the role of blood platelets
   - the types of hemotoxins and their effects

3. Describe the liver, including the following:
   - seven principal functions of the liver
   - the clinical symptoms of toxic hepatitis
   - the most common types of hepatotoxins
   - medical diagnosis of toxic hepatitis

4. Describe the functions of the heart, including the following:
   - the role of the pacemaker in synchronizing the heart beat
   - the effect of cardiotoxins on the patterns of the heart beat

5. Describe the kidney, including the following:
   - the principal functions of the kidneys
   - the clinical symptoms of renal failure
   - the most common types of nephrotoxins

6. Describe the immune system, including the following:
   - the general functions of the immune system
   - the functional components of the immune system and their roles in the immune response
   - the effects of immunotoxins on the immune response

7. Explain how certain metals might negatively impact the kidney's function and how these effects might cause anemia and the appearance of adverse symptoms related to nerve cells and the heart.

8 Reproductive Toxins, Mutagens, and Carcinogens

Overview

This chapter introduces the mechanisms of actions of mutagens, carcinogens, and teratogens. Although these agents can affect all body systems, the reproductive systems are at particular risk. Therefore, the general organization and functions are provided for the male and female reproductive systems. Several specific classes of toxic substances are identified which are known teratogens, mutagens, and carcinogens.

Mutagens

The command, control, communication, and intelligence (C^3I), of the cell is run by the nucleic acids. These come in two forms, depending upon the associated sugars, ribose (RNA) and deoxyribose (DNA) (Table 8-1). Substances that mess with the C^3I can play havoc with the cell; these substances are known as mutagens. **Mutagens** are responsible for mutations. A **mutation** is a change in the nucleic acid sequence, which changes the genetic code of a cell, resulting in some message change. A **codon** is a sequence of 3 DNA base pairs.

Double Strand DNA		DNA	Single Strand RNA	
A	–	T	A	
G	–	C	G	Codon 1
C	–	G	C	
T	–	A	U	
A	–	T	A	Codon 2
C	–	G	C	
A	–	T	A	
G	–	C	G	Codon 3
T	–	A	U	
T	–			
A	–			Codon 4
C	–			

(Fill in the correct sequence)

A Frame Shift Mutation with one missing pair				
A	–	T	A	
G	–	C	G	Codon 1
Missing Base Pair				
T	–	A	U	
A	–	T	A	Codon 2
C	–	G	C	
A	–	T	A	
G	–	C	G	Codon 3
T	–	X	Y	
		(A point mutation)		
T	–			
(Thymine dimer)				Codon 4

Table 8-1

DNA is a double stranded molecule. The principal bases are Adenine (A), Thymine (T), Guanine (G), Cytosine (C) and Uracil (U). For DNA they pair A-T and G-C. For RNA the T is replaced by U. Three DNA base pairs represent a codon, which codes for a particular amino acid

Functional Groups

The properties of organic compounds are highly influenced by their functional groups. Hydrocarbons are molecules made out of hydrogen and carbon, such as oil. Adding a hydroxyl group turns a gas such as ethane into an alcohol like ethanol, grain alcohol, which is found in beer, wine, and whiskey. Carbonyl groups such as aldehydes, ketones, and carboxyl are found in sugars. A carboxyl group is an organic acid, bestowing acidic properties on a molecule. Amines, which are basic, bestow alkaline properties, and are linked with carboxyl groups to create amino acids. Phosphates are used for energy storage, ATP, and for hooking molecules together such as with DNA and RNA (Figure 8-1).

Figure 8-1

Protein Synthesis. The information contained in DNA is read by a strand of messenger RNA (mRNA). The mRNA leaves the nucleus through a pore in the nucleus and travels to the rough Endoplasmic Reticulum (rER) where protein is synthesized. The **rough** *in rER is a unit called the Ribosome. It reads the mRNA and attaches a fitting transfer RNA (tRNA) group with an associated amino acid. A string of amino acids is called a polypeptide chain. A large polypeptide is what is known as a protein. One gene, or segment of DNA, codes for one protein. If the process is disturbed no protein, or a dysfunctional protein, may be produced. Such a mutation may be good, bad, or indifferent. The general probability is that it will be bad.*

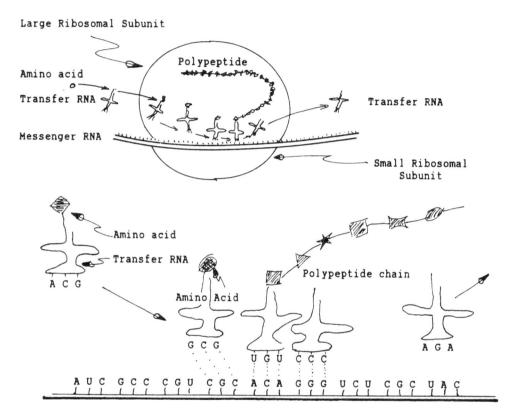

Codons and Information

One codon codes for a particular amino acid or set of actions. The alteration of the codon, mutation, generally causes a change, or substitution, of amino acids. The net effect of any message alteration to C^3I must be assessed on an individual basis. Mutations can be good, bad, or indifferent. (Figure 8-2) Evolution is an example of a possible beneficial aspect. Cancer and teratogenic effects tend to be negative. The majority of mutations tend to have no effect or impact.

Genetic toxicology is the study of the physical and chemical substances which induce changes in the hereditary process. Changes may occur either to the body (**somatic**), or to the heritage and **progeny** (**germ**). The somatic cell mutation theory of cancer hypothesizes that mutations cause cells to change such that they undergo rapid, uncontrollable, and uninhibited division, multiplying to the general detriment of the organism. The germ cell mutation passes bad information or

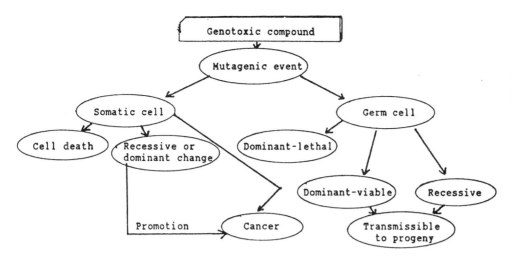

Figure 8-2

Possible consequences of mutagenic event in somatic and germinal cells.

misinformation to the progeny, creating monsters, **terata**, or causing cancer. **Teratogens** are substances that create **terata**. Thus, somatic cell mutations affect the organism while germ mutations affect the progeny, or future generations.

Both physical and chemical mutagens occur naturally in the environment. Darwin's theory of evolution depends upon mutations for the diversity of species and for the variety of life-forms that we know today. Coupled with natural selection, often termed survival of the fittest, we have a plausible explanation of how life has evolved to the point where we observe it today. Radiation passing through the atmosphere and free radicals (reactive chemical species) are the major factors of natural mutagenesis. Benzpyrene is a naturally occurring and potent **carcinogen**, (a mutagen that causes cancer), which is omnipresent in our environment. It is a by-product of the burning, or cooking, of any organic mate-

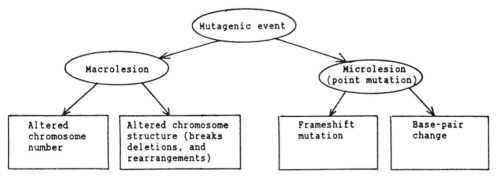

Figure 8-3

Types of mutagenic changes (Source: Brusick, Principles of General Toxicology, New York: Plenum Press, 1980).

rial. Thus, mutagens are a two-edged sword: mutations encourage evolution, allowing us to evolve to where we are today, while simultaneously, cutting down many in the midst of life. (Figure 8-3)

Three basic events may cause a transmissible change in the nucleic acids:

1. Infidelity in DNA replication
2. Point mutations
3. Chromosomal aberrations

Infidelity in DNA replication or the copying of a message may result from a variety of causes. There may be an inaccurate initiation of replication as may happen when the read does not start in the correct place. There may also be a failure of the transcription enzymes to accurately read the DNA. Therefore, the

Reproductive Toxins, Mutagens, and Carcinogens   87

information passed is not faithful and, thus, bad. Interruptions of the transcription process may also occur. This is caused by agents that intercalate themselves within the DNA molecule or between the DNA and the transcription enzymes. Such interruptions result in missing chunks of information. Bad information disrupts the C³I functions and tends to negatively impact the cell.

Point mutations may be either changes in the base-pairs or frameshift mutations. Base-pair changes result from either transition or transversion of the DNA base pairs. In these cases, the number of bases is unchanged but the sequence is altered. Such an event results in the corruption of a single codon. A frameshift mutation results from the insertion or deletion of one or more base pairs from the linear sequence of the DNA. This results in the displacement of the transcription process by the corresponding number of bases and, thus, causes an alteration of the subsequent codons.

Chromosomal aberrations may be present in the form of breaks or gaps in the chromatids, exchanges of corresponding segments between arms of a chromosome, or asymmetrical interchanges between chromosomes. Chromosomal aberrations passed on through germ cells can have dire consequences, such as embryotoxicity, congenital malformations, growth retardation, or mental retardation.

Carcinogens

Cancer is a process by which cells undergo some basic change that allows them to grow without limit and without contact inhibition. **Malignant** cancers exhibit these characteristics:

1. **Cell proliferation**: growth much more rapid than that of their normal parent cells
2. **Loss of differentiation**: reversion to more primitive cell line status with the loss of some of the features typical of their parent cells
3. **Metastasis**: invasion and destruction of adjacent tissues that spread to distant locations where they establish secondary cancer foci

Benign cancer tumors are not malignant. Generally, this is because they have not metastasized. Whether or not benign tumors have the potential to become malignant is the subject of much debate. The early stages of both malignant and benign tumor development is similar. It is hypothesized that, if an organism lived

Figure 8-4

Mutagenic event ramifications

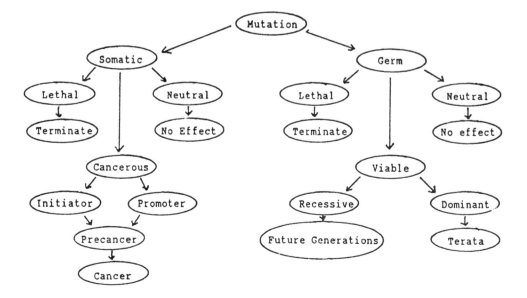

long enough, benign tumors would eventually become malignant. Regulatory agencies have resolved the issue by classifying all chemicals that produce **neoplasm** (benign or malignant) as carcinogens.

Cancer is a multifaceted process that may evolve in any number of ways. Chemical-induced carcinogenesis is hypothesized to involve at least three distinct stages. The first stage is **initiation**. This leads to a permanent genetic alteration, or mutation, in the cell. This cell is capable of **phenotypic** expression in the somatic cell and changes; i.e., the cell loses some of the characteristics of the normal cell. The second stage is a **promotional** stage, where physiological and biochemical changes facilitate the growth and expression of the initiated cell. It appears to demonstrate a dose-response relationship. The third stage is the **progressional** stage, which like the initiation stage, is thought to be irreversible. It is characterized by rapid cell growth, invasion of normal tissue, increased frequency of metastases, lack of response to environmental factors, and lack of cellular differentiation. Unrestrained growth and expression of a cell line is cancer. (Figure 8-4)

This stage distinction is used to explain two observable aspects of carcinogenesis. The first is the dormant nature of cancer—it is believed that the manifestation of neoplasms occurs years, or even decades, after the exposure phase. The second is the belief that there are a number of different mechanisms by which chemicals may increase the incidence, or expression, of tumors.

Carcinogens have been divided into two broad groups, genotoxic and epigenetic. **Genotoxic** carcinogens are those substances that act by directly altering DNA or genetic expressions (initiators). Some examples of these are the organics—nitrosamines, benzanthracene, epoxides, dimethyl sulfate, and nitrosoureas—and the inorganic metals—cadmium, chromium, and nickel. (See Figure 8-5) **Epigenetic** carcinogens are substances not involved in the direct interaction with genetic material, promoters. Examples of these are asbestos, estrogens and androgens, azathioprine, ethanol, solvents, catechol, phorbol esters, and tetrachloroethylene.

There are four basic types of cancers:

1. **Leukemias**: cancers associated with the WBC and their stem lines
2. **Lymphomas**: cancers associated with the lymphatic system, such as Hodgkin's disease
3. **Sarcomas**: cancers associated with connective and muscle tissues
4. **Carcinomas**: cancers associated with epithelial tissues—ectodermal, mesodermal, or endodermal

In terms of industrial toxicology, however, it is advantageous to look at cancers associated with particular organs or organ systems. (Figure 8-6) Some of these are described below.

Lung Cancer

Lung cancers are technically bronchogenic carcinomas and pulmonary adenocarcinomas. The smoking of tobacco is considered to be the primary cause of bronchogenic carcinomas. Metal fumes of cadmium, nickel, and chromium are cancer promoters. Pneumoconiosis, silicosis, asbestosis, and anthoconiosis are known promoters of cancer.

Urinary/Bladder Cancer

Cancers of the bladder are typically epithelial carcinomas. Prolonged contact between the concentrated toxon and the transitional epithelium occurs during

Figure 8-5

Representative carcinogenic compounds

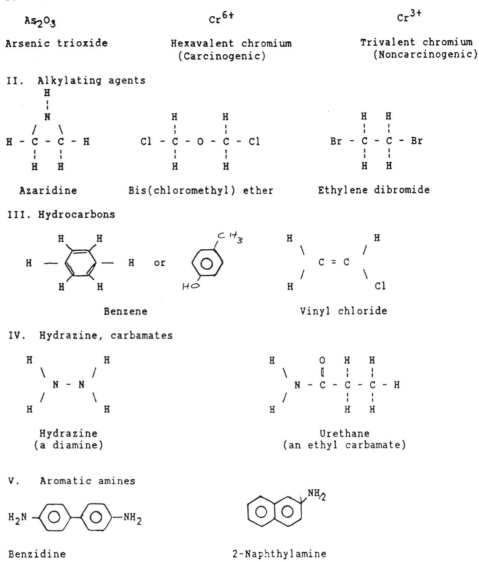

I. Metals

As₂O₃ — Arsenic trioxide

Cr⁶⁺ — Hexavalent chromium (Carcinogenic)

Cr³⁺ — Trivalent chromium (Noncarcinogenic)

II. Alkylating agents

Azaridine

Bis(chloromethyl) ether

Ethylene dibromide

III. Hydrocarbons

Benzene

Vinyl chloride

IV. Hydrazine, carbamates

Hydrazine (a diamine)

Urethane (an ethyl carbamate)

V. Aromatic amines

Benzidine

2-Naphthylamine

VI. Unsaturated nitriles

Acrylonitrile

urine storage. Polycyclic aromatic hydrocarbons and the active metabolites of benzidine are stored in the bladder.

Liver Cancer

Hepatocellular (hepato = liver) carcinomas are characterized by hepatomegaly with multiple scattered nodules and large malignant tumors throughout the liver. Many nontoxic chemicals, the procarcinogens, are metabolized into reactive carcinogenic chemicals. An example of this is benzidine, which is metabolized to an active form in the liver. Polycyclic aromatic hydrocarbons have been associated with liver carcinomas. Cancers of the biliary tract are cholangiocarcinomas.

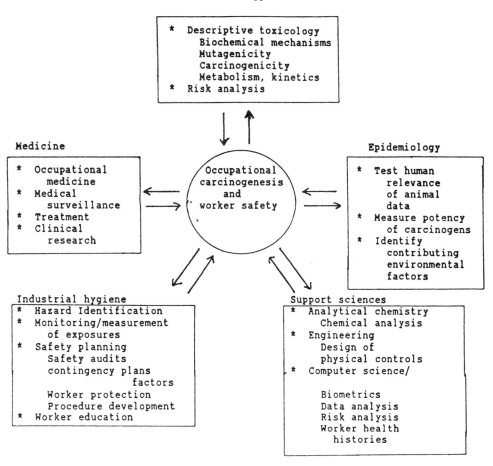

Toxicology

* Descriptive toxicology
  Biochemical mechanisms
  Mutagenicity
  Carcinogenicity
  Metabolism, kinetics
* Risk analysis

Occupational carcinogenesis and worker safety

Medicine

* Occupational medicine
* Medical surveillance
* Treatment
* Clinical research

Epidemiology

* Test human relevance of animal data
* Measure potency of carcinogens
* Identify contributing environmental factors

Industrial hygiene

* Hazard Identification
* Monitoring/measurement of exposures
* Safety planning
  Safety audits
  contingency plans
                factors
  Worker protection
  Procedure development
* Worker education

Support sciences

* Analytical chemistry
  Chemical analysis
* Engineering
  Design of physical controls
* Computer science/

  Biometrics
  Data analysis
  Risk analysis
  Worker health histories

Figure 8-6

Identifying and reducing chemical carcinogens requires an interdisciplinary approach, in which health professions interface with other scientific disciplines.

Leukemia

Leukemia is frequently caused by chemically induced bone marrow changes. Chemical inducers of leukemia include benzene, chloramphenicol, and phenylbutazone.

Skin Cancer

Epidermal carcinomas are of various types. Squamous cell carcinomas are commonly associated with exposure to ultraviolet light. Contact with petroleum products may cause promotion, and photoactivation may occur. Tobacco products can promote carcinomas within the mouth. Polycyclic aromatic hydrocarbons have been associated with scrotal cancers.

Threshold and Additive Effects

The existence of thresholds for chemical carcinogens is an area of excited scientific debate. The existence of thresholds for chemical carcinogens continues to be rejected because their existence cannot be experimentally demonstrated. Part of this debate is due to the inability to accurately extrapolate dose-effect curves to zero dose. The existence of many natural mutagenic agents and events create difficulties in establishing true control groups.

Additive effects imply that exposure to chemical carcinogens are additive over the life-span of the organism. Carcinogenic substances are subject to the same bioaccumulation, transformation, and excretion principles previously discussed. However, if there exists no threshold, and a risk is assumed for any absorbed dose, then the probability of cancer induction would be additive.

There are practical limits that may be used for all carcinogens. For example, if exposure to a carcinogen is small enough to reduce cancer incidence to one in a trillion, the level of exposure is of no practical significance, since there are not that many people in the world. If the dose can be related to induction period, such as when larger doses lead to shorter induction periods, decreasing the dose to induction periods greater than 100 years, extends the induction period past the average human life-span.

While not all lung cancer is a result of smoking, it has been estimated, by many experts, that up to 90% of lung cancer deaths are directly attributable to smoking. There are numerous instances where smoking and occupational exposure have a synergistic interaction. Generally, smokers have a risk of dying from lung cancer of about 10 to 15 times that of nonsmokers. Nonsmokers exposed to asbestos have about a fivefold increase in risk of dying from lung cancer as compared to the general public. Smokers exposed to asbestos run a risk 50 to 70 times that of the general populace in contracting lung cancer. For more risk evaluations, see Chapter 16.

Cancer Treatment

Treatment of cancer is very difficult. The basic problem is the need to rid the organism of every single cell. For comparison, an antibiotic that is 99.9999% effective would leave one bacterium in a million alive. Typically, the body's immune system can deal with this odd bacterium. However, this is not generally true for cancer cells; one cancer cell can **metastasize** to re-establish itself. An interesting hypothesis is that tumor cells secrete an agent which inhibits metastasized cells from expressing themselves (forming tumors). When the tumor is removed, the inhibition is removed and the metastasized cells subsequently develop into tumors.

There are three classical types of cancer treatments. The first is surgery, which is effective at removing centralized tumors. In the second treatment, chemotherapy, the difficult challenge is to kill the cancer cells without killing the organism. Chemotherapy functions by killing rapidly proliferating, dividing cells. Cancer cells constitute such a population. Unfortunately, so too do skin, hair, bone, blood, intestinal, and digestive cells. These cell populations are also significantly impacted by chemotherapy, which thus tends to have disastrous side effects. The third treatment is radiotherapy. Again, the difficulty is to eradicate the cancer cells without killing the patient.

There are a number of natural products—herbals and extracts—which are touted as have anticancer properties. These may have agents which initiate cancer cell apoptosis, inhibit metastasized cancer and tumor cells, or have antineogenic properties. Currently the general prognosis (+5 years) of classical therapy is poor; the best defense is prevention and mitigation. (Figure 8-7)

Reproductive System

The reproductive system consists of the gonads and a series of tubes which allow egress of the sperm or eggs. Sexual development begins at about seven weeks after conception. Differentiation into male or female is due to hormonal influences. The male system is determined by 85 days and the male sex organs are

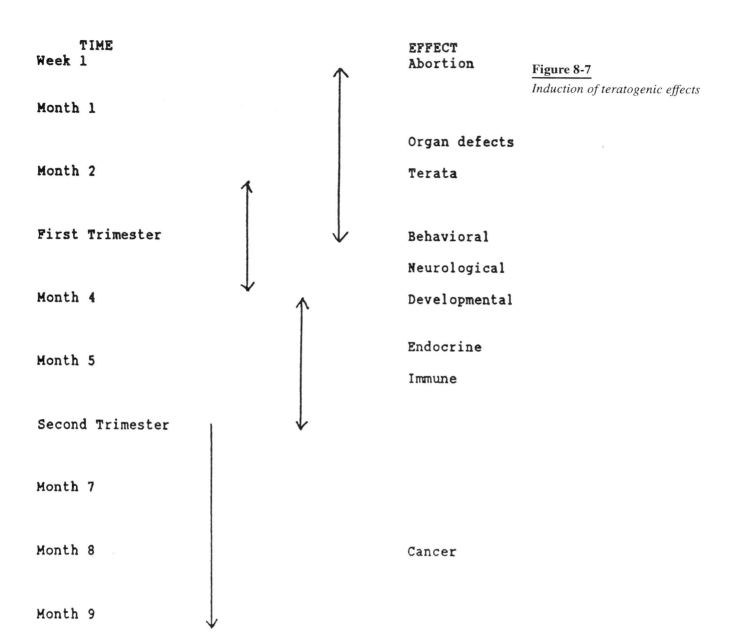

TIME		EFFECT	
Week 1		Abortion	**Figure 8-7**
			Induction of teratogenic effects
Month 1			
		Organ defects	
Month 2		Terata	
First Trimester		Behavioral	
		Neurological	
Month 4		Developmental	
Month 5		Endocrine	
		Immune	
Second Trimester			
Month 7			
Month 8		Cancer	
Month 9			

fully developed by seven months. Male germ cells, sperm, are not produced until puberty. Female sexuality is completely defined in about 100 days, and the sex organs are fully formed at about five months. In contrast to male development, in which no sperm are produced until puberty, the female fetus develops all the ova she will ever have before birth. At birth there are 300,000 to 400,000 ova per ovary. At puberty, there are between 150,000 and 200,000, and at age 30, there are about 25,000. About 400 mature ova are released over a lifetime.

Lead decreases the fertility of male workers and increases the rate of spontaneous abortion in their wives. Cadmium, nickel, and methyl mercury cause testicular damage. Halogenated pesticides have been linked to adverse responses of the male reproductive system. Organic solvents such as toluene, benzene, and xylene have been reported to cause low sperm counts, abnormal sperm, and varying degrees of infertility. Dinitrotoluene causes reduced sperm counts and abnormal sperm morphology. Mutations to the male germ cells can cause infertility or be transmitted to offspring.

Agents affecting the female ova, (e.g., radiation) can mutate the ova or cause infertility. Hormonal agents, including birth control pills, alter the female hormonal cycle, ovulation, and menstruation. Most of the adverse effects occur directly to the developing fetus during pregnancy.

The developing embryo is exquisitely sensitive to perturbations. Toxic exposure during the first week of pregnancy prevents proper implantation and causes embryonic death. During the first three months of pregnancy, organogenesis occurs. The nature of birth defects and terata observed during organogenesis all depend upon the exact period of exposure. Thalidomide has a particular effect on limb bud formation (phocomelia). Toxic exposure during the third and fourth months of pregnancy leads to neurological, developmental, and behavioral defects. During the fifth and sixth months of pregnancy, exposure leads to endocrine and immune system dysfunction. After seven months of pregnancy, exposure leads to the development of cancer.

The embryo is more susceptible to physical and chemical teratogens than the mother. Actinomycin D inhibits the formation of RNA. Colchicine inhibits cell division. Heavy metals inhibit enzymatic activity. Vitamin deficiencies tend to be teratogenic. There are a variety of transplacental carcinogens. Diethylstilbestrol (DES) causes adenocarcinoma of the vagina and scrotal seminoma. Tobacco smoke, alcohol, hallucinogenic drugs, and mineral deficiencies have been linked to birth defects. Alkylating agents, anesthetic gases, methyl mercury, organophosphates, carbamates, and acute hypoxia are all agents considered to be toxic to the female reproductive system and potential teratogens.

As is typical of the response to exposure, birth defects follow the dose-response relationship for a population. However, it is not possible to assign individual risk from such a generalized relationship as when an expectant mother cannot assume to be safe just because she is below average for exposure. Typically, the dose-response relationships are exceedingly steep, arguing for the concept of a threshold dose. Unfortunately, individual toxon effects cannot be eliminated. Because of our relative ignorance about the causes of most birth defects, and the extreme sensitivity of developing embryos to toxic chemicals, strict exposure avoidance is recommended for pregnant women.

Study Exercises

1. Describe these effects of mutagens:
   - the differences in consequences between somatic cell mutations and reproductive cell mutations
   - the differences between genetic point mutations and chromosomal alterations

2. Describe each of the following:
   - the characteristics of cancer
   - the difference between benign and malignant tumors
   - the difference between initiator and promoter types of carcinogens
   - the carcinogens associated with lung cancer
   - the carcinogens associated with urinary bladder cancer
   - the carcinogens associated with liver cancer
   - the carcinogens associated with leukemia
   - the carcinogens associated with skin cancer
   - the normal rate of cancer in the population
   - treatments of cancer

3. Describe these effects of fetal toxins:
   - the differences between embryotoxicity, congenital malformations, growth retardation, and mental retardation
   - the effects of prenatal carcinogens

4. Describe these functions of the reproductive system:
   - the principal functions of the male reproductive system and the potential effects of toxic substances
   - the principal functions of the female reproductive system and the potential effects of toxic substances

5. Explain how addition and deletion of nucleotides result in spontaneous mutations.

6. Explain a mechanism of action that may account for how teratogenic agents cause their effects.

7. How does the mechanism of action for a genotoxic carcinogen differ from that of an epigenetic carcinogen?

8. Identify several lifestyle or dietary factors that may contribute to the development of cancer.

9

Survey of Common Hazardous Agents
Toxic Substances

Overview

This chapter on chemical hazards is designed to provide information on, and the toxicology associated with, the principal chemical groups. The material is oriented towards those substances most commonly released. Included are corrosives, metals, organic solvents, organic compounds, and pesticides.

Chemical Hazards

Many hazardous and toxic substances are used by industry, in businesses, and around the house. The material presented in this chapter is generic in nature and deals specifically with those materials that have the greatest risk for exposure in industrial settings.

Metals

Most of the commonly used metals in industry can be toxic if absorbed. Metals tend to cause enzyme inhibition, bind to cofactors or vitamins, substitute for essential metals, and/or cause metal imbalances. Potentially toxic metals include aluminum, arsenic, beryllium, bismuth, cadmium, chromium, cobalt, copper, iron, lead, manganese, mercury, nickel, selenium, strontium, thallium, tin, and zinc.

Arsenic

Arsenic exists primarily in trivalent (man-made) and pentavalent (natural and in some wood preservatives) forms, either as an organic or inorganic. All forms of arsenic are toxic. Symptoms of chronic arsenic poisoning include general feelings of illness and fatigue, with stomach and intestinal distress. Hyperpigmentation (dark spots on the skin) and Mee's lines (dark bands on the fingernails and toenails) may develop. Hyperkeratosis (thickening of the skin) on the palms and soles are classically associated with chronic arsenic poisoning. Three types of neuropathy have been reported: sensory, movement and control of muscles, and CNS (dementia). Arsenic is a human carcinogen, leading to increased risk of skin and lung cancer.

Beryllium

Beryllium is widely used in the space industry because of its light weight and high strength. Its use for fluorophors in CRTs has been discontinued. Beryllium appears to inhibit certain magnesium-activated enzymes. Soluble beryllium salts are directly irritating to the skin and mucous membranes and induce acute pneumonitis with pulmonary edema. Chronic pulmonary granulomatosis develops as a result of hypersensitivity to beryllium in the tissues. Onset of symptoms occurs 2 to 5 weeks after an acute exposure. Chronic symptoms, such as dyspnea, may begin 3 months to 11 years after first exposure. Dust and fumes from beryllium processes must be rigidly controlled. No beryllium is allowable in air.

Cadmium

Cadmium use is widespread as a plating metal because of its resistance to oxidation. It is also used in nickel-cadmium (NiCad) batteries. Cadmium-plated refrigerator racks converted for use in homemade barbecues can result in serious poisoning. Cadmium fumes, if inhaled, can produce rapid, and sometimes fatal, pulmonary edema; chronic exposure can lead to fibrosis or emphysema. Chronic absorption leads to liver and kidney damage, which is only slowly reversible due to cadmium's extremely long half-life in the body, 10 to 30 years. Occupational exposures thus accumulate over a lifetime.

Chromium

Chromium is used in chemical synthesis, steel making, electroplating, and leather tanning, and as a radiator antirustant. The toxicity of chromium compounds depends on the valence state of the metal. The exposure limit for metal dust and chromium salts of valence 2 or 3 is 0.5 mg/m^3, while compounds of valence 6, have an exposure limit of 0.05 mg/m^3. Chromium and chromates are irritating and destructive to all cells of the body. Acute poisoning results in hemorrhagic nephritis. Chronic poisoning, leads to incapacitating eczematous dermatitis with edema and ulcerations, which heal slowly. Conjunctivitis, lacrimation, and acute hepatitis with jaundice have been observed. Chromium VI is considered a carcinogen. Lung cancer incidence is increased 15 fold over normal in workers exposed to Chromium VI.

Lead

See Chapter 10.

Mercury

Mercury exists as an element (metallic) with inorganic and organic forms. Each form is associated with its own set of symptoms. Symptoms of chronic elemental mercury poisoning are related primarily to subtle neuropsychiatric changes (such as excessive shyness, insomnia, emotional instability, irritability, and mood swings into depressions). These symptoms are typically slow to develop, making diagnosis difficult. Mercury vapor is in the monatomic state and is lipophilic; therefore, it diffuses through the lipid membranes of cells. It is transferred to brain cells where it is oxidized to Hg^{+2} to produce toxic effects. Poisoning from inorganic mercury salts like mercuric chloride results in extensive kidney damage. Mercurous chloride (calomel) is less toxic than the mercuric form and is usually clinically insignificant. Organic forms of mercury affect the CNS, primarily the sensory-motor system, resulting in decreases in hearing, tunnel vision, and altered sense of touch and pain. Organic mercurials like methylmercury can cause severe birth defects producing mental retardation in infants of mothers exposed during the later stages in pregnancy. Seed grain treated with methylmercury to inhibit mold growth and seafood contaminated with organic mercury in Japan have resulted in epidemic poisoning.

Zinc

At higher doses, zinc ingestion causes vomiting, nausea, and diarrhea. Inhalation of zinc oxide fumes may produce **metal fume fever**. Chest pain, coughing, chills, and fever appear 4 to 8 hours after exposure and persist for 1 to 2 days. Zinc chloride is corrosive and causes ulceration of exposed skin. Solder fumes containing zinc chloride may cause irritation of the mucous membranes and, at higher doses, pulmonary edema and respiratory collapse.

Gases

Chlorine

Chlorine is a reactive gas commonly used in a variety of industrial processes. It is frequently stored as a liquid under high pressure. Chlorine will react with water to form a strong acid. Exposure causes local effects. Both simple asphyxiation and chemical asphyxiation are of concern.

Ammonia

Ammonia is a reactive gas commonly used in a variety of industrial processes. It is frequently stored as a liquid under high pressure. Ammonia will react with water to form a strong base. Exposure causes local effects. Both simple asphyxiation and chemical asphyxiation are of concern.

Ozone

Ozone is a colorless gas. Its odor can be detected at 10 ppb. It is very irritating to all mucous membranes. Nasal and throat irritation, congestion in the chest, headaches, vertigo, and pulmonary edema may result from inhalation. Sensitivity is increased in asthmatics. It may cause chromosomal damage by some of the same mechanisms as radiation.

Nitrogen Oxides

Nitrous oxide, also known as laughing gas, is a weak narcotic not normally found in industrial settings. Nitrogen dioxide and nitrogen tetroxide are always found together at normal environmental temperatures and are mildly irritating to lung tissue. Pulmonary edema may occur. Nitric oxide inhalation has been linked to CNS paralysis and anoxia from methemoglobin formation. Acute inhalation poisoning (50 to 300 ppm of nitrogen dioxide) leads to progressive weakness, dyspnea, cough, and cyanosis (1 to 3 weeks after exposure). Recovery from the acute phase requires 1 to 6 months and emphysematous changes persist.

Phosgene

Phosgene COCl$_2$, is a reactive lung irritant. Only a relatively small portion of phosgene hydrolyzes in the respiratory passages, but in the terminal end of the alveoli complete hydrolysis occurs with irritant effects on the epithelium. This results in increasing edema until up to 50% of the plasma has accumulated in the lungs, causing drowning. It is produced by mixing a chlorine compound with ammonia hydroxide.

Carbon Disulfide

Carbon disulfide is used as a solvent. Exposure at concentrations from 100 to 1000 ppm causes symptoms progressing from restlessness, irritation of the mucous membranes, blurred vision, nausea, vomiting, headache, unconsciousness, and paralysis of respiration. Chronic exposure causes bizarre sensations in the extremities followed by sensory loss and muscular weakness. There is increased risk of coronary heart disease and increased incidence of abortions, sterility, and amenorrhea in women.

Hydrogen Cyanide

Hydrogen cyanide gas, or the ingestion of cyanide salts, leads to cyanide poisoning. Cyanide inhibits the cytochrome oxidase enzymes of the mitochondria preventing ATP production. Because oxygen utilization in the tissues is essentially blocked, venous blood may be as bright red as arterial blood, imparting a flushed appearance to the skin and mucous membranes. Cyanide has a characteristic bitter-almonds odor which can readily aid in diagnosis. However, a significant percent of the population is genetically incapable of detecting this odor. Therapeutic treatment must be initiated immediately to be lifesaving. Cyanide kits should be readily available in any industry or business which uses cyanide-based products. IDLH is 50 ppm.

Hydrogen Sulfide

Hydrogen sulfide is a common component of the natural decay of organic matter high in sulfur and is used and produced in many industrial processes. Hydrogen sulfide has a strong, unpleasant, characteristic odor of rotten eggs. Olfactory fatigue, the loss of the ability to smell hydrogen sulfide, occurs rapidly. Therefore, although the odor appears to disappear quickly, toxic amounts of the gas may remain present. As with cyanide, it inhibits the cytochrome oxidases causing cellular asphyxia. It irritates the mucous membranes and causes conjunctivitis, headache, nausea, cough, and dizziness at 50 to 100 ppm. IDLH is 300 ppm, at which levels pulmonary edema occurs.

Sulfur Oxides

Sulfur oxides are produced by oxidizing sulfur, typically by burning sulfur-containing products in the air. Automobile exhaust, coal and oil burning, smelters, refineries, paper manufacturing, and a variety of industrial processes produce SO$_2$ and SO$_3$ as air pollutants. They are major components of acid rain. They are irritants to the eye and upper respiratory track. High doses may cause pneumonitis and pulmonary edema. Respiratory rate may increase at levels above 1 ppm. A sharp odor or taste may be noticed at 3 to 5 ppm. Eye and throat irritation occurs at 8 to 12 ppm. Choking, cough, and bronchial constriction occur at 10 to 50 ppm. IDLH is 100 ppm. As with nitrogen oxides, additional symptoms may develop 2 to 3 weeks after exposure due to pulmonary fibrosis and the destruction of bronchial epithelia with concomitant edema.

Corrosives

Strong acids and bases are corrosive upon local contact. They will immediately attack any tissue with which they come in contact. Acids tend to cause an immediate burning sensation and are easier to wash off with water. Bases, caustics, and alkalies, tend to do damage before the burning is felt and tend to be difficult to remove by washing. Many corrosives, like acids, are formed by hydrating a gas: oleum (sulfur trioxide) forms sulfuric acid when mixed with water. Reaction of a metal with water (e.g., sodium, potassium, and calcium), can create a base (NaOH, KOH, or CaOH). These are inorganic acids and bases. A large variety of organic acids and bases exist; they tend to be less corrosive than the inorganic corrosives.

Solvents and Organic Compounds

Three different types of toxicological hazards may be associated with these compounds.

- **Dermatitis**
  Organic solvents dissolve the fats from the skin such that repeated contact will result in a dry, red, itchy skin that may become quite uncomfortable. These skin changes also tend to increase the permeability of the skin to solvents and other toxic chemicals, thereby enhancing the potential for skin absorption.

- **Aspiration pneumonitis**
  Aspiration of solvents into the lungs may result in rapid and severe pulmonary damage. The low viscosity and surface tension of organic solvents, water immiscible, may lead to aspiration of solvents into the lungs with vomiting. Thus, vomiting should not be induced when organic solvents have been ingested.

- **CNS depression**
  Inhalation of organic vapors may cause a progression from an initial high to lethargy, dizziness, staggering, and other signs of drunkenness, coma, and death. Intoxication from solvent exposure is not unlike that from drinking alcohol, and abuse of solvent (e.g., glue sniffing and huffing) is widespread among youth. Unfortunately, the dosage is not easily controlled when vapors are inhaled; overdoses are common, as are recurrent side effects. Repeated heavy exposures to solvents may lead to prolonged and potentially permanent damage to the brain. Subtle changes in higher brain functions can cause sleep disturbances, anxiety, short-term memory loss, and behavioral changes.

Saturated Aliphatic Hydrocarbons

Paraffin hydrocarbons are typically straight-chain hydrocarbons derived from petroleum. They include methane, ethane, propane, butane, pentane, hexane, heptane, and octane. Methane and ethane are pharmacologically inert but are simple asphyxiants. The hydrocarbons above ethane are CNS depressants. Vapors are mildly irritating to mucous membranes.

Unsaturated Aliphatic Hydrocarbons

Olefins and diolefins are formed as by-products of the cracking of petroleum fractions. They include ethylene, propylene, butadiene, and isoprene. These tend to be simple asphyxiants and/or weak anesthetics.

Alicyclic Hydrocarbons

The cycloparaffins or cycloalkalines are saturated or unsaturated cyclic hydrocarbons having the chemical properties of the aliphatic series. These include cyclohexane, methylcyclohexane, decaline, dicyclopentadiene, and turpentines. Their toxic effects resemble those of acyclic hydrocarbons (anesthesia and CNS depression) and include irritation to the mucous membranes.

Aromatic Hydrocarbons

Aromatic hydrocarbons are chemicals composed of one or more benzene rings. The single ring compounds include benzene, toluene, styrene, ethylbenzene, xylene, and naphthalene. Diphenyl and polyphenol compounds and polynuclear aromatics (polycyclic aromatic hydrocarbons - PAH) are included in this class. The aromatics are far more irritating than the aliphatics. They act as primary irritants to the mucous membranes, causing pulmonary edema, pneumonitis, and hemorrhaging in the respiratory system. They may also cause CNS depression and narcosis. Many of these have been linked to organ (liver) damage and carcinogenesis.

Aliphatic Halogenated Hydrocarbons

Aliphatic halogenated hydrocarbons include methyl chloride, methylene chloride, chloroform, bromoform, carbon tetrachloride, ethyl chloride, methyl chloroform, vinyl chloride, vinylidene chloride, and numerous other compounds. Typical toxicological actions include renal and hepatic damage, possible cardiac sensitization, and carcinogenesis. Numerous instances have been reported where the startle response, which is associated with a large and rapid release of adrenaline from the adrenal gland, has precipitated a heart attack in individuals who were breathing large quantities of chlorinated hydrocarbons (e.g., from glue sniffing and huffing).

Cyclic Halogenated Hydrocarbons

Cyclic halogenated hydrocarbons include chlorobenzene, dichlorobenzene, dieldrin, and other compounds. Typical toxicological actions include renal and hepatic damage, and carcinogenesis.

Alcohols

Alcohols are hydrocarbons in which one or more hydrogen atoms are substituted by hydroxyl groups. Alcohols are used extensively as solvents and thinners. These include methanol, ethanol, propanol, isopropanol, isobutanol, amyl alcohols, and allyl alcohols. Toxicity increases progressively from ethyl to amyl, and then, as the higher alcohols become less soluble in body fluids, there is a decreasing toxicity. Generally, there are some narcotic effects such as irritation to the upper respiratory tract, local muscle spasm, tremors, headaches, pulmonary edema, diarrhea, convulsions, vomiting, and liver and kidney involvement. A developing fetus exposed to alcohol through its mother can be born with a condition called Fetal Alcohol Syndrome (FAS). An infant with this condition is faced with substantial nervous system disorders and characteristic birth defects, particularly those of the head and face.

Glycols and Derivatives

Glycols are doubly substituted alcohols. These include ethylene glycol, diethylene glycol, polypropylene glycols, and ethylene glycol monoethyl ether. In general, they are less acutely toxic than the monohydroxy alcohols and are not significantly irritating to the eyes or skin. Their vapor pressures are so low that toxic concentrations are unusual at room temperature. Symptoms include narcosis, upper respiratory tract irritation, pulmonary edema, vomiting, headache, and tremors. Ethylene glycol is the compound of primary concern; a single oral dose of 100 ml is lethal. It is metabolized to oxalate, which is toxic to the kidneys and may cause renal failure. Ethanol can be used as a competitive inhibitor.

Aromatic Alcohols

Phenols have the ability to denature and precipitate proteins, and are cytotoxic. Phenols include phenol and cresol, which are readily absorbed by all routes of exposure. They are quite corrosive as well as being CNS depressants. Dihydroxy (resorcinol, hydroquinone, and catechol), trihydroxy, and chlorinated phenols have increased toxicity and, frequently, the ability to reduce blood oxygen content.

Ethers

Ethers are very effective anesthetics, a property that increases with the size of the molecule. These include methyl ether, ethyl ether, isopropyl ether, chloromethyl ether, and chloromethyl methyl ether. Ethers irritate the mucous membrane with possible associated pulmonary edema, vomiting, headaches, and nausea. Halogenated ethers can cause severe irritation to the skin, eyes, and lungs; they are potent alkylating agents and several are known carcinogens.

Epoxies

Epoxy compounds are cyclic ethers and include epichlorohydrin, ethylene oxide, and propylene oxide. Epoxides tend to cause nausea and vomiting, are irritating to the eyes, skin, and respiratory tract, and act as sensitizers. Pulmonary edema may occur as well as liver, kidney, and lung damage. Ethylene oxide is a known carcinogen.

Ketones

Ketones are generally very chemically stable. Industrial ketones include acetone, methyl ethyl ketone, methyl isobutyl ketone, and diisobutyl ketone. Exposure to high concentrations of ketones produces narcosis, headaches, nausea, vomiting, dizziness, loss of coordination, and unconsciousness. They are CNS depressants, but cause irritation to the eyes and respiratory passages at lower doses. However, exposure at lower concentrations may impair judgment and result in death from respiratory failure.

Aldehydes

Aldehydes are one of the most important classes of industrial chemicals. These include acetaldehyde, acrolein, formaldehyde, furfural, and chloral (chloral hydrate). Typically, they are strong irritants to the mucous membranes, causing edema, bronchitis, bronchopneumonia, pulmonary sensitization, asthmatic attacks, and narcosis. While they are generally toxic, they are not carcinogenic. Chloral is a trisubstituted acetaldehyde derivative and acts as a CNS depressant (Mickey Finn).

Organic Acids and Anhydrides

Carboxylic acids may be aliphatic or aromatic. Acid anhydrides are acid derivatives. They include acetic acid, acetic anhydride, formic acid, oxalic acid, and phthalic anhydride. Since they are primarily irritants, as are the corrosives, sensitization is common with the anhydrides. Bronchopneumonia, pulmonary edema, and kidney damage have been reported.

Esters

Esters are formed by replacing the carboxylic acid hydrogen with an organic grouping. They include acetate, ethyl silicate, ethyl formate, and methyl formate. Simple aliphatic esters are characterized by anesthesia and irritation. Halogenated esters are characterized by lacrimation, vesication, and lung irritation. Phosphate esters may cause cumulative organic damage to the nervous system and neuropathy. Most aliphatic and aromatic esters used as plasticizers are physiologically inert.

Amines

Amine-substituted chemicals are among the most toxic solvents of organic chemicals. They are strong irritants and act as strong corrosives. They include methylamine, dimethylamine, trimethylamine, ethylamine, diethylamine, triethylamine, propylamine, butylamine, allylamine, and cyclohexylamine. They are easily and well absorbed by all routes. They are toxic to all tissues in which they are absorbed and adversely affect a number of organs. They produce methemoglobin, cause sensitization, and are aromatic. A number of other amines are carcinogenic.

Amides

Above formamide, the simple amides are solid; they are generally not hazardous by the usual routes of exposure and are rapidly hydrolyzed. The unsaturated amides are very toxic and have pronounced CNS, liver, and kidney toxicities.

Biocides

Biocides are commonly used in industry, business, and around the house. They include algicides, fungicides, herbicides, insecticides, nematocides, molluscides, and rodenticides. Biocides present a unique problem because they are specifically designed for the purpose of killing biological organisms and because they are directly applied in massive quantities to the environment. Occupational hazards associated with biocides are manifold. Formulators and applicators are generally at highest risk, followed by those involved in manufacturing and packaging. Migrant farm workers are particularly at risk due to insufficient training in the safe use of certain chemicals and their lack of familiarity with regulations. Additionally, many migrant workers lack ready access to good medical care.

Historically inorganic substances such as lead arsenate, copper sulfate, and sulfur were used as pesticides. In 1938, Dr. Paul Mueller introduced dichlorodiphenyltrichloroethane (DDT), which was effective in eliminating malaria-carrying mosquitoes. The **organochlorine** insecticides were followed by the **organophosphate** insecticides (a spin-off of nerve gas research) and, subsequently, the **carbamates**. These related organics constitute the vast majority of all chemical biocides.

Organophosphate pesticides are extremely effective and toxic to mammals. Thus, they are of particular concern to man, especially handlers and applicators. The basic mechanism of action is the binding to, and the inactivation of, acetylcholinesterase. Because the brain, the autonomic nervous system, and the neuromuscular junctions all utilize acetylcholinesterase, signs and symptoms of organophosphate poisoning are quite broad.

Symptoms of organophosphate poisoning include diarrhea, excessive salivation, constriction of the pupils, congestion in the lungs, difficulty in breathing, and loss of bladder control. Effected are skeletal muscles (twitching, weakness, and paralysis), the autonomic nervous system (inhibition), smooth muscle contraction, (cramping, and fluid into the lung), and the CNS (tremors, delirium, confusion, slurred speech, disequilibrium, loss of coordination, and convulsions). Organophosphates are rapidly absorbed through the skin, lungs, and digestive tract. Blood tests can determine whether exposure to organophosphates is causing significant cholinesterase inhibition. Specific and effective antidotes are available.

Carbamate pesticides are similar to organophosphate pesticides; however, they tend to be readily reversible. Therefore, they are generally regarded as far less toxic than organochlorides and organophosphates. (Aldicarb is, however, highly toxic.) Sevin and Baygon are the most widely used insecticides in this class and are the principal ingredients in many household products, flea collars, and pet shampoos. Acute symptoms include light-headedness, nausea, vomiting, sweating, blurred vision, salivation, muscular weakness, and convulsions. Atropine is the only effective treatment for severe cases.

Organochlorine pesticides are less acutely toxic than the organophosphates, and more chronically toxic. DDT was the first synthetic insecticide to find widespread use, was considered relatively safe to humans, yet was later banned due to adverse environmental impact. Lindane, dieldrin, endrin, and chlordane are all of this class. These chemicals are quite persistent in the environment and, for this reason, have been phased out of use. They are tightly regulated by the EPA. Methoxychlor is now commonly used because of its low toxicity to mammals and relatively low persistence in the environment.

Generally, the organochlorine chemicals serve as CNS stimulants and produce disruptions in CNS function, leading to convulsions, coma, and death. Chlorinated cyclodiene pesticides produce convulsions, birth defects, and are toxic to the fetus. Kepone, used to control fire ants, accumulates in fat and in the liver. When excreted in bile, it is rapidly reabsorbed from the digestive tract and recycled to the liver. Many of the organochlorines produce liver cancer in laboratory animals. Several of these compounds, along with other halogenated hydrocarbons and dioxins, have been identified as having hormonelike qualities and are under investigation for their possible role in promoting hormone-induced cancers, including breast cancer.

Chlorophenol compounds are used to kill broadleaf weeds. They are weakly toxic to mammals; death occurs from ventricular fibrillation. They have been associated with miscarriages; however, the vast majority of animal data and human epidemiological studies indicate that such effects are highly unlikely at the doses encountered from normal use of these compounds. There is controversy over purported chronic effects of 2,4–D and 2,4,5–T, because of the potential for trace contamination with Dioxins, which are very toxic. Chloracne is a symptom of dioxin exposure.

Dinitrophenols are used in weed control. Acute human symptoms include nausea, hot flashes, rapid breathing, sweating, tachycardia, and coma. These compounds uncouple oxidative phosphorylation in the mitochondrial cytochrome system. Atropine is contraindicated.

Paraquat is a broad spectrum herbicide (weed killer) that is the most toxic of the commonly used herbicides. It is considerably more toxic than many insecticides. Paraquat produces delayed-onset lung damage, similar to emphysema, which is frequently fatal. Acute symptoms such as gastrointestinal distress, nausea, vomiting, and malaise may subside within a day after exposure, and the prognosis may appear to be positive, only to have the patient readmitted a week or two later with progressive failure of the lungs. In addition, it causes liver and kidney damage. The lethal dose is less than 50 ppm; the mode of action is unknown.

The principal rodenticide is warfarin, which is essentially an anticoagulant or blood thinner. As with other coumarin anticoagulants, large doses of vitamin K are antidotal. Zinc phosphide, or other forms of phosphorous, are used agriculturally; they produce gastrointestinal damage leading to bloody vomitus, diarrhea, and sloughing of large sections of intestinal cells. Sodium fluoroacetate (compound 1080) is used to control ground squirrels, coyotes, and other mammalian pests. This chemical effectively and irreversibly blocks cellular respiration and therefore is toxic to all oxygen-breathing organisms. Its use has been strictly curtailed by the EPA because of its effects on wildlife.

Pentachlorophenol (PCP) is a widely used fungicide and wood preservative. PCP interferes with the ability of the mitochondria to produce ATP. The body consumes carbohydrates and oxygen in an effort to phosphorylate ADP, but the reaction is decoupled and the energy is released as heat. Consequently, the principal symptom and toxic effect of PCP poisoning is fever, which may be confused with a severe case of the flu. PCP is slowly eliminated from the body. Therefore, occupational exposures may result in the accumulation of PCP over extended periods of time. Chronic exposure may result in what initially appears to be an acute episode, whereas symptoms have actually developed slowly.

Study Exercises

1. Identify the most commonly released chemicals associated with human injury or death.

2. Describe the principal toxicity of chlorine.

3. Describe the principal toxicity of ammonia.

4. Describe the principal toxicity of corrosives.

5. Describe the principal toxicity of the following metals:
   - arsenic
   - mercury
   - cadmium
   - beryllium
   - chromium VI

6. Describe the principal toxicity of methane.

7. Describe the principal toxicity of gasoline.

8. Describe the principal toxicity of methylene chloride.

9. Describe the principal toxicity of trichloroethane.

10. Describe the principal toxicity of the freons.

11. Describe the principal toxicity of methanol.

12. Describe the principal toxicity of glycol ether.

13. Describe the principal toxicity of epoxy compounds.

14. Describe the principal toxicity of acetone.

15. Describe the principal toxicity of benzene, toluene, and xylene.

16. Describe the principal toxicity of the polycyclic aromatic compounds.

17. Describe the principal toxicity of the organochlorine insecticides.

18. Describe the principal toxicity of organophosphate insecticides.

19. Describe the principal toxicity of carbamate insecticides.

20. List five hazardous or toxic materials for which you are at risk at home or in the workplace. Develop a brief scenario showing how one might be exposed, the route of absorption, expected adverse effects, and a treatment plan.

21. What are the primary target tissues of solvents and their major toxic effects? How does this relate to "sniffing" and "huffing"?

22. Why should you not mix bleach and ammonia hydroxide to create a super cleaning solution?

23. Give two examples that illustrate why smell is not a reliable means of determining the presence or absence of toxic substances.

24. Exposure to which pesticides might result in leg cramps, twitching, and weakness, confusion, and slurred speech?

10 Survey of Common Hazardous Agents
Physical and Biological Hazards

Overview

This chapter provides information about physical and biological hazards. These hazards include the pollution of indoor air. The principal toxicities associated with asbestos, radon, and the criteria air pollutants are examined and discussed. The most common indoor air pollutants are identified and assessed. The subject of infectious wastes is introduced, defined, and discussed. The pertinent regulations for California, the most restrictive in the nation, are reviewed.

Physical and Biological Hazards

This chapter deals with nonchemical etiologic agents. (**Etiology** deals with the causes or origins of diseases.) Here, we examine a variety of physical and biological hazards, their sources, and prevention. We also explore a number of situations consisting of mixed hazards, such as chemical-physical and chemical-biological hazards.

Radiation

Radiation is the transfer of energy through space (i.e., over a distance). This implies that there exists a space around any radiation emitter which is full of energy. This energy may be hazardous to your health. The degree of the hazard depends upon the type and quantity of radiation. Radiation may be either low energy and nonionizing, or high energy and ionizing.

An example of nonionizing radiation is the electromagnetic field created in the vicinity of a power line. The risks associated with such EM fields has yet to be determined. Preliminary scientific investigations suggest that such radiation can affect plants and animals.

Ionizing radiation consists of high-energy photons and particles. Photons of greater energy than those of ultraviolet rays, including X rays, have detrimental effects. Particles include electrons (beta radiation), protons, neutrons, helium nuclei (alpha particles), and heavier ions. These all have the capacity of causing cellular and tissue damage. Therefore, exposure to these hazards should be minimized. Chapter 16, "Risk Assessment," deals with exposure to radiation sources such as X-ray machines, airline flights, and altitude variations such as those caused by living in the mountains.

Each of the sources of ionizing radiation has a characteristic sphere of influence. The size of this sphere can be decreased by the use of shielding. Proper shielding prevents the radiation from penetrating beyond the shield, thus allowing personnel access to the area immediately adjacent to the shielding. Shielded radiation sources may be safely transported. Care must be exercised to ensure that shielding is not broached, thereby releasing the radiation.

Nuclear power plants require the use of radioactive materials. Occupational exposure may occur during any of the many activities of the industry. The mining of the uranium ore and the processing of it for use in the reactor is risky. Workers are exposed to radon and radon daughter products in the mining and processing of uranium ore. Nuclear reactor operators receive routine exposure which, while it is much less than medical or industrial exposures, on average, is greater than background levels of exposure. The removal of the spent nuclear fuel for disposal, transportation, and storage is hazardous. Occasionally, workers receive dangerous exposure, as in the case of an accident or system failure. Nuclear accidents at both Three Mile Island and Chernobyl exposed workers to excessive or lethal levels of radiation. In the transport of nuclear fuels and high-level waste, occupational exposure does occur. The radioactive waste cleanup and disposal process may expose workers to radiation. These risks need to be fully disclosed and accurately calculated for both workers and community. This process has prevented the establishment of "permanent" storage sites, e.g., Ward Valley in California. These are problems that currently are being considered by the federal government.

The activity of a radioactive substance is measured by the rate at which it decays. One curie (Ci) of any radioactive material represents a decay rate of 3.7×10^{10} disintegrations per second. Useful units are the microcurie, 3.7×10^4, and

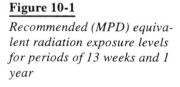

Figure 10-1

Recommended (MPD) equivalent radiation exposure levels for periods of 13 weeks and 1 year

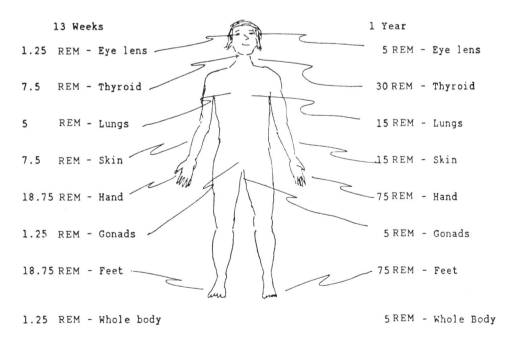

13 Weeks

1.25 REM - Eye lens
7.5 REM - Thyroid
5 REM - Lungs
7.5 REM - Skin
18.75 REM - Hand
1.25 REM - Gonads
18.75 REM - Feet

1.25 REM - Whole body

1 Year

5 REM - Eye lens
30 REM - Thyroid
15 REM - Lungs
15 REM - Skin
75 REM - Hand
5 REM - Gonads
75 REM - Feet

5 REM - Whole Body

Quality	Units
Activity-disintegration of radioactivity per second.	Curie = 3.7 X 10^{10} disintegration/sec Becquerel = 1 disintegration/sec
Exposure-charge liberated by ionizing radiation per unit of air.	Roentgen (R) = 2.584 x 10^{-4} coulomb/kg = 1 esu/cm^3 @ **STP**
Absorbed Dose-energy absorbed from ionizing radiation.	Gray (Gy) = 1 joule/kg 1 Gy = 100 rad (older term) 1 rad = 100 ergs/gm
Dosage-Equivalent type of radiation energy absorbed from ionization radiation per tissue mass per unit time.	Sievert (Sv) 1 Sv = 100 rem (older term) radiation equivalent to man (rem)

Table 10-1

Units used to describe radioactivity

picocurie, 3.7×10^{-2} disintegrations per second. One nanocurie is equal to 37 becquerel (Bq), where one bq equals one disintegration per second.

The **roentgen** (R) is the unit of exposure for radiation. Represented as 1 electrostatic unit (esu) per cc of dry air, it is the unit used to measure the amount of radiation, or ionization, present in the air. This very small ionization (0.000258 coulomb) charge per kilogram of air is used as the basis for standardizing and calibrating equipment which detects radiation. Due to safety practices, the dosage, the amount actually absorbed from exposure, is small. Therefore, the dose rate is usually much less than one roentgen per hour (1 R/hr.). Thus, subdivisions of one one-thousandths of a R ($\frac{1}{1000}$ R), milliroentgen (mR), and one one-millionth of a R ($\frac{1}{1,000,000}$ R), microroentgen (μR).

Roentgen is an exposure dose. Absorbed dose is measured as the **gray (Gy)**, which is equal to 100 **rad**. One rad is 100 ergs absorbed per gram of any substance. For workers who deal with radioactive materials or radiation-producing machines, absorbed dose is an important concept. A person may be exposed to radiation, but not all of the radiation contacts the body, and not all radiation irradiating the body is absorbed by the tissue. In the case of X rays and gamma rays, an exposure of 1R would result in an absorbed dose, in soft tissues, of an amount equal to about 0.01Gy. The gray is the official international unit which corresponds to an absorption of energy of 1 joule/kg. We approximate the conversion because *soft tissue* is a vague term encompassing what actually has variable characteristics, since the body is neither homogenous nor consistent.

The effects of different types of ionizing radiation on biological systems varies; i.e., they have a different biological effectiveness. The **Relative Biological Effectiveness (RBE)** is a measure of this variability. The **rem** is a unit of dose equivalent. The absorbed dose, rad, times the RBE for a particular type of ionizing radiation, provides a measure of the dose equivalent in rem. The annual whole-body **Maximum Permissible Dose (MPD)** is 5 rem. Figure 10-1 displays Maximum Permissible Dose (MPD) for various organs for 13 weeks and one year.

Biological Hazards

Biological hazards encompass the great diversity of life forms. They range from the very small, viruses, which can only be imaged by the use of an electron microscope, to the very large, protozoa, which may be seen with the naked eye. Life forms have several common properties, including the ability to grow and reproduce. Humans serve as a food source for a variety of life forms. Because such colonization and growth is generally detrimental to us, these etiologic agents cause disease.

Viruses are the simplest life form. They are parasitic in that they cannot live independently of their host. They colonize a living cell and subvert its processes to their own uses. There are no easy cures for viral infections. Vaccinations and the human immune system have the capacity to deal with, and thus prevent, some viral invasions.

Bacteria are complete functional units. However, they have frequently evolved to specialize in living within the human body. Our immune system, through vaccinations and immunization, has the capacity to deal with a variety of bacterial species. Antibiotics have been developed that can, in most cases, control the growth of bacteria. However, the indiscriminant use of these agents has encouraged the mutation of resistant bacterial populations. Consequently, we are seeing more viral and unstoppable infections.

Molds and fungi generally do not colonize living cells. However, they can stimulate the immune system, a situation known as an allergy (e.g., hay fever and asthma). They can colonize dead tissue such as on the outer layer of the body, skin, and nails (e.g., athlete's foot). By colonizing and degrading the outer protective surface layers, these etiologic agents can cause a number of problems. These problems range from cracking skin, bleeding, and the associated pain, to the creation of an entrance for other agents, such as bacteria, thus causing infection. Several chemical agents are available to combat molds and fungi and their effects.

Protozoa are the simplest of animals. They, and a variety of other tiny animals, are capable of colonizing humans. Some live in the digestive tract (e.g., amoeba and giardia, or beaver fever), some live on the surface (e.g., lice), and some live in the tissues (e.g., trichina which causes trichinosis). Most protozoan infestations can be prevented by the proper cooking of food and water. Some can be treated medically. The best prevention is to eliminate exposure.

Asbestos

Asbestos is a generic term for a large and diverse group of naturally occurring substances. These substances are hydrated silicates that have crystallized in a fibrous format. The various types of asbestos fibers differ in their chemical composition, morphology, and durability. The rodlike amphiboles, which account for less than 10% of the world's production of asbestos, are a frequent contaminant of other forms of asbestos and appear to penetrate the cells of the peripheral lung. Fibers longer than 8 microns, and less than 0.25 microns in diameter, have the most marked carcinogenic promotion potential. The legal definition of a fiber, as promulgated by the EPA and other U.S. regulatory agencies, is one that possesses an aspect ratio greater than 3 to 1.

Occupational exposure to asbestos has been linked to four types of disorders:

- **Asbestosis**
  This is a pulmonary, interstitial fibrosis with excessive deposition of collagen, causing progressive lung stiffening and impaired gas exchange.

- **Lung cancers**
  Tumors arising in the tracheal, bronchial, and alveolar epithelium have been found in asbestos workers who are smokers. These tumors are only rarely found in nonsmokers and, in most cases, 20 or more years have passed after the initial asbestos exposure.

- **Mesotheliomas of the pleura, pericardium, and the peritoneum**
  These are diffuse malignancies arising from mesothelial or underlying

mesenchymal cells. The time between diagnosis and initial occupational exposure commonly exceeds 30 years.

- **Benign changes in the pleura**
  These rarely cause functional impairment. They include pleural effusions, fibrosis, and plaques.

Currently, asbestos is incorporated into cement construction materials, (e.g., roofing, shingles, and cement pipes), friction materials (e.g., brake linings and clutch pads), jointing and gaskets, asphalt coats, and sealants. In the past, asbestos was commonly used for insulation and heat protection. Some 733,000 public and commercial buildings now require abatement. About 20% of buildings contain asbestos construction materials (ACM).

The risk is highest in miners and workers involved in the manufacturing and removal of ACMs. If undisturbed, asbestos in buildings represents a relatively low risk. Exposure may be either by inhalation or ingestion. Impaled cells produce a mixture of fibroblast growth factor and active oxygen species, which are believed to be causative agents of both asbestosis and asbestos-related malignancies. Of the 23 agents designated as Group 1 human carcinogens by the International Agency for Research on Cancer (IARC), only asbestos and conjugated estrogens were nongenotoxic.

The airborne PEL for asbestos in the U.S. workplace is 0.2 fibers per cubic centimeter. The levels of airborne asbestos in buildings, even with damaged ACM, are magnitudes lower, 1/100 th, than workplace PELs. With few exceptions, the type of asbestos fiber found predominantly in buildings is chrysotile, most probably not associated with medical problems. Long, thin asbestos fibers are rarely found in building air samples. It is now believed that the risk associated with asbestos in buildings is minuscule and is not a health risk in the nonoccupational environment.

Lead

As a heavy metal, lead has significant effects on the nervous system. Principal neurological symptoms can include retardation, convulsions, comas, and death. Low levels, persisting during childhood, slow a child's normal mental and physical development and cause learning and behavioral problems. The Agency of Toxic Substances and Disease Registry (ATSDR) reports long-lasting impacts on intelligence, motor control, hearing, and emotional development of children who have levels of lead in the body that are not associated with obvious symptoms. Lead also adversely affects the cardiovascular system and the kidneys.

Inhalation and ingestion are the major routes of exposure for both children and adults. Lead exists in both an inorganic and organic form. Most inorganic lead salts are poorly absorbed from the digestive tract, on the order of 5 to 7% in adults, but as high as 40% in children. The reason for this physiological difference is unknown, but it partially explains why childhood lead poisoning is much more common than adult poisoning. Once in the body, lead is distributed via the bloodstream to RBCs, soft tissue, and bone. Nearly 90% of the human total body burden of lead is found in bone; the biological half-life of lead in bone is 10 to 20 years. Lead in the body is eliminated very slowly by the kidneys and gastrointestinal tract. Lead serves no useful purpose in the body. The resulting damage from lead poisoning may be permanent, and, in some cases, fatal.

Over the past 20 years, the Centers for Disease Control (CDC) has responded to emerging knowledge about the effect of low-level lead exposure in children by progressively lowering the blood lead levels said to warrant medical intervention. In 1970, the level was 60 micrograms per decaliter. By October 1991, the intervention level was revised downwards to 10 micrograms per decaliter, and the single, all-purpose definition of childhood lead poisoning was replaced by a multitier approach to followup. The multitier approach emphasizes the implementation of primary prevention activities as blood lead levels of concern are lowered. One such activity includes the elimination of lead hazards before children are poisoned. CDC now states that "the goal of all lead poisoning prevention activities should be to reduce children's blood lead levels below ten micrograms per decaliter." A home inspection and remediation should be done for children with blood lead levels of 15 to 19 micrograms per decaliter. Under no circumstances should 10 micrograms per decaliter be regarded as a harmless level of blood lead.

Lead in our environment comes from a variety of sources. Until the EPA banned leaded gasoline, this was one of the largest sources. Lead is still pervasive on the ground due to auto exhaust. Lead in water is due to lead pipes, lead solder, and lead in brass. Proposition 65, the California Safe Drinking Water and Toxic Enforcement Act of 1986, mandated the removal of lead from water supplies. The largest remaining source of lead pollution is dust from lead paint. Lead was used extensively in paint before 1950, and continued in use, to a lesser extent, until 1978. Therefore, pre-1978 construction is at risk from lead paint. In San Francisco alone there are currently between 45,000 and 60,000 lead-poisoned children. Over 250,000 units are at risk from lead paint. Good housekeeping and personal hygiene are essential in reducing the risk from lead poisoning.

Radon

The EPA recommends action be taken to reduce risk when radon levels inside a building or room reach 4 picocuries. A picocurie is a trillionth of a curie. Exposure to four picocuries is equivalent to smoking ten cigarettes a day. The risk of getting lung cancer at an exposure of 4 picocuries/l of air is 1 in 1000 over a 70-year lifetime exposure. Acting on a mandate from Congress, the EPA is working on changing its recommendations to urge action when radon levels reach 0.2 to 0.7 picocuries/l of air. There are no other proven health effects from exposure to radon except an increased risk of lung cancer. It is an alpha radiation emitter, which cannot penetrate skin, so it does not affect internal organs other than the lungs.

Indoor radon concentrations can be reduced by applying natural ventilation, forced ventilation, covering exposed earth, sealing cracks and openings, using drain tile suction, block wall ventilation, and sub-slab suction methods. Safety standards are being recommended by states, professional and trade organizations, and the EPA. All lack the force of law. There is likely to be further research and debate before a national standard emerges.

Indoor Air Pollution

Many adverse health effects have been attributed to pollution of indoor air. Indoor air pollution is frequently referred to as tight building syndrome and sick building syndrome and is due to inadequate ventilation and the use of modern building materials. Beginning in the early 1970s, complaints by office workers about poor indoor air quality began to appear in significant numbers. The State of California's office building in Sacramento, the Bateson Building, which opened in 1981, is an example. Originally intended as a model for energy conservation, the building made people who worked there ill. Eighty percent of the employees on one floor of the building had one or more symptoms that they attributed to the building environment. The symptoms included ailments of the upper and lower respiratory tracts, nausea, itching or burning eyes, sinus problems, skin irritation, dizziness, and fainting spells.

Indoor air pollution is composed of airborne gases, vapors, particles, odors, and exhaled air. These are typically released inside of the building by the outgassing of structural materials, clothing, people, office products, office furnishing, and smoking. Other pollutants include chemicals and solvents from office reproduction equipment, combustion products from stoves and furnaces, and fungal and microorganismal growth in wet areas of the building or in its ventilation system. Occasionally, polluted air enters the fresh air ventilation system intakes.

Of 446 episodes of tight building syndrome investigated by NIOSH (*Indoor Air Quality,* Cincinnati: NIOSH, 1989) the causes were determined to be as follows:

1. Inadequate ventilation, 52%
2. Contamination from inside the building, 17%
3. Contamination from outside the building, 11%
4. Microbiological contamination, 5%
5. Contamination from the building fabric, 3%

In 1976, there was an outbreak of Legionella bacteria in Philadelphia. This resulted in 182 reported cases of illness and 29 deaths due to acute respiratory infections. Prevention of this type of bacterial infection depends on the elimination of potential sources of microbial contamination of building intake air, control of microbial growth within the building, and control of water accumulation in ventilation systems, on carpets, and on furnishing.

The use of synthetic materials in construction and furniture, and the use of organic chemicals in office supplies and equipment, have added to the burden of pollution on indoor air. Formaldehyde is used extensively in the plastics, adhesives, and polymer industries. At even a few ppm, the gas is extremely irritating to the skin, eyes, and mucous membranes, and it is a strong sensitizer. Symptoms include skin rashes, headaches, puffy eyes, and respiratory problems. Animal studies give no evidence that formaldehyde causes cancer in sites other than nasal passages, and epidemiological studies have not linked human exposure to cancer incidence. The allergenic properties of formaldehyde are of much greater significance to the general public than its potential carcinogenicity.

Carbon monoxide (CO) is found in the exhaust gases from fuel-driven motors, poorly ventilated fireplaces, and coal and natural gas used for cooking and heating. Poor heating, ventilation, and air conditioning (HVAC) design can lead to the mixing of exhaust gases and CO with the intake air. Carbon monoxide poisoning can cause oxygen deficiency in the tissues. (About 5 to 10% of the hemoglobin of heavy smokers remains in the form of carboxyhemoglobin. This percentage, close to or higher than the 8%, is reached by exposure to a concentration of 50 ppm, the TLV, of carbon monoxide. Passive smoking by children, born and unborn, in a smoking family, and by nonsmokers inhaling the smoke waste of their smoking friends, can lead to appreciable carboxyhemoglobin levels in the blood.)

Medical Wastes

Medical waste is, in general, any waste generated at any medical or health care facility, including veterinary facilities, clinics, and laboratories. It encompasses both infectious and noninfectious wastes. Infectious waste definitions vary and include biohazardous, health-services hazardous, pathological, biological, and infectious wastes. They may be generalized into four classes:

1. **Pathological waste**: materials from surgery and labs
2. **Infectious waste**: materials contacting patients
3. **Hazardous waste**: defined by RCRA
4. **Solid wastes**: defined by RCRA

Infectious diseases occur as a result of interaction between an infectious agent (pathogen) and a susceptible host. Medical wastes are a source of such pathogens. Interaction may occur in several ways including infection or intoxication. Intoxication refers to the induction of a disease by the production of a toxin by a pathogen. *Clostridium botulinum* causes botulism in this manner. Infection is the more common form of disease induction.

Infectious waste is any material contaminated with a type of microorganism, bacteria, mold, parasite, or virus, which normally causes, or significantly contributes to the cause of, increased morbidity or mortality of human beings.

Communicable diseases are worth noting here. Examples of some low-level communicable diseases are the common cold, influenza, or other diseases not representing a significant danger to nonimmuno-compromised persons. Of special concern are the highly communicable diseases. These are diseases such as those caused by organisms classified by the federal Centers for Disease Control as Biosafety Level IV organisms. Such organisms merit special precautions to protect staff, patients, and other persons from infections.

Infectious medical wastes, potentially infectious wastes, and suspected infectious wastes must all be treated as infectious wastes. The testing of medical wastes is contraindicated. Each medical waste must be treated as a separate entity, and any waste contaminated by an infectious waste must be also treated as infectious waste and treated accordingly. Infectious medical wastes are generally classified as follows:

- Sharps: needles, blades, and syringes
- Obstetrical waste
- Renal dialysis wastes
- Surgical dressings
- Contaminated disposable items
- Drapes
- Body parts
- Pathological items
- Chemicals
- Bacteriological cultures
- Urine and feces
- Blood and blood products
- Animal remains
- Biological specimens
- Miscellaneous infected material such as dressings and swabs

Solid waste may constitute up to 70% of medical wastes and may be segregated at the source if not contaminated with pathogenic, infectious, and hazardous wastes. Studies have shown that viruses may live in a waste station for up to eight days.

Anthrax is a disease affecting the skin, the lungs, or the digestive tract. It is caused by contacting or inhaling spores of *Bacillus anthraces*. The spores are highly resistant and may persist in the soil or in the hide of infected animals years after they have died.

Salmonellosis, or Salmonella poisoning, is a disease which affects the digestive tract. It is caused by the *Salmonella typhi* bacteria. It is spread by the contamination of food and water supply.

Tuberculosis (TB) is a disease which affects the lungs of humans and animals. It is caused by *Mycobacterium bovis* and *tuberculosis*. It is transmitted through respiratory secretions, feces, and milk.

Brucellosis is a zoonotic disease that produces pneumonia-like effects in humans. Outbreaks have been linked with contaminated water.

Medical and Infectious Waste Management
(California Medical Waste Management Act of 1990)

California has enacted a regulatory program for medical and infectious waste management. The specific regulatory requirements apply to clinical and research laboratories, biotechnology facilities, hospitals, nursing centers, clinics, and medical waste transporters and management firms. The requirements vary based on the size of the facility and amount of waste generated or handled, and the nature of the waste-related activity. There is no federal counterpart to this state program. However, future federal regulatory activity involving medical wastes is probable.

Medical wastes are generally defined to include biohazardous wastes, sharps waste, and other wastes. Medical waste does not include radioactive waste, household waste, waste that is commonly found in medical facilities but is not biohazardous, waste generated from normal agricultural and livestock practices, noninfectious waste containing microbiological cultures used in food processing and biotechnology, human bodily fluids (such as saliva and vomit), and bodily waste unless they contain fluid blood and hazardous wastes.

Biohazardous waste includes various types of laboratory waste, waste containing microbiologic specimens sent for laboratory analysis, human tissue or other surgical specimens potentially contaminated with contagious infectious agents, animal tissues, fluids and carcasses similarly suspected of being contaminated, waste containing fluid blood or blood products, and waste contaminated by quarantined humans or animals with highly communicable diseases.

Sharps waste includes hypodermic needles, syringes, blades, broken glass items such as pipettes, and other instruments or materials with acute corners or protuberances capable of cutting or piercing and which are contaminated with biohazardous waste.

The regulatory program covers medical waste generators, medical waste haulers, and medical waste treatment facilities.

- Medical waste generators are divided into small-quantity generators (less than 200 pounds of medical waste per month) and large-quantity generators (more than 200 pounds of medical waste per month).

- Medical waste haulers are persons who transport medical wastes.

- Treatment facilities are those that treat medical wastes by incineration, steam sterilization, or microwave technology, and may include any other process designated by the DHS as a process or technique employed to eliminate the disease-causing potential of medical waste. Treatment facilities are regulated in accordance with whether they are onsite treatment facilities or offsite treatment facilities. Some treatment facilities had been regulated as hazardous waste treatment facilities under the Title 22 hazardous waste regulations. The Medical Waste Management Act establishes alternative permitting for medical waste treatment facilities.

Containment and storage requirements for medical waste apply to all generators, haulers, transfer, and treatment facilities. Medical and biohazardous waste must be contained at the point of origin in a red biohazard bag, conspicuously labeled as biohazardous waste or bearing the international symbol biohazardous waste and the word *Biohazard*. The bags must be tied to prevent leakage and then placed in rigid covered containers for storage or transportation. The wastes must not be removed from the bag or disposed of until treatment is complete.

Bagged waste may be stored under refrigeration for up to 90 days before treatment, or seven days without refrigeration. Small quantity generators may store wastes, if properly contained, without refrigeration. Rigid containers may be reused if sanitized. Other containers may not be reused. Sharps waste must be placed in a sealed sharps container and may be enclosed in bags with other wastes.

Accumulation areas used to store medical waste containers must be secure and marked with warning signs in English:

CAUTION—BIOHAZARDOUS WASTE STORAGE AREA—
UNAUTHORIZED PERSONS KEEP OUT

and in Spanish:

CUIDADO—ZONA DE RESIDUOS—BIOLOGICOS PELIGROSAS—
PROHIBIDA LA ENTRADA A PERSONAS NO AUTORIZADAS

Languages, in addition to English, must be used if the generator or enforcement agency determine them to be appropriate. Warning signs shall be readily legible during daylight from a distance of a least 25 feet (8 m).

The use of trash chutes, compactors, and grinders to process medical waste is prohibited unless the waste has previously been treated or it is part of the treatment process.

Radioactive Medical Waste

Radioactive medical wastes are regulated by the United States Nuclear Regulatory Commission (NRC), the Department of Transportation (DOT), EPA, the Department of Environmental Protection (DEP), and the Department of Health (DOH). The packaging and transportation of radioactive wastes is regulated by the NRC and the DOT, (10 CFR 20; 49 CFR 173, 174, 175, 176, and 177).

Biological radioactive waste consists of pathogenic and infectious waste (including carcasses) and all ancillary equipment (syringes, test- tubes). This waste should be placed in a 30 gal container, lined with 4 mil plastic liner, and placed upright within a DOT-approved 55 gal drum. The annular space within the container should be filled with approved absorbent to a capacity of 200% of the liquid contained.

California Biohazard Regulations

Large-Quantity Generators

Large-quantity generators are required to register with the enforcement agency (DHS or county). Shared facilities in the same building, property, or adjacent facilities may register as a single generator. The registration must be renewed annually (90 days before it expires) and updated if any material change occurs in the operation. A medical waste management plan also must be filed with the enforcement agency on specified forms. The waste management plan must include the following information:

1. The name of the person filing the plan
2. The business address of the person
3. The type of business
4. The type and estimated monthly quantity of medical waste generated
5. The type of treatment used on site, if applicable (For generators with onsite medical waste treatment facilities, including incinerators or steam sterilizers or other treatment facilities as determined by the enforcement agency, the treatment capacity of the onsite treatment facility)
6. The name and business address of the registered medical waste hauler used by the generator to have untreated medical waste removed for treatment, if applicable
7. The name and business address of the registered hazardous waste hauler service provided by the building management to which the building tenants may subscribe or are required by the building management to subscribe, if applicable
8. The name and business address of the offsite medical waste treatment facility to which the medical waste is being hauled, if applicable
9. An emergency action plan complying with regulations adopted by the department
10. A statement certifying that the information provided is complete and accurate

Other regulations governing large-quantity generators include these:

- Treatment and tracking records must be maintained for three years.
- Large quantity generators are subject to at least annual inspections by the enforcement agency.
- Containment and storage of medical waste must meet the requirements described for all medical waste generators.
- Registration and annual permit fees are established by the DHS and the counties.

Small Quantity Generators

Small-quantity generators are required to register if the small quantity generator treats medical wastes on site using steam sterilization, incubation, or microwave technology. Small-quantity generators required to register must file a medical waste management plan with the local agency, which includes the following information on forms specified by the agency:

1. The name of the person filing the plan
2. The business address of the person
3. The type of business
4. The types and estimated monthly quantity of medical waste generated
5. The type of treatment used on site
6. The name and business address of the medical waste treatment facility used by the generator for backup treatment and disposal of waste for which the onsite treatment method is not appropriate due to hazardous or radioactive characteristics, and the name of the registered hazardous waste

hauler used by the generator to have untreated medical waste removed for treatment and disposal

7. A statement indicating that the generator is hauling the medical waste generated in his or her business under the limited-quantity hauling exemption (described later) and the name and business address of the offsite medical waste treatment facility to which the medical waste is being hauled, if applicable

8. The name and business address of the registered medical waste hauler service provided by the building management to which the building tenants may subscribe or are required by the building management to subscribe, if applicable

9. A statement certifying that the information provided is complete and accurate

Other regulations governing small-quantity generators include these:

• Treatment and tracking records must be maintained for three years.

• Containment and storage of medical waste must meet the requirements described for all medical waste generators.

• Registration and biannual permit fees are established by the DHS and the counties (currently $100).

Small quantity generators that do not treat medical wastes on site are not required to register. Such small quantity generators are required to maintain the following records in their files for two years:

1. An information document stating how the generator contains stores, treats, and disposes of any medical waste generated through any act or process of the generator

2. Records of any medical waste transported offsite for treatment and disposal, including the quantity of waste transported, the date transported, and the name of the registered hazardous waste hauler or individual hauling the waste

Medical Waste Haulers

Medical waste haulers are required to be registered. However, a medical waste hauler that generates less than 20 pounds of medical waste per week, transports less than 20 pounds at any one time, and maintains the documentation required of small-quantity generators, may transport the waste under a limited quantity hauling exemption. The exempt hauler must maintain a medical waste tracking document.

Medical waste haulers who are not exempt must comply with the act's requirements, which include these:

1. Registration with the enforcement agency

2. Issuance of a registration certificate by the agency

3. Transportation of wastes to treatment facilities or transfer stations

4. Transportation of wastes in leak-resistant and fully enclosed rigid containers in vehicle compartments separate from other wastes being transported

5. Issuance of permits to transfer facilities by the enforcement agency and with annual inspections (The permit fee for transfer facilities is $500.)

6. Proper protective clothing for personnel (Persons loading or unloading containers of medical wastes must be provided with and are required to

wear clean, protective gloves, coveralls, lab coats, or other protective clothing.)

7. Maintenance of tracking documents for all wastes removed for treatment or disposal for three years:

The generator must receive a copy of this document which includes the following information:

1. The name, address, and telephone number of the hauler
2. The type and quantity of medical waste transported
3. The name of the generator
4. The name, address, telephone number, and signature of an authorized representative of the permitted facility receiving the waste

Any hazardous waste hauler or generator transporting medical waste in a vehicle must possess a tracking document while transporting the waste. The tracking document must be shown upon demand to any enforcement agency personnel or an officer of the California Highway Patrol. If the waste is transported by rail, vessel, or air, the railroad, vessel operator, or airline shall enter on the shipping papers any information about the waste which the enforcement agency may require. Medical waste transported out of state shall be consigned to a medical waste facility in the receiving state. If there is not a permitted facility in the receiving state, or if medical waste is crossing an international border, the waste must be treated to render it safe before it is transported.

Medical Waste Treatment Facilities

Medical waste treatment facilities must be permitted. Offsite facilities are permitted and inspected by DHS. Onsite treatment facilities are permitted and inspected by the enforcement agency as mandated by federal, state, or local regulations.

The treatment facility registration and permit application must include the following information:

1. The name of the applicant
2. The business address of the applicant
3. The type of treatment provided, the treatment capacity of the facility, a characterization of the waste treated at this facility, and the estimated average monthly quantity of waste treated at this facility
4. A disclosure statement about the ownership and regulatory history of the facility
5. Evidence satisfying the department or local enforcement agency requirements that the operator of the medical waste treatment facility has the ability to comply with the regulations adopted pursuant to the Medical Waste Management Act
6. Any other information required by the DHS or the enforcement agency for the administration or enforcement of regulations

The decision to grant the permit must be made within 120 days of the application. The permit is effective for five years, unless revoked by the DHS or enforcement agency for cause.

Other regulatory requirements include recordkeeping for three years regarding the facility and its capacity, operating records, and tracking documents for wastes received and treated.

The annual permit fees for an offsite treatment facility are $10,000 for an autoclave and $15,000 for an incinerator or other approved technology. DHS may charge the applicant $100 per hour, up to a maximum of $50,000, to process the application or as otherwise provided in the department's regulations. Permit fees and application processing fees for onsite treatment facilities are established by the local enforcement agency.

Fines start at $1,000 per violation and increase to $10,000 per day per violation. The penalty for the first offense is a misdemeanor with a $2,000 fine and a year in the county jail. Subsequent and serious violations are felonies with a minimum fine of $5,000, a maximum fine of $25,000, and 1 to 3 years in state prison.

Study Exercises

1. Define and identify etiologic agents.

2. Distinguish between ionizing radiation and nonionizing radiation.

3. Describe and distinguish virus, bacteria, mold, fungus, protozoa, and parasite.

4. Describe the principal toxicity of asbestos.

5. Describe the principal toxicity of lead.

6. Describe the principal toxicity of radon.

7. Describe the principal toxicity of the criteria air pollutants.

8. Describe the principal types of indoor air pollutants.

9. Describe the principal sources of infectious wastes.

10. Identify three different organs affected by lead exposure and the symptoms associated with such exposure.

11. Explain why children especially should not be exposed to lead. What are the manifestations of childhood lead poisoning?

12. Explain how radiation may result in mutations, teratogenic effects, or cancer.

13. Why does irradiating food keep it from spoiling? Would the handling or consumption of such materials subsequent to irradiation be hazardous to your health?

11 Industrial Hygiene and Occupational Health Hazards

Arleen Goldberg, CIH

Contributing Author

Overview

Occupational health hazards are those hazards present in a working environment that pose a risk to personnel. It is crucial to develop an awareness of all the different types of hazards that may routinely exist in the workplace. The purpose of this chapter is to provide an overview of occupational and health hazards. The reader will understand the need for involvement in occupational health and the rationale for industrial hygiene concepts. It is also critical to understand what hazards may occur and to develop a repertoire of specific examples for the various hazard categories.

Industrial Hygiene

Industrial hygiene is the field of study concerned with the anticipation, recognition, evaluation, and control of hazards arising in the workplace. Of primary concern in this field is the health of workers. The American Industrial Hygiene Association (AIHA) defines industrial hygiene as "that science and art devoted to the anticipation, recognition, evaluation, and control of those environmental factors or stresses arising in or from the workplace, which may cause sickness, impair health and well-being, or cause significant discomfort among workers or among the citizens of the community." In addition, the training and education of the workers, management, and the citizens of the community is of growing importance to the organization.

Anticipation and recognition of environmental hazards and stresses associated with work is of prime importance to the industrial hygienist. The magnitude of the problem is determined from a review of the scientific literature, followed by documented quantitative measurement techniques. Once the magnitude of the problem is determined, it is necessary to devise and recommend methods to minimize or remove the hazard.

Training and education are of critical importance. Employers and employees, as well as the potentially affected citizens around the facility, must be informed of the potential health hazards associated with the facility or industry. It is important that employees learn to recognize potential hazards, develop the means of preventing these hazards, and respond to hazards or emergency situations that may arise.

AAIH Code of Ethics

The Code of Ethics of the American Academy of Industrial Hygiene (AAIH) lists these primary responsibilities of the industrial hygienist:

1. Protect the health of employees.
2. Maintain an objective attitude toward the recognition, evaluation, and control of health hazards regardless of external influences.
3. Counsel employees regarding health hazards and necessary precautions to avoid adverse health effects.
4. Respect confidences, advise honestly, and report findings and recommendations accurately.
5. Act responsibly in the application of industrial hygiene principles toward the attainment of healthful working environments.
6. Hold responsibilities to the employer subservient to the protection of the employee health.
7. Monitor and analyze the extent of exposure in the workplace.
8. Evaluate the administrative, workplace, and engineering controls used to control the hazards in the workplace.
9. Quantify the environmental stresses that may endanger life and health, accelerate the aging process, or cause significant discomfort in the workplace.
10. Provide an expert opinion as to the degree or amount of risk posed by the environmental stress.
11. Develop corrective measures in order to control health hazards by eliminating or minimizing exposure.

Industrial hygienists may have backgrounds in chemistry, biology, physics, or engineering. They have either acquired master's degrees or have demonstrated equivalent levels of expertise, based on extended experiences in evaluating the health effects of chemical and physical agents in the workplace. The industrial hygiene technician or technologist aids the industrial hygienist in his or her work. The role of the industrial hygienist in the occupational setting is something of a cross between a detective and a coach. Like good detectives, they seek out hazards lurking in the workplace; like a good coach, the industrial hygienist will help workers to avoid those hazards.

Management and the industrial hygienist must communicate effectively to create an optimal program for a healthy workplace. The industrial hygienist must exercise a variety of skills to successfully bridge the employee–employer gap and to implement the requisite safety measures and programs. Thus, the industrial hygienist is like a coach communicating with both the owners and the players to achieve the end goal of winning the game. In order to succeed, the IH must win the respect of the players, make them fully aware of the opposition, and train and prepare them well before they go into the game. Employees must respect the industrial hygienist and fully understand the hazards associated with the workplace, including knowing what actions and equipment are necessary for worker protection while still enabling them to get the job done.

As a good detective, the industrial hygienist spends a lot of time on preliminary research, gathering information about processes or similar facilities prior to the time any investigation actually begins. This allows the IH to know what clues to look for during the investigation and helps her solve the case.

The industrial hygienist routinely interacts with many other professionals in detecting hazards to worker health. Key occupational health team members include the occupational physician, the occupational nurse, the toxicologist, the health physicist, and safety professionals. Safety professionals study work operations and potential hazards in order to make recommendations to minimize injuries (as opposed to illnesses) in the workplace.

Hazards

Hazards lead to an unsafe workplace or cause environmental stress. Hazards can be categorized into four groups. Before we look at these categories, let's decide exactly what a hazard is.

Student Exercise

The dictionary says a hazard is _____

_____

Your definition of a hazard is _____

_____

What do you think is a potential hazard? _____

_____

Is this different from a hazard? _____

_____

Chemical Hazards

Chemical agents in the workplace may be in the form of mists, vapors, gases, or solids (Chapter 14), and subject to either inhalation or skin absorption. Common occasions for hazard exposure to chemical substances include the following:

- Points of hazardous material measurement or sampling
- Transfer points for hazardous materials
- Packaging and unpackaging of hazardous materials
- Process operations involving hazardous materials, such as welding, machining, plating, spray coating, and using solvents
- Maintenance activities
- Process upsets and releases

(For more information on chemical hazards, review the material on toxicology.)

Physical Hazards

Physical hazards may include the following:

- Fire and explosion due to ignition, chemical reaction, or the sudden release of a pressurized gas such as the rupture of a gas cylinder or storage tank

- Oxygen deficiency, once the oxygen in the atmosphere is consumed, or when air is displaced by some other gas, like methane or carbon dioxide

- Temperature extremes leading to heat stress, stroke, or prolonged exposure to cold

- Noise that startles, annoys, or distracts workers
  (Continued exposure to loud noise can cause temporary or permanent hearing loss by damaging the hearing organs.)

- Electrical hazards, including frayed or bare wires, improperly grounded electrical work, wetting of normally dry work areas, and catastrophic damage to electrical lines, such as those that occur after earthquakes

- General physical safety issues
  (Even common objects can often constitute hazards. Ladders and scaffolding, cat walks, and access to elevated structures pose risks and require extreme caution. Holes and ditches, slippery surfaces, steep grades, uneven terrain, trenches, and unstable structures can be hazardous. Sharp objects, precariously positioned objects, and jagged edges may cause injury and damage.)

- Ionizing radiation
  (Discussed in Chapter 10, this is invisible and undetectable by human senses and can be deadly. Areas where exposure may occur should be avoided.)

- Electromagnetic radiation
  (This has been the subject of much research and discussion. Monitors and televisions, CRT's, and the areas around high tension lines may impose risk.)

Ergonomic Hazards

These include improper lifting, poor seating arrangements, repeated motion (especially in awkward positions), improper keyboarding, improper monitor positioning, poor lighting, and poorly designed tools and work areas.

Biological Hazards

These include contamination by biological organisms such as molds, fungi, bacteria, viruses, plants (such as poison oak), and protozoa. (See Chapter 10)

Healthful Working Conditions

In 1970, the Occupational Safety and Health Act (OSHA) was passed to "assure, so far as possible, every working man and woman in the nation safe and healthful working conditions." The U. S. Department of Labor has the responsibility for promulgating and enforcing occupational safety and health standards. Therefore, almost all employers are required to put in place some basic elements of an industrial hygiene program.

OSHA attempts to assure safe and healthful working conditions for workers. It is the responsibility of OSHA's compliance officers (inspectors) to enforce the law. It is also the responsibility of safety and health professionals in the various facilities to assist employers in adhering to the Occupational Safety and Health Act. To accomplish these objectives, it is necessary to have an understanding of the structure of the regulations enforcing OSHA.

Lets review the following definitions relating to this chapter:

Health

The World Health Organization states that "health is a state of complete mental, physical, and social well-being." Most attention is focused on the physical well-being. The mental and social well-being have recently begun to receive more attention.

Curative Medicine

In this country, a large group of professionals are involved in actively addressing public health problems. In general, physicians, nurses, and medical facilities are concerned with treatment in the area of public health medical problems. First aid is an elementary form of curative medicine.

Preventive Medicine

Preventive medicine seeks to prevent illness. This is accomplished with education, training, hazard elimination, establishing improved work practices, and establishing a medical surveillance program. One intention of the OSHA is to attempt to prevent workers from suffering illness and injury by providing mandatory standards for workplaces.

Industrial Hygiene

Industrial hygiene is the term used for the practice of maintaining a safe and healthful work environment. The industrial hygienist is the professional concerned with preventive medicine at the worksite.

Epidemiology

This is literally the study of epidemics—large outbreaks of infectious disease. The epidemiologist is a professional involved in the study of disease and its spread in human populations. The epidemiologist is interested in studying the host (person affected), the agent (the infectious substance), the vector (how it was transmitted to the victim or host), and the environment (circumstances or situations in which the event took place). The epidemiologist's work includes reviewing medical records, analyzing samples taken in the workplace, in-plant observations, and searching other historical data sources such as death certificates or payroll records. (See Chapter 16)

Occupational Stress

An occupational stress is work related to physical or mental tension. It is typically caused by exposure to harmful agents or conditions in the work environment. Humans resist stress to a large degree, but when a person's resistance wanes, occupational illness or disease can occur.

Ergonomics

The study of the man-machine relationship. It includes both physical stresses and psychological stresses. The proper design of a forklift seat, or the design of a keyboard or automatic hand scanner involve ergonomics. The purpose is to make the worker more comfortable, improve performance, and reduce risk of injury or accident.

Occupational Disease

A health condition is defined as an occupational disease when it is caused by the work environment. Examples include mercury and lead poisoning which have been associated with the trades of metal smelting, felt making, and house painting.

Regulations and Standards

We live in an age of exploding industrial technology. Safety and health professionals, as well as employees and employers, must be able to anticipate and recognize hazards in the workplace. Standards have been written into federal, state, and local regulations to aid in achieving a safe and healthful workplace. Just as the coach knows his rule book, the industrial hygienist must be aware of the regulations and standards in the workplace and try to achieve adherence to requirements.

Several decades ago, the American Conference of Governmental Industrial Hygienists (ACGIH) started setting acceptable levels for occupational hazard exposure. Initially, these levels were called Maximum Allowable Concentrations (MACs). Now they are called **Threshold Limit Values** (TLVs). These exposure limits are continually being revised and re-evaluated based on epidemiological analysis of actual reported illnesses in work populations. These levels are not regulatory limits, but guidelines to further aid the industrial hygiene professional in achieving his or her goal.

Legally enforceable standards dealing with agent minimization, work practices, and administrative controls are covered by federal and state regulations. (See OSHA's General Industry Standards, 29 CFR 1910 000.) In California, the standards are referred to as Safety Orders.

Agent minimization standards set the maximum **Permissible Exposure Limits (PEL)** of a contaminant allowed in the workplace. For example, trichloroethylene has a permissible exposure limit of 100 ppm, meaning 100 parts of trichloroethylene for every million parts of air is the maximum permitted in any workplace.

Examples of work standards include improving sanitation and hygiene practices, changing work habits, and making other changes in employee work practices and job performance. An example of an administrative control is limiting the access or number of employees in areas where they may be exposed to hazards.

As we said at the beginning of this chapter, the four main duties of industrial hygienists and their staff are anticipation, recognition, evaluation, and control of environmental factors in the workplace. Other professionals frequently utilize a four-part methodology consisting of assessment, analysis, implementation, and evaluation. There are many similarities among these processes. The particular methodology used depends upon the goals and objectives of the process. It is very important to establish a pattern of use, to use it consistently, and to continuously recycle the sequence through the chain of events. Industrial hygienists view their role as being cyclic—moving from anticipation to recognition, to evaluation, to control implementation, then back again to anticipation. The wheel goes round and round.

Anticipation and Recognition

The hygienist may investigate a facility during routine operation or after recognition of some type of problem or incident. Either way, the overall process is similar.

The first task of hygienists is investigative. They anticipate by researching the particular process or type of facility so as to be able to enter the facility with a basic understanding of its function. They will anticipate what might be found by "blueskying," conducting a broad review, before coming on site. A knowledge of sources, bibliographies, and research skills is necessary. There is a need to review regulations and standards that generally cover this type of operation especially those that may be specific or unique to this process or facility.

Once the background investigative work is done, the hygienist is ready to enter the facility and start the real detective work. Normally, the first step in visiting the facility is to review records that are found on site. This would include records of injury and illness in the workplace, written respiratory or hearing conservation programs, written standard operation procedures, and other documentation that might be useful or required by the applicable standards, orders, rules, and regulations that were reviewed prior to arrival on the site.

After the hygienist is familiar with the written documents for the facility, it is time to begin with the formal assessment phase of the process. This step commonly starts with the **"walk-through."** As the name implies, you walk through each phase of the operation. During the walk-through, you typically need to draw a **plot plan** and mark the location of all employees, indicate their proximity to sources of potential hazards, and the duration of their workshift that is spent in this location.

The circumstances in the work situation are identified, and the work patterns reviewed (stationary versus mobile). Clues to hazards may be identified in the immediate and adjacent work areas, e.g., the condition of employee clothing, the cleanliness of the work area, and the proximity of additional hazards. Their senses, experience, and expertise are used in the initial, assessment walk-through, to help the IH detective gather clues and data, the evidence. As with any investigator, it is critical to take comprehensive and extensive notes during this stage. Depending on the situation, you may need to use a tape recorder, video recorder, or camera, as well as simple measuring devices to aid in this initial assessment.

Further information and evidence is obtained by questioning the employees, management, supervisors, and other personnel. Interpersonal skills must be highly developed in order to effectively question workers of various educational backgrounds and to discover the facts. Care must be taken to ensure confidentiality and to maintain the employees' sense of dignity. Although difficult, it is essential that the hygienist remain impartial and not lead the individual to potentially false or erroneous conclusions.

No two worksites are the same. Slight differences in the workplace may be significant in determining whether stress conditions prevail. Once you have seen a probable cause for concern, it is imperative to document the exposure. Therefore, samples must be taken, the evidence collected, notes made, and documentation completed for later evaluation. Subjective estimates of stress are often useful in overall analysis, even if they cannot be quantified at a later time.

After the walk-through is complete, the industrial hygienist will review all records and data compiled, identify areas of concern that warrant additional sampling, and develop a sampling plan that details the type and number of samples to be taken.

Selection of correct sampling techniques is critical to a sampling plan. (See Chapter 12) Sampling techniques must be selected based on the levels of accuracy available or required by current regulations. The sampling technique can be affected by the method of analysis being utilized by the laboratory. Direct reading instruments work well for some contaminants but may not offer a low enough detection level for others. Direct reading devices give an instantaneous answer. However, direct readings only evaluate one point in time while the relevant level may be one averaged over an entire 8-hour work shift.

Sampling includes either capturing the air containing suspected contaminants or extracting the contaminant from the air. The former require use of some type of container and a way to draw the air into the container. Vacutainers (bags or cans which have had all the air removed) may simply be unsealed in the work area and the air from the area of concern drawn into the container. Other containers require the use of a pump connected to the container via tubing to draw air from the area into the container. These containers are then transported to a laboratory for analysis.

While collecting the air itself is often sufficient for securing samples from a fixed sample location—called an **area or general area sample**—the hygienist often wants to know what the concentration is in and around the area where the employees are breathing, known as a **breathing zone sample**. These samples must be taken within a one-foot radius from the end of the employee's nose.

Instead of collecting the air, along with the contaminants, we may filter the contaminants out from the air, collecting them on a filter or in sampling media. To accomplish this, the sampling media is placed in a tube which is attached to a sampling pump. The end of the tube is attached to the employee's collar while the pump is attached to the belt. The assembly stays with the employee for a full shift, thus revealing the exposure averaged over the work period. After collection, the sampling filter is sent to a laboratory for analysis.

For chemical-specific sampling, an experienced industrial hygienist seeks the advice and cooperation of laboratory chemists. They are resources who can provide information on the exact sample media, sample collection and preservation techniques, holding times, and other specifics that must be used so as not to affect the laboratory analysis. They should also be involved in a discussion of laboratory limitations, such as equipment or analytical method, detection limits, and interference factors.

All instrumentation used, including collection pumps and direct reading instrumentation, must be calibrated before and after use to confirm the accuracy of data collected. The calibration of the instruments involved, and the quality of the data, is the responsibility of the industrial hygienist. The work is often conducted by the technician under the hygienist's direction.

Many things must be considered in order to select the proper type of sampling devices as well as to assure the proper location of sampling devices. During sample collection, the industrial hygienist and the technician keep copious notes regarding everything that transpires in and around the location where sampling is occurring to aid in data evaluation later. Work duties, proximity to various types of equipment, duration of duties, and the location of other workers are all crucial clues for workplace detectives. Great care is used during the entire process, and the procedures are covered in more detail in Chapter 12.

Evaluation

After all samples are analyzed in the laboratory and the data is received, this information, along with all the field notes from the sample collection activities and the walk-through, is reviewed in detail. Decisions must then be made to either conduct additional investigation or document the evidence of concern and develop possible alternative solutions. Once a hazard has been identified and documented, it is time to research alternative solutions and develop feasible methods of control. Our ultimate objective is to reduce or eliminate the hazard.

Hazard Control

How does the industrial hygienist "coach" control hazards in the workplace? Just as in a ball game, he may call time-out, remove a particular player or players, change the lineup, or bring in new plays.

The best control is eliminating the hazard. This can be done by eliminating the need for the process causing it, or by substituting something else in its place. We could isolate the hazard of the work by moving the hazard, or putting up a wall between the hazard and the employee. This is called **isolation**.

We can modify our process and change some of the equipment (**engineering controls**), change the way we handle a particular process or chemical, or limit the amount of time our employees spend in a particular area. While evaluating the effects of the control measures, or after finding the control measures are not completely successful, we can place the employee in protective clothing (PPE). This should be considered only as a temporary solution. (Use of personal protective equipment is the subject of Chapter 13.)

All of these are control alternatives. An iterative process is often necessary to identify the best alternative in a particular situation. This involves active input from all of the members of the team as well as fine tuning intervention activities.

The issues of analysis and implementation of controls for occupational stress agents are the topics of Chapter 14.

Implementation is perhaps the most difficult step in ensuring workplace safety and health. Even when it is the law, it is often difficult to sell the necessary measures to both employers and employees. Both management and workers must agree to buy the industrial hygienist's product if the workplace is to be safe.

Managers, supervisors, and workers must become aware of what serious health concerns are involved in their industry. They must be convinced that the hygienist's knowledge, estimates, and measurements are valid. They must accept the fact that they are at risk. They must believe that it is possible to mitigate this risk and realize that it is in their best interest to comply with the suggested measures.

The management level, which is the key to industrial operation, is in direct supervision of workers. Supervision puts management policy into practice. Company policies, such as mandatory use of respirators and special maintenance procedures, must be applied in an expeditious and even-handed manner. The supervisors are frequently the key to implementation because they are directly responsible for the operation.

Study Problem

Sam is in the garage using a very volatile methylene chloride based "Strip Away Anything Paint Remover" to remove 7 layers of old paint from Aunt Augusta's old bureau. Sam doesn't like restrictive clothing while he is working so he has on a t-shirt and shorts with tennis shoes. However, it is rather cool outside so he has two heaters on in the garage. The garage door is closed.

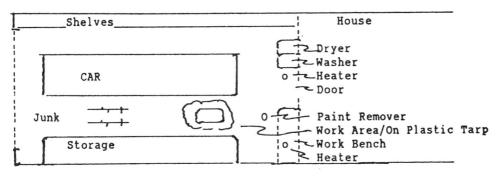

Figure 11-1

Floor plan of Sam's garage

Sam just closed the door to the house because Susie yelled that he was smelling up the kitchen and the 2-year-old twins were trying to join him in the garage. Since the house is small, the garage is packed with stored items, and there is limited space to move around. Sam's nose was itching about a half hour ago, but has since stopped.

What hazards do you think exist in Figure 11-1?

What controls would you suggest?

Sources of Technical Information on Toxic Substances

Environmental Protection Agency (EPA)
Office of Health and Environmental Assessment (OHEA) (RD-689)
401 M Street, SW, Room 3703
Washington, DC 20460
(202) 382-7345

Environmental Protection Agency (EPA)
Center for Environmental Research Information (CERI)
Office of Research and Development
26 West Martin Luther King Drive
Cincinnati, OH 45268
(513) 569-7562

Environmental Protection Agency (EPA)
Region IX
(415) 974-8131

National Institute for Occupational Safety and Health (NIOSH)
Information Dissemination, C-13
4676 Columbia Parkway
Cincinnati, OH 45226
(513) 533-8287

NIOSH Registry of Toxic Effects of Chemical Substances (RTECS)
U. S. Government Printing Office
Superintendent of Documents
Washington, DC 20402
(202) 783-3238 (213) 894-5841

American Conference of Governmental Industrial Hygienists (ACGIH)
6500 Glenway Ave., Bldg. D-5
Cincinnati, OH 45211
(513) 661-7881

American Industrial Hygiene Association (AIHA)
P. O. Box 8390
345 White Pond Drive
Akron, OH 44320
(216) 873-2442

Study Exercises

1. Identify the four facets of the practice of industrial hygiene.

2. Review the AAIH Code of Ethics and note which responsibilities fulfill the industrial hygienist's detective role and which reflect the role of the coach.

3. Define *hazard*.

4. Distinguish between chemical, physical, and biological hazards.
   Give examples of each.

5. Identify four common occasions of hazardous material exposure.

6. Identify the following:
   - HESIS
   - EPA
   - OHEA
   - NIOSH
   - RTECS
   - ACGIH
   - AIHA
   - ABIH
   - CA DHS RCHAS
   - CA DHS TSCD

12 Monitoring of Harmful Agents

Arleen Goldberg, CIH
Contributing Author

Overview

The intent of this chapter is to describe and characterize the various types of air monitoring instruments that are commonly used in the field. First, we will discuss the needs and implications of air monitoring, or sampling. Next, the advantages and disadvantages of the different air monitoring devices are delineated. Also, the availability of equipment and instruments and their usage is explained. Numerous examples are provided.

Air Monitoring

Monitoring air quality is crucial for ensuring worker health and safety. Air monitoring should be conducted at all work sites where there is a potential exposure of workers to hazardous materials. Such sites include industrial, waste cleanup, transfer, disposal, and emergency response activities. Identification and quantification of any and all possible hazards is a vital element of any health and safety program.

Sampling and the Sampling Plan

Each sample is acquired at a particular location at a specific time. By its nature, sampling can only provide a snap shot of the real situation. From these snap shots, it is necessary to develop an overall picture. In space, it is possible to create the big picture, a mosaic of snap shots. The more snap shots taken, the easier it is to compile the big picture, and the more accurate the rendition of the big picture. Often, it is necessary to evaluate the change in conditions over a period of time, the motion picture. Sequences of samples over a period of time provide for a time-lapsed motion picture. The greater the number of samples, the greater the resolution of our image of the situation and the more accurate our model of the situation. Unfortunately, sampling is very expensive; therefore, the objective of the sampling plan is to minimize the number of samples to minimize cost. The trick is to determine the minimum number of samples required to develop an accurate picture of the total situation.

Because of the cost, time, effort, and importance of sampling, sampling planning is a critical activity. Good planning of the sampling activity is essential. Therefore, before proceeding into the field, a considerable amount of work must be performed. To start, a preliminary assessment should be conducted. The initial assessment objectives are as follows:

1. Applicable regulations
2. Site characterization
3. Source characterization
4. Potential site contamination characterization
5. Health and safety planning

Sampling methodology can be divided into a number of steps. By walking through the steps, the sampling plan can be developed. Following the plan allows for the development of the picture. As the picture emerges, that plan is modified to fill in the blanks not revealed elsewhere.

The first step, conducted during the walk-through inspection of the facility, is mapping the site in time and space. A site plan provides the physical layout. Overlaid on the map, in time, are a number a variables. Wind and air currents, responsible for the dispersion of airborne contaminants, should be noted. Also important are the locations of personnel on the site and traffic patterns of both personnel and vehicles. A picture of where and when events are occurring on site is produced from this layout.

The next step is to identify contaminant **sources**. Sources may be either point sources, originating from a single point, or diffuse, coming from a general area. Sampling is required to identify these sources. In sampling along a line towards a suspected point source, the concentration of the contaminant will increase. If the contamination is from a diffuse source, samples taken from throughout the site should be relatively uniform.

After the source is determined, the third step is to identify contaminant **sinks**, i.e., where are the contaminants going? These may be point sinks, ventilation hoods, doors, windows, or diffuse sinks, which send the contaminants up into the air. Again, sampling can be used to identify these sinks. In general, contaminants will flow from source to sink. This flow is actually a detectable air current. For example, smoke tubes can be used to trace air currents. The other sink is that of personnel. Workers on site inhale the air. If information on the flow and concentration of contaminants is mapped, then concentrations around workers can be estimated. Personal monitoring can be used to determine actual worker exposure.

Plumes are areas of contamination downwind from a source (flow from a source to a sink). As the contaminant flows, it will diffuse. Generally, the width of a plume increases downwind from a source. If a centerline, taken upwind to the source is established, sampling on either side of the centerline provides information on the **dispersion**, or lateral diffusion, of the contamination within the plume.

Once the big picture is obtained, steps can be implemented to mitigate the effects of the contamination and protect the health and safety of the workers. Chapter 13 covers the use of Personal Protective Equipment (PPE), which can be employed immediately to protect personnel on site. Chapter 14 explores control methodologies which can be used to reduce or eliminate contamination and hazards.

There are many different terms used to refer to the types of samples collected. The most common terms will refer to samples taken on an individual employee or in a specific location. Samples taken on employees are referred to as **Breathing Zone** sampling or personnel sampling. Samples taken in a specific location may be referred to as **General Area** samples, zone samples, work zone samples, prevalent level samples, source samples, or simply, area samples. Source samples denote samples taken at the point of discharge or emission of some contaminant. General area or zone samples are taken in a room or area within a facility and may take into account multiple sources. These would represent conditions of potential exposure levels in the workplace.

The level of airborne contaminant is measured by collecting some of the air and analyzing it. How, when, and where the sample is taken can have significant impacts on the results. Documentation must be maintained on each sample taken, including any assumptions or estimations made in taking the sample. Collection technique and media can also significantly impact the results. Sample analysis techniques and procedures affect the results; knowledge of these are critical for successful interpretation of results. All of this information is required before the results can be fully evaluated. The skills of the detective are truly useful here.

Human senses are not sufficient to detect and evaluate all possible health hazards. Although some hazards such as dust, vapors, mists, fumes, and gases can be sensed, sensing is only partial and it is not quantitative. Certainly, any sensed hazard should be avoided, but, just because no hazards were sensed, does not mean the environment is safe. There are numerous lethal hazards which humans cannot sense. Human senses will also fatigue and degrade. Many gases can initially be smelled. However, over a period of time, the worker grows used to the odor, and although the gas is no longer smelled, it has not gone away. Gases which build up over a period of time while the worker is on site may not be detected at all. However, another worker walking on to the site would immediately detect the odor. There is great variability among humans as to what can be smelled and what levels can be detected. Using your senses is necessary, but not sufficient, for hazard detection.

Instruments and Equipment

A variety of measuring instruments are available for air monitoring and sampling. These can be divided into **Direct Reading** and **Indirect Reading** devices. Direct reading devices give an immediate indication of the presence of the hazard they are designed to detect. They are useful for early warning of hazards. They may be used to identify any situation of **Immediate Danger to Life and Health (IDLH)** conditions, e.g., oxygen deficiency, explosive or flammable atmospheres, toxic levels of some vapors, gases, and aerosols (carbon monoxide), and radioactive environments. Although some direct reading instruments are very sensitive and able to detect contaminants in concentrations down to the ppm level, they tend not to be as sensitive as a number of available laboratory analytic techniques. Therefore, to detect the low-level contaminants which may contribute to long-term exposure and its effects, indirect reading devices are utilized. Risk assessment of the workplace requires the use of both direct and indirect devices for air monitoring.

Direct Reading Instruments

Direct reading devices are the principal tool of initial site characterization. The data obtained by direct reading devices can be used to institute appropriate protective measures (evacuation or the selection of PPE), to determine appropriate equipment for further monitoring, and to develop sampling and analytic protocols. Direct reading devices tend to be very specific, measuring only a specific class of chemicals. Mixtures of chemicals may confuse the instruments and give false readings. Therefore, the validity of the results in response to unknown contaminants is suspect. The validity of the results is highly dependent upon the instrument's calibration. Devices should be calibrated before, during, and after sampling. **Calibration** is the process of adjusting the instrument readout so that it corresponds to the actual concentration. Instruments which are not calibrated give **Relative Responses**, such as when the meter deflection is proportional to the concentration. However, the absolute concentration is unknown. This information should be logged into the documentation of the mission.

Indicator Tubes

Indicator tubes tend to be colormetric; i.e., they change color in the presence of a specific substance. A glass tube is filled with material impregnated with an indicating chemical. The tube is connected to a piston or bellows-type pump. A known volume of contaminated air is pulled through the tube at a predetermined rate by the pump. Indicator tubes are available for a variety of hazardous materials, as they may be confused by mixtures. Accurate reading requires both an accurate determination of the amount of air pumped through the tube and the matching of the final color, or color position, to a standard. These are often referred to as "length of stain" indicators. Manufacturers report error factors of up to 50% for some tubes. Tubes have a limited shelf-life, are temperature sensitive, and frequently humidity sensitive. Indicator tubes are single-use instruments.

Indicator Badges

Indicator badges are designed for personal monitoring. They are also colormetric and have the same type of advantages and limitations as indicator tubes. However, rather than pumping air through the indicator material, they rely upon casual contact with air and the diffusion of the substance upon their surface.

Oxygen Monitors

Oxygen meters measure the percentage of oxygen in the air. They generally rely on an electrochemical sensor to measure the partial pressure of oxygen. They must be calibrated to compensate for altitude and barometric pressure. Some gases, e.g., oxidants (ozone), can affect the readings. Carbon dioxide can poison the detector. As always, successful operation requires a basic understanding of the operating principles and procedures for the specific instrument.

Combustible Gas Indicators

Combustible gas meters measure the concentrations of combustible gases and vapors. Generally, they contain a platinum filament which burns the combustible gas or vapor, so that the increased heat is measured. The instrument needs to be calibrated for a specific gas, as different gases can cause different readings at the same concentration. The filament can be damaged by certain compounds. It does not provide a valid reading under oxygen deficient conditions. Again, successful operation requires an understanding of the operating principles and procedures for the specific instrument.

Hydrocarbon Detectors

Photoionization detectors (PID) can detect many organic and some inorganic gases and vapors. These detectors ionize molecules using UV radiation to produce a current that is proportional to the number of ions. They do not detect methane or compounds with ionization potentials greater than the probe's energy level. Different gases cause different readings, as PIDs are sensitive to calibration of the gas used. Successful operation requires an understanding of the operating principles and procedures for the specific instrument.

The portable infrared (IR) spectrophotometer can detect many gases and vapors. It passes different IR radiation through the specimen. It is necessary to make repeated passes to achieve reliable results. It is not approved for use in a potentially flammable or explosive atmosphere, or where there is interference by water vapor and carbon dioxide. Successful operation requires an understanding of the operating principles and procedures for the specific instrument.

Radiation Meter

The Geiger-Mueller tube (Geiger counter), an ion and proportional detector tube, uses gas-filled tubes with a filament or a wire in the tube. Ionizing radiation passing through the tube leaves a trail which allows current to flow across to the filament. These tubes are not effective at detecting low-energy alpha or beta particles. They are relatively simple to use, however, interpretation of the data requires some experience.

Scintillation detection devices contain a solid, crystalline material, which flashes when struck by ionizing radiation. The light flashes are measured and counted by the instrument, giving an indication of the presence of alpha or gamma radiation.

Radiation meters are generally calibrated in milliroentgens/hour (mR/Hr) or in Rads or millirads (mR). Roentgens measure the presence of radiation, while rads are used to indicate absorbed dose. (See Chapter 10 for a more detailed discussion of these terms.)

Indirect Detectors

The most accurate method for assessing any air contaminant is to collect samples and have them analyzed by a reputable laboratory. But there are two distinct disadvantages to this method. The first is cost and the second is time. Having a laboratory analyze a large number of samples can be very expensive. Generally, laboratories charge a premium for quick turnaround. Onsite laboratories can reduce the turnaround time; however, they are expensive to operate. In emergencies, there is usually not sufficient time available to wait for a laboratory analysis.

Indirect methods require a four-step process. The first step is to collect a sample at the site; this will be covered in more detail. The second step is to transport the sample to the lab. Because many samples may be taken and transported together, it is imperative to accurately document the process. Information concerning the sample must be logged, the sample must be labeled and packaged for transport, and the package should be labeled. The third step is the laboratory analysis. Typically, a report is generated containing the findings of the analysis.

The fourth step is to review and evaluate the findings. It is best to log all of this information in a notebook. Since the notebook is a legal document and may be required in court for the defense and explanation of actions implemented, it is imperative that it be complete and accurate.

Several pieces of equipment are required for taking samples. The first is a standard industrial hygiene pump that has calibrated air flow. Depending on the contaminant and collection time wanted, different flow rates are desirable. For example, aerosol samples should be taken at a relatively high flow rate (2 liters/minute). Second, a flexible tube is used to connect the pump to the filter assembly. The tube needs to be large enough in diameter to accommodate the airflow at the desired flow rate. The third item is the collection assembly, which consists of some type of collection media and a holder. This assembly holds the media. The different types of collection media range from filters to various types of solid sorbent.

Solid Filters

A variety of different types of filters are available. Filters are selected based upon the material to be collected. Commonly used vapor filters include activated carbon, porous polymers, and polar sorbents. Each has advantages for collecting particular types of materials.

Activated carbon is useful in collecting vapors of materials with a boiling point above 0°C. Most odorous organic substances, such as solvent vapors are among those materials well adsorbed.

Porous polymers like Tenax or Chromosob (low ash polyvinyl chlorides) are good for collecting high-molecular-weight hydrocarbons, organophosphorous compounds, and a number of pesticides.

Polar sorbents such as silica gel are good for collecting organic vapors exhibiting relatively high dipole moments (e.g., aromatic amines).

Filters can frequently be put to double duty. For example, a filter can be weighed to determine total particulate or nuisance dust and then sent to the lab for destructive or nondestructive analysis for other contaminants (e.g., metals, vapors, or gases).

Particulate Filters

Atomic absorption (AA) is a laboratory analytical technique that has the capability of identifying most metals. This is a destructive test; special filters, AA filters, are used for these tests. The AA filter has a 0.8 micron pore size. The pore size refers to the average diameter of the holes, or pores, in the filter; they are about 150 microns thick. These filters have a very high collection efficiency and are not readily attacked by dilute acids or bases. They are very effective in collecting particulates. The AA filter does not depend on the flow rate or impact of the particle on the filter for collection.

AA filters are able to collect particulates much smaller than the 0.8 micron pore size for two reasons. First, the pores are not a straight shot through the filter. Particles become caught in the twists and turns of the pores. Second, there is an electrostatic attraction to most particulates. Therefore, the particles stick to the surfaces of the filter. The efficiency of the filter tends to increase after a few minutes, because a very fine layer of dust builds up on the surface and tends to act as an additional filtering medium.

Polyvinyl chloride (PVC) filters work well for gravimetric dust, liquid mist, and respirable silica. This type of filter has an average pore size of 5 microns. The larger pore size works well when sampling for particles with a wide variety of diameters. With the larger pore size, there is much less likelihood of the filter plugging up and reducing the flow rate. Just as with AA filters, these filters will capture particles that are much smaller than average pore size. Another advantage of the PVC filter is that it does not readily absorb moisture from the air and thus is more easily weighed with accuracy.

Liquid Media

Some contaminants may be collected by pulling air through a container called an impinger, which is full of a liquid solution such as dilute acid or water. The liquid is subsequently analyzed in the laboratory.

For information on environmental sampling, see *Sampling and Monitoring of Environmental Contaminants,* Richard C. Barth and Karl Topper, McGraw-Hill Inc. (1993). ISBN 0-07-005153-4.

SAMPLING TERMINOLOGY AND CALCULATIONS

Documentation

OSHA regulations, in-house program directives, and sampling procedure details determine your specific recordkeeping requirements. For accurate documentation, it is necessary to verify sample collection and to maintain a chain of custody for all samples. It is critical to document the entire sample collection process; therefore, it is important to arrange for this procedure before sampling begins. To prevent tampering and, thus, nonverifiable data, the technician must be at the site throughout the entire sample collection period. Accurate documentation of field sampling is crucial. In the event of controversy, field notes, including sampling data, are part of the permanent record and, thus, can be subpoenaed by court order.

Extreme caution is advised; the technician should note usual and unusual occurrences, work practices, or procedures that may contribute to worker exposure. Document routine operations and verify the flow rate for each of the sampling pumps during rounds, with a notation in the sampling log as to the current time. Record data on when pumps were operable, the sampling rate throughout the sample collection, and alterations to the flow rates caused by tubing disconnection, temporary battery failure, overloading of sample collection media, or tampering. Since the electricity, or a piece of machinery, can fail, there can be interruptions in production, or there can be a breakdown in the ventilation (exhaust) equipment. Constant monitoring is therefore critical. As any of these events will affect normal exposure readings, a sampling made under any of these conditions may represent a worst-case scenario, rather than a normal workday.

Personal Sampling: Explanation to Employee

The manager should always briefly explain to the employee what the monitoring process is, and why it is being performed. (If English is not a worker's primary language, arrangement should be made in advance to have an interpreter present.) Emphasis should be placed on the health benefits of workplace safety, wearing personal sampling devices, and worker cooperation and assistance. The worker can be the employer's best ally or worst enemy. It is imperative that time be taken to encourage worker participation and ensure that all their questions and concerns be addressed.

Refusal to participate is usually indicative of management's inability to effectively communicate with the worker. Benefits to the worker, should be emphasized; however, management should not confront the worker. Supervisors should not place themselves in a yes/no situation. Instead, it might be better to hold the discussion at a different time. That the employee's participation is required as a condition of employment is best downplayed. Instead, a peer might explain to the worker the benefits of personal sampling. If the company has a safety committee or a union, the worker should talk to the employee representative. If all else fails, the supervisor can follow the worker around with the sampling equipment to collect samples.

Sample Result Interpretation

How sampling is conducted affects the usefulness of the data. For instance, if one sample has been collected over an entire eight-hour work shift, (a **full-shift sample**), there is no way to determine peak concentrations, low concentrations, or to characterize what concentration levels the employee was exposed to during the workday. But with a **Time Weighted Average (TWA)** sample, we have an average of air quality experienced over the entire day.

A TWA is obtained by collecting a sample for a full shift. This may be accomplished by collecting just one sample or by collecting several samples and averaging the results. To determine the TWA, it is necessary to calculate the sampling time and the flow rates. This data is used to calculate the volume of air collected. To obtain concentration, one divides the weight of the collected sample by the total volume of air passed through the collector; this value is the TWA.

To determine other parameters (e.g., exposure during a particular operation or specific time of day), it is necessary to collect multiple samples throughout the day. Together, these represent the employee's full worshift exposure; individually they provide more data about specific time periods. Such an approach allows a more accurate characterization of employee exposure. When the standards call for the TWA, all of the filters collected for the same individual are simply averaged.

Example One: **Full shift**

Four filters were collected over a full shift (480 minutes).

Sample 1 was collected for 120 minutes; the concentration was calculated to be 6 mg per cubic meter.

Sample 2 was collected for 120 minutes; the concentration was calculated to be 4 mg per cubic meter.

Sample 3 was collected for one hour (60 minutes); the concentration was calculated to be 1 mg per cubic meter.

Sample 4 was collected for 180 minutes; the concentration was calculated to be 4 mg per cubic meter.

The samples are averaged as follows:

$$TWA = \frac{C1T1 + C2T2 + C3T3 + C4T4}{480 \text{ minutes}}$$

(C = Concentration, T = Sampling Time)

$$\frac{\begin{array}{r} 6 \text{ mg per cubic meter} \times 120 \text{ minutes} \\ + \ 4 \text{ mg per cubic meter} \times 120 \text{ minutes} \\ + \ 1 \text{ mg per cubic meter} \times \ \ 60 \text{ minutes} \\ + \ 4 \text{ mg per cubic meter} \times 180 \text{ minutes} \end{array}}{480 \text{ minutes}}$$

$$\text{TWA} = \frac{720 + 480 + 60 + 720}{480 \text{ minutes}} = \frac{1980}{480 \text{ min}} = 4.1 \text{ mg/m}^3$$

Where: mg/m^3 = milligrams per cubic meter

Now, let us assume, this material had a ceiling concentration of 5 mg/m^3 and a **Permissible Exposure Level (PEL)** of 4.5 mg/m^3. Looking at the individual results, notice that the ceiling limit was exceeded during collection of Sample 1. After calculating the TWA, notice that the TWA was not exceeded. However, using a single full-shift sample collection methodology, the excursion past the ceiling concentration would not have been noted. In this example all of the worker's shift was sampled and accounted for.

Example Two: Unsampled time during the work shift with zero exposure

Taking the previous problem, assume for Sample 4, the time of collection was 150 minutes. The total time sampled is now only 450 minutes. Therefore, there are 30 minutes of unsampled time. This 30 minutes must be accounted for in our TWA calculation. There are two ways to approach this situation. First, we can assume that the concentration during that time was zero. In other words, we add a Sample 5, with zero concentration, for 30 minutes. Based on OSHA's internal guidance for their compliance officers in the *Industrial Hygiene Field Operations Manual,* a TWA must be determined from a sampling time that exceeds 7 hours.

Calculate the TWA for Example Two using the formula given for Example One. Use Sample 5 with a 30 minutes duration and zero concentration.

Example Three: Unsampled time during the work shift, with potential exposure

The second manner to approach the situation of Example Two, is to make a professional judgment regarding the concentration during the unsampled time. Based upon the available information and documentation, you must make the assumption that the concentration of the unsampled period is the same, or essentially the same, as one of the sampled time periods.

In Example One, assume that we have documentation that the unsampled period was time an employee spent at a stationary operation, generating a known quantity of material, at a constant rate. During the entire time the employee was in the area, the exposure was at the same rate as during Sample 3. Therefore, we assume that the concentration for our 30 minute interval (Sample 5), instead of being zero, is the same as the concentration for Sample 3 or 1 mg/m^3.

Calculate the TWA for this situation, using Sample 1, 2, 3 and their respective concentrations from Example One, and Sample 4, from Example Two, and Sample 5 (one-half hour exposure at a rate of 1 mg/m^3 per hour) in your calculations.

Work Breaks

When evaluating an employee's personal full-shift exposure to airborne contaminants and noise, it is OSHA's intent that the employee carry the sampling equipment on his or her body at all times during the work shift. This is necessary in order to account for all movements and fluctuations in contaminant levels. If an employee leaves the worksite for lunch, or to run an errand, the technician or industrial hygienist must account for this time. In most cases, unless the supervisor can accompany the worker, the sampling assembly should be removed when the worker leaves the site and reattached when he or she returns. If lunch is taken on site, the technician must use professional judgment about whether to continue monitoring or to assume zero concentration for this time period. It is always easier to leave the sampling device on, if possible. Then there is no question as to whether exposure occurred or not.

If the workshift is eight hours, sampling must be conducted for at least seven hours for a valid full-shift sample. If you are only concerned with a single job operation, which occurs for a specific time period, you might only sample for that time. However, in such a situation, you are not obtaining a full-shift sample. Rather, you are obtaining a **grab** or a **short-term** sample.

Example Four: Total time

An employee is working in a glass manufacturing facility, performing a finishing grinding operation. The employee will be monitored for respirable silica dust. The following is a detailed documentation of the shift times of the employee:

TIME	SAMPLE
7:02 AM	Off — Employee starts work
7:10 AM	On — Start sampling pump at 1.5 lpm
8:10 AM	On — Pump Check—OK
9:10 AM	On — Pump check—OK
9:20 AM	On — Employee leaves job for restroom
9:30 AM	On — Employee returns to job
9:35 AM	On — Employee returns to work
10:30 AM	On — Pump check—OK
11:00 AM	Off — Pump OFF—Lunch in clean lunchroom
11:33 AM	On — Pump ON—Returned to work
12:30 PM	On — Pump Check—OK
1:30 PM	On — Break in lunchroom
1:45 PM	On — Return to work
2:10 PM	On — Restroom
2:20 PM	On — Return to work
3:20 PM	Pump OFF
3:30 PM	End of workday

The employee was not sampled during the lunch break. Sampling was conducted during the break periods, in the restroom, and during the coffee breaks in the lunchroom. These activities are included in workday activities. The time engaged at nonexposure functions resulted in ZERO exposure.

Work Shift Time: Before Lunch	238 min.
Times Sampled Before Lunch	230 min.
Unsampled Lunch Time	33 min.
(Time Out of Work Area)	(10 min.)
Work Shift Time After Lunch	237 min.
Time Sampled After Lunch	227 min.
(Time out of Work Area)	(25 min.)
Total Time Sampled	457 min.
Total Time Out of Area	35 min.

Example Five: **Calculation of volume**

The total volume of air sampled, expressed in liters, equals the flow rate in liters per minute multiplied by the total sampling time in minutes. (One cubic meter, m^3, is equivalent to 1000 liters.)

We conduct sampling for 480 minutes at a flow-rate of 2.0 liters per minute (lpm). Total volume sampled would be 960 liters or 0.96 m^3 of air.

Exercise

Part A: Go back to Example 4. Assume that the flow rate was 1.5 liters per minute. What would be the total volume sampled in liters, and cubic meters?

Part B: Continuing with Example 4, assume that the pump should have been calibrated at 1.2 liters per minute, but in the field, we found that the flow rate for the morning sampling was 1.1 liters per minute, while the afternoon samples were at 1.3 lpm.

   1. Calculate the volume sampled assuming the pump had been sampling at a rate of 1.2 lpm all day long.

   2. Calculate the actual volumes sampled in the morning and the afternoon using the 1.1 lpm and the 1.3 lpm flow rates.

   3. What, if any, effects do you think this level of variability might have on the final sample result, if we had submitted the total volume for the whole day based upon the 1.2 lpm value, versus the actual values you calculated?

Example Six: **Exposure time calculations**

The Safety Officer is taking a field sample in a carbon-black bagging operation. The sampling periods are as follows:

Sampling Period 1

| 8:01 am | Start pump |
| 11:20 am | Stop pump for employee (lunch break) |

Sampling Period 2

| 12:15 pm | Start pump |
| 4:34 pm | Stop pump; end sampling |

Calculate the total sampling time.

Example Seven: **Calculation of volume**

Air Sampled

$$\text{Flow Rate} \times \text{Total Sampling Time} = \text{Total Volume of Air Sampled}$$

The flow rate is known (from pump calibration). The sampling time is also known. By multiplying flow rate (lpm) by sample time, we get the total liters of air sampled.

$$\text{Total liters} = \text{lpm} \times \text{minutes} = \text{liters}$$

We convert total liters to cubic meters. There are 1000 liters in one m^3, so we divide the total volume (lpm × minutes) in liters by 1000 liters.

$$\frac{\text{liters} \times \text{m}}{1000 \text{ liters}} = \text{ volume in } m^3$$

Data: Total dust, sampling at 2.0 lpm

Sample time: 480 minutes

Example Eight: **Determination of sample weight**

To find the total sample weight, it is necessary to subtract the filter pre-weight from the post-weight.

Example	Data
Filter pre-weight:	9.785 mg
Filter post-weight:	16.505 mg
	16.505 mg
	−9.785 mg
	6.720 mg = Sample Weight

Example Nine: **Calculation of airborne concentration sampling for total dust or metals using one filter**

The two quantities needed for this calculation are the sample weight and the volume of air sampled.

Concentration = Sample weight / volume of air samples

Data

Sample weight: 7.438 mg
Time: 350 min.
Volume of air sampled: 0.70 m^3

Concentration = 7.438 mg/ 0.70 m^3 = 10.63 mg/ m^3

Example Ten: **Calculation for the TWA when more than one filter is needed**

Calculate the concentration for each sample period. Then multiply each sample concentration by the time for that filter. Sum these values. Finally, divide by the total time.

$$\text{TWA} = \text{Concentration (C1)} \times \text{Time (T1)} + (C2 \times T2) + \ldots (Cn \times Tn)/\text{Total time}$$

Data

Sample	1	2	3	4
Filter No.	1	2	3	4
Time (min.)	97	117	114	101
Flow Rate (lpm)	2.0	2.0	2.0	2.0
Weight (mg)	2.392	4.361	4.915	2.489

Example Eleven: **Short shift**

If, for some reason (short shift or incomplete sample), the entire sample is less than 7 hours, it is assumed that during the unsampled time (remainder of the 8-hour period) the concentration was zero. This period of zero concentration is included in the TWA calculation.

Filter 1 Time: 175 min.
Concentration: 5.32 mg/m^3

Filter 2 Time: 100 min.
Concentration: 7.89 mg/m^3

Time of Sample = 175 min. + 100 min.
= 275 min.

Unsampled Time = 480 = 275 min.
= 205 min.

$$\text{TWA} = \frac{(5.32 \times 175) + (7.89 \times 100) + (0 \times 205)}{175 + 100 + 205}$$

$$= \frac{931 + 789 + 0}{480} = \frac{1720}{480}$$

$$\text{TWA} = 3.58 \text{ mg/m}^3$$

If the entire sample is between 7 and 8 hours, but less than a full 8 hours, assume a concentration of zero for the unsampled time.

In some cases, given sufficient information, exposure can be estimated for the unsampled period, such as when an 8-hour operation is continuous and the concentration of the substance of concern does not vary. This is likely when the nature of the process is known to not vary substantially throughout the shift or day. The employee, by virtue of the nature of the job, may be assumed to be exposed continuously to essentially the same concentration. Therefore, assume that the exposure for the unsampled time is similar to that measured for the sampled time. In this situation, calculate a TWA by dividing the sample results by the actual time sampled. Then compare the resulting TWA with the 8-hour standard.

Note: Carefully document the rationale and all assumptions regarding unsampled exposure periods.

Silica Sampling

After taking a dust sample containing silica, two calculations are necessary. First, the TWA concentration of respirable dust must be determined. This is accomplished following the same procedure as that for total dust concentrations. Second, the PEL must be found. This is accomplished by using the percentages of silica in the respirable dust sample.

If more than one form of silica is present, (i.e., quartz, cristobalite, or tridymite) the proportions and composition of the mixture must be considered.

The formula for determining PEL from percentage of silica can be found in Table Z-3 of 29 CFR 1910.1000.

$$\text{Quartz (respirable) PEL} = 10 \text{ mg/m}^3$$

$$\text{Cristobalite PEL} = \frac{10 \text{ mg/m}^3}{2}$$

(½ the PEL value of the quartz standard)

$$\text{Tridymite PEL} = \frac{10 \text{ mg/m}^3}{2}$$

(½ the PEL value of the quartz standard)

Applying the mixture formula in 1910.1000(d)(2)(i), the PEL for the mixture is as follows:

$$\text{PEL} = (10 \text{ mg/m}^3)/(\% \text{ quartz}) + 2(\% \text{ cristobalite} + 2(\% \text{ tridymite}))$$

Insert the percentages, which are obtained from the laboratory analysis, into the formula. Calculate the PEL and compare it with the concentration calculated for the respirable dust sample.

Example Twelve: Respirable silica dust, one filter for entire sample

Data

Flow rate:	1.7 lpm
Sample time:	480 min.
Sample weight:	1.602 mg
Quartz:	17 %
Cristobalite:	8 %
Tridymite:	0 %

a. What is the respirable dust concentration?

b. What is the PEL for respirable dust from this sample?

If more than one filter is used for a silica sample, the lab analysis will indicate the percentages of quartz, cristobalite, and tridymite for each filter. It is then necessary to combine the percentages for each type of silica into overall percentages for the entire sample. The total weight of quartz on the first filter is the weight of dust on that filter times the percentage of that weight which is quartz. The total weight of quartz on the second filter is the weight of dust on that filter times the percentage of that weight which is quartz. The same is true for additional filters and additional components such as cristobalite.

Example Thirteen: **Multiple filters with respirable quartz dust**

Filter 1: Post-weight   2.500 mg
        Pre-weight   2.200 mg
                   0.300 mg = total weight of dust on filter

       Lab analysis: 15% quartz
                      7% cristobalite

                 $1 \times 15\%$ = percent which is quartz

       Total weight of quartz on filter 1:

$$0.15\% \times 0.300 \text{ mg} = 0.045 \text{ mg}$$

Repeat this procedure for all the filters used, add up the total weights of quartz on all the filters, and this is the total weight of quartz for the entire sample.

The same procedure is used to find the total weight of cristobalite.

Percentage of quartz for the entire sample:

$$\text{Total weight quartz} = \frac{\text{quartz weight filter 1} + \text{quartz weight filter 2} + ...}{\text{Total weight of all samples}}$$

Percentage of cristobalite in the entire sample is calculated in the same manner.

These percentages (for the entire sample) are the percentages used in the PEL formula.

Data

Flow rate:	1.7 lpm
Period 1:	220 min.
Period 2:	241 min.
Weight (filter 1):	2.198 mg
Weight (filter 2):	2.404 mg

Lab Analysis

Filter No.	1	2
Quartz	14%	20%
Cristobalite	5%	0%

a. What is the time-weighted average concentration, assuming zero exposure during the unsampled time?

b. What is the PEL for respirable dust containing quartz and cristobalite for this sample?

Homework Problems

Problem Set 1

Problem 1

Look at the following time analysis of a worker during sampling.

Time	Samples	
6:55 AM	Off	Start work
7:03 AM	On	Start sampling pump at 2.0 lpm
7:45 AM	On	Check pump—OK
8:45 AM	On	Pump check—OK
9:03 AM	On	Employee leaves job for restroom
9:12 AM	On	Employee returns
9:40 AM	Off	Changed sample media,
10:00 AM		Coffee break in the clean-air lunchroom
10:15 AM	On	Returns to work
10:45 AM	On	Pump check—OK
11:30 AM	Off	Lunch in clean lunchroom; pump turned off sample media changed
12:05 PM	On	Returned to work, pump turned on
1:15 PM	On	Check pump—OK
2:15 PM	Off	Break in lunchroom, sample media changed
2:30 PM	On	Return to work
2:45 PM	On	Restroom
2:50 PM	On	Return to work
3:20 PM	Off	Pump off; end sampling rate 2.0 lpm
3:30 PM		End of workday

Sample Results = 5 ,7 12, 22 milligrams per cubic meter respectively.

a. Calculate the appropriate exposure time and state all assumptions and justifications.

b. Based on concentrations given for each sample, calculate the worker's TWA.

Essay Questions

1. Does the time have to be recorded precisely or could we just "guestimate" and round to the nearest five-minute interval? Why or why not?

2. What are the advantages and disadvantages of taking one full-shift sample as in the case of Problem Set 1 and Problem 2, as shown below?

3. Would it be advisable to use military time when sampling? Why?

Problem 2

You are sampling three operators as noted below:

Operator One

7:01 am	Start pump
10:20 am	Stop pump; change sample media
11:15 am	Start pump
3:45 pm	Stop pump

Operator Two

7:45 am	Start Pump
3:50 pm	Stop Pump

Operator Three

8:00 am	Start pump
11:30 am	Stop pump
11:30 am	Lunch break
12:00 pm	Start pump
3:30 pm	Stop pump; end sampling

a. What is the sampling time for each of the above?

b. Given a sample time of 480 minutes and a flow-rate of 1.5 liters per minute, calculate total volume in liters and cubic meters.

c. Given a sample time of 453 minutes and a flow rate of 2.4 lpm, what is the total volume in liters and cubic meters?

d. Assume the sample weights for the previous three operators were 5, 7, and 10 milligrams of diethyl amine. Calculate the concentration in milligrams per cubic meter and ppm (mw = 73).

Problem 3

Assume that the contaminant we were previously monitoring the three operators for was for tetracyclic death. We sent our sample media to the laboratory.

Sample results

Operator One:	7 mg/m^3 and 15 mg/m^3 for Samples 1 and 2
Operator Two:	2 mg/m^3
Operator Three:	10 mg/m^3 and 14 mg/m^3

Assume the flow-rate for sampling of Operators 1, 2, and 3 was 2.5, 1.2, and 0.5 liters per minute respectively.

a. Calculate the sample time.

b. Calculate sample concentrations for each individual sample.

c. Calculate the TWA for each.

d. Given a PEL of 12 mg/cubic meter and a ceiling concentration of 15 mg/cubic meter, determine if each of these is in compliance.

Problem 4

Multiple collection media

On some occasions more than one sample is collected to represent an employee's daily or full-shift exposure. Using the examples from before, calculate the concentration for each sample. Then multiply each by its time and divide by the total time.

Sample number	1	2	3	4
Sample Time (min.)	100	100	108	110
Sample Flow rate (lpm)	1.5	1.5	1.5	1.5

a. Air volume in liters

b. Air volume in cubic meters

c. Concentration in mg/cubic meter
(Remember: Conc* = concentration of each individual sample)

$$\text{TWA} = \frac{\text{Conc* of \#1} \times \text{time of \#1} + \text{Conc \#2} \times \text{time \#2} + \text{Conc \#3} \times \text{time \#3, etc.}}{480 \text{ minutes}}$$

d. Is all of the shift time accounted for in your calculations above?

Problem Set 2

Problem 1

Sample Period 1
7:30 am	Start pump
11:45 am	Stop pump for employee

(lunch break)

Sample Period 2
12:30 pm	Start pump
3:25 pm	Stop pump; end sampling

What is the total sampling time?

Problem 2

Data: Total dust sampling at 2.0 lpm

Sample time: 453 minutes

What is the total volume of air sampled (in m^3)?

Problem 3

The Safety Officer is taking a total dust sample for a nuisance dust such as gypsum in a drywall manufacturing plant.

Data

Flow rate = 2.0 lpm

Sample Period 1

8:00 am	Start pump
11:12 am	Stop pump; employee lunch break

Sample Period 2

12:15 pm	Start pump
4:34 pm	Stop pump; end sampling

Filter post weight: 21.324 mg

Filter pre-weight: 10.942 mg

What is the concentration of airborne gypsum?

Problem 4

The Safety Officer is inspecting a large waste-paper shredder and baler located in a paper mill. There is no visible dust resulting from the process. However, when a forklift passed overhead on the second floor, dust was seen coming down from shaken rafters. The Safety Officer, therefore, decides to sample the shredder operator for TWA exposure for cellulose paper fiber. These are considered nuisance dusts.

Data

Flow rate = 2.0 lpm

Period 1:	7:32 am	Start pump
	10:45 am	Stop pump; change filter due to loading
Period 2:	10:50 am	Start pump
	12:00 pm	Stop pump; employee lunch break
Period 3:	12:30 pm	Start pump
	1:30 pm	Stop pump; change filter
Period 4:	1:35 pm	Start pump
	3:45 pm	Stop pump; ending sample

Filter No.	1	2	3
Post-weight	14.753	12.491	11.897(mg)
Pre-weight	9.342	10.211	8.423 (mg)

What is the TWA?

a. If the concentration for the unsampled time is not known?

b. Assuming that you have enough information to say that the unsampled time is essentially the same concentration as the sampled time.

Problem 5

Data

Flow rate:	1.7 lpm
Sample time:	439 min.
Sample weight:	1.703 mg
Lab analysis:	Quartz 12 %
	Cristobalite 2 %

a. What is the respirable dust TWA concentration, assuming zero exposure during the unsampled time?

b. What is the PEL for respirable dust containing quartz and cristobalite for this sample?

Problem 6

Sample Period 1

7:05 am	Start pump
8:56 am	Stop pump; remove and replace filter

Sample Period 2

9:02 am	Start pump
11:00 am	Stop pump; remove and replace filter

Sample Period 3

12:15 pm	start pump
3:12 pm	stop pump; end sampling

Filter No.	1	2	3
Post-weight	13.357	14.538	12.496 mg
Pre-weight	12.734	13.215	11.914 mg

Lab Analysis	1	2	3
Quartz	15%	18%	16%
Cristobalite	7%	6%	5%

a. Determine the TWA or equivalent exposure. (Assume zero exposure during the unsampled time.)

b. Is the TWA in excess of the PEL for respirable dust for this sample?

Study Exercises

1. Explain the principal uses of air monitoring for toxic contaminants.

2. List the limitations of relying on sensory awareness for monitoring the presence of toxic substances.

3. Describe the advantages and disadvantages of using Indirect Air Monitoring Instruments, including these:

   • Low-flow air sampling pumps
   • High-flow air sampling pumps
   • Passive dosimeter badges
   • Radiation film badges

4. Describe the advantages and disadvantages of using Direct-reading Air Monitoring Instruments, including these:

   • Indicator tubes
   • Indicator badges
   • Oxygen monitors
   • Combustible Gas Indicator (CGI, explosimeters)
   • Hydrocarbon Detector (PID, HNU)
   • Radiation meter (Geiger-Mueller Counter)

5. When collecting a field sample, what might be the advantages of collecting four samples—two during the morning and two during the afternoon—and averaging the results, versus simply taking one sample?

6. Describe and develop a full-shift (8-hour) sampling table annotated with assumptions and estimations required.

7. Identify limitations of sampling table data.

8. Calculate worker shift exposure (TWA) from a sampling table.

13 Exposure Limits and Personal Protective Equipment

Arleen Goldberg, CIH
Contributing Author

Overview

This chapter explores the need, use, and limitations of personal protective equipment (PPE). PPE is used to reduce worker exposure to potentially hazardous materials when other methods are inadequate. Included in this chapter is an analysis of those factors that should be considered in the selection of respiratory protective equipment, protective clothing, and head, eye, ear, hand, and foot protection.

Personal Protective Equipment (PPE)

Workers are often required to perform tasks in hazardous environments. The function of **personal protective equipment (PPE)** is to protect the worker from exposure to the hazardous environment in these situations. The regulations covering these situations are found in 29 CFR 1910. Training in the proper use of PPE is required prior to use in a hazardous environment.

There are three primary routes of entry (ROE), as discussed in earlier chapters. These are inhalation, direct skin contact, and ingestion. However, when evaluating PPE, a fourth category—injection—must be added. Respirators are designed to protect the lungs, gastrointestinal tract, and eyes against airborne hazards. Protective clothing is designed to protect the skin from contact with hazards. The ingestion of hazardous materials is avoided by the adherence to good personal hygiene practices, i.e., refraining from eating, smoking, or drinking on the job and the use of proper decontamination procedures upon exit from the hazardous environment. We protect against injection hazards by the use of heavy gloves and boots. The example that conjures up the most vivid mental picture is that of stepping on a rusty nail in a dirty industrial facility.

Levels of Protection

Equipment to protect the body against contact with known or anticipated hazards has been divided into four categories according to the degree of protection that is afforded.

Level A Should be worn when the highest level of respiratory, skin, and eye protection is required.

Level B Should be selected when the highest level of respiratory protection is required, but a lower level of skin protection is acceptable.

Level C Should be selected when the type of airborne substance is known, the concentration is measured, and the criteria for using air purifying respirators are met.

Level D Should NOT be worn on any site with respiratory or skin hazards. PPE at this level is primarily work clothing providing minimal protection.

Selection Criteria

The criteria for selection of the proper level of protection are these:
(See also Figure 13-1)

1. Type and measured concentration of the chemical or biological substance in ambient atmosphere and its toxicity.
2. Potential or measured exposure to substances in air, splashes of liquids, or other direct contact with material due to work being performed.
3. The scope and degree of physical hazard likely to be encountered.

The following information should be assessed and analyzed in order to evaluate the correct level of protection to be implemented. In situations when insufficient data is provided, such as when responding to an unknown spill, full protection, or Level A, should be selected.

- identity of the material
- its state (i.e., solid, liquid, or gas)
- its probable location in the environment (e.g., will it float on water or sink in air?)
- Threshold Limit Value (TLV)
- Immediately Dangerous to Life and Health (IDLH) concentration
- concentration at the site
- duration and relative chances of exposure during the planned activities
- Routes of Entry (ROEs) of the material into the body
- warning properties or early symptoms of exposure and whether these effects are delayed
- probability of a fire, explosion, or violent reaction occurring while personnel are in proximity
- adverse reactivities (e.g., water-reactive materials should not be exposed to water)

No single combination of PPE is capable of protecting against all hazards. The use of PPE can also create a spectrum of significant worker hazards. These include heat stress, physical and psychological stress, impaired vision, loss of mobility, and degradation of communications. It must be remembered that the

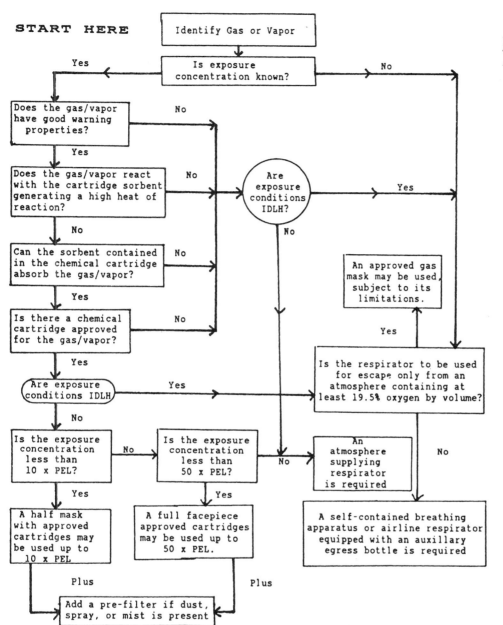

START HERE

Identify Gas or Vapor

Is exposure concentration known?

Yes / No

Does the gas/vapor have good warning properties?

Does the gas/vapor react with the cartridge sorbent generating a high heat of reaction?

Can the sorbent contained in the chemical cartridge absorb the gas/vapor?

Is there a chemical cartridge approved for the gas/vapor?

Are exposure conditions IDLH?

Are exposure conditions IDLH?

Is the exposure concentration less than 10 x PEL?

Is the exposure concentration less than 50 x PEL?

A half mask with approved cartridges may be used up to 10 x PEL

A full facepiece approved cartridges may be used up to 50 x PEL.

An approved gas mask may be used, subject to its limitations.

Is the respirator to be used for escape only from an atmosphere containing at least 19.5% oxygen by volume?

An atmosphere supplying respirator is required

A self-contained breathing apparatus or airline respirator equipped with an auxiliary egress bottle is required

Add a pre-filter if dust, spray, or mist is present

Plus

Figure 13-1

Decision chart for PPE

greater the level of protection, the greater the associated risks. Therefore, for any given situation, the lowest level of protection adequate for protection should be selected.

Selection Criteria

Due to the multiplicity of hazards that may exist in a given operation, care and diligence is required for the selection of the proper respirator. (See Table 13-1) This selection is made more difficult by the diversity of available respirators. Each type of respirator has its own limitations, areas of application, and operational and maintenance requirements. The proper selection of a respirator for any given situation first requires an assessment of the situation and subsequently an analysis of the following factors:

Table 13-1

Types of personal protective clothing required (USCG, Chemical Hazards Response Information System)

1 - Organic Vapor Canister	8 - Slicker Suit
2 - Air-Supplied Mask	9 - Rubber Suit
3 - Rubber Gloves	10 - Safety Helmet
4 - Chemical Safety Goggles	11 - Self-Contained Breathing Apparatus
5 - Face Splash Shield	12 - Acid Goggles
6 - Plastic Gloves	13 - Impervious Gloves
7 - Rubber Boots	

HAZARDOUS MATERIAL	1	2	3	4	5	6	7	8	9	10	11	12	13
Acetone	1	2	3	4	5								
Acetone Cyanohydrin		2	3	4	5	6	7	8		10			
Acrylonitrile		2	3	4	5	6	7	8		10			
Aerozine-50	1	2	3		5		7				11		
Anhydrous Ammonia			3	4			7				11		
Butadiene Inhibited			3	4			7		9		11		
Chlorine				4							11		
Ethyl Acrylate Inhibited		2										12	13
Ethylene Oxide		2		4	5		7	8					
Hydrazine Anhydrous		2				6					11		
Hydrocyanic Acid		2	3	4									
Hydrogen, Liquid				4	5						11		
Isobutane LPG				4							11		
Methyl Alcohol		2	3	4									
Methyl Bromide			3					8			11		
Methylhydrazine			3	4	5						11		
Nitrogen Tetroxide Liquid		2	3	4	5						11		
Oxygen, Liquid				4	5						11		
Propane LPG											11		
Propylene LPG	1	2		4	5								
Sodium Hydrosulfide Solution			3				7		9		11		
Sodium Hydroxide Solution	1		3	4	5		7						
Styrene Monomer Inhibited	1	2	3		5	6	7						
Toluene	1	2		4	5	6							
Unsymmetrical Dimethylhydrazine			3	4	5				9				
Vinyl Acetate	1	2	3	4	5	6							
Vinyl Chloride	1		3	4			7				11		

- nature of the hazard
- extent of the hazard
- work requirements and conditions
- characteristics and limitations of available respirators

Chemical and physical properties, toxicity, and concentration of hazardous materials are important considerations in respirator selection. Only respirators that provide an independent, breathable atmosphere should be used in oxygen-deficient atmospheres. In situations where the concentration of oxygen is deficient or the toxicity of material present in the atmosphere is high, use only those respirators listed as suitable for respiratory protection against oxygen deficiency. If in doubt, then it is necessary to proceed to the next level of protection. Normally, a SCBA, or airline respirator with auxiliary self-contained air supply, is used under these conditions.

IDLH situations require special care. If it is probable that an atmosphere immediately dangerous to life or health may occur, then both the normally expected inward leakage and the reliability of the respirator are primary considerations. In an oxygen-deficient atmosphere, with no toxic materials, inward leakage is normally not a problem, unless the leakage exceeds a small percentage. It is essential in highly toxic atmospheres that inward leakage be avoided.

Respirators

Several hundred respirators have been tested and certified under Bureau of Mines and NIOSH schedules. The 30 CFR Part 11 approval tests, as well as the old Bureau of Mines tests, were designed to approve respirators for protection against specific hazards or groups of hazards. Selecting the correct respirator for protection against a particular hazard requires a thorough knowledge of the types available. Choosing the best respirator from among the hundreds available is a formidable task. One methodology entails evaluating a respirator in terms of its components, and subsequently, assembling it as a system.

Respirator Classifications and Configurations

A way of categorizing respirators is by the physical configuration of the inlet covering. The respiratory inlet covering serves as an impervious barrier against the contaminated atmosphere and as a framework for attaching air-purifying or atmosphere-supplying elements.

Tight-fitting coverings are usually called facepieces. They are made of flexible molded rubber or plastic. Rubber, or woven elastic, headstraps are attached at two to six points. The straps buckle together at the back of the head, or sometimes, they may consist of a continuous loop of material.

Facepieces are available in three basic configurations:

- The first, called a quarter mask, covers the mouth and nose. The lower sealing surface rests between the chin and mouth. Good protection may be obtained with a quarter mask, but it is more easily dislodged than other types.

- The second type, a half mask, fits over the nose and under the chin. Half masks generally seal more reliably than quarter masks; thus, they are preferred for use against more toxic materials.

- The third type, a full facepiece, covers from the hair line to below the chin. These provide the greatest protection and seal the most reliably. In this configuration the lenses or eyepieces are built-in.

Note: the lenses or eyepieces incorporated into any respirator must meet impact and penetration requirements.

A special, tight-fitting respirator increasingly in use is the single-use, disposable type. It is shaped much like the half or quarter mask. The air purifier is permanently attached to the facepiece, or the entire facepiece, including the mouthpiece and nose clamp, is made of filter material. The mouthpiece is held in the teeth—the lips seal around it—and a clamp closes the nostrils. The air purifying elements may be either permanent or replaceable. These small devices are easily carried in a pocket and are designed primarily for intermittent use, such as in an emergency escape. They do not provide eye protection.

Loose-Fitting Respirators

Loose-fitting respirators include hoods, helmets, suits, and blouses. Loose-fitting respirators enclose at least the head, neck, and shoulders. The enclosure contains perforated rigid or flexible tubing, which allows clean compressed air to be distributed around the breathing zone. A light flexible device covering the head, neck, and shoulders is called a hood. Helmets are constructed of a rigid material. Blouses cover the entire torso, while suits encapsulate the entire body.

Chin-Style Respirators

Chin-style gas masks typically have a canister rigidly attached to a full facepiece. The useful lifetime is less than that of a front- or back-mounted canister, owing to its smaller sorbent volume. However, the lifetime is usually greater than that of chemical cartridges.

Escape Masks

Escape masks are gas masks for use during escape from atmospheres immediately hazardous to life and health. Gas masks are approved only if equipped with a full facepiece. Mouthpiece respirators can also be used for escape purposes.

Respirator Classification
Methods of Operations

Different methods exist for classifying respirators. First, they are classified based on their method of operation or how they work. Next, different configurations such as how they are constructed or what they look like will be explored.

Air Purifying Respirators

The first method to be considered for the operation of respirators is air purifying. As the name implies, this type of device purifies or cleanses the existing onsite air. This purification is accomplished by drawing air through filtration elements attached to the unit. The filtering elements are designed to remove a specific contaminant or contaminants. The physical configuration choices are quarter mask, half face, full face, helmet, or hood configuration. These configurations will be examined in more detail.

Many different air purifying elements are available. These different elements tailor the respirators for protection against specific contaminants. Particulate-removing elements that intercept particles such as dusts, mists, and fumes form one subclass of air purifying respirators.

Respirators that remove particulates are referred to as dust, fume, or mist respirators. All dust, fume, and mist respirators protect in the same manner: they remove and retain the particulate before it can be inhaled. All particulate-removing respirators use a filter made of fibrous material that removes the contaminant. As you inhale, air is drawn through a cartridge filled with the filtering media, and particles drawn into the filter are then trapped by the fibers. Whether or not a single particle will be trapped depends on factors such as size, velocity, and the composition and shape of particle and fiber.

Efficiency

No filter is absolutely efficient in removing all sizes of particles. Although a 100% efficient filtration is possible, it would be unusable as it would be almost impossible to breathe through this type of filter. As efficiency increases, breathing resistance increases. As particulate material collects on the fibers, making the openings between them smaller, breathing resistance increases. The filter becomes more efficient, but breathing becomes much more difficult.

High-efficiency dust, fume, and mist filters are designed to protect against particulate contaminants with a permissible exposure limit less than 0.05 mg per m^3. These filters are at least 99.97% efficient against 0.3 micron particles. A

Normal oxygen content of atmospheric air is 19.5 to 23.5%; the balance is predominately nitrogen.

The water content of compressed air required for a particular grade can vary from saturated to dry depending upon the intended use. If a specific water limit is required, it should be specified as a limiting dewpoint (expressed in temperature °F at one atmosphere absolute pressure) or concentration in ppm (v/v).

There should be no odor.

Table 13-2

Characteristics of Grade D and better breathing air (Compressed Gas Association, Inc., Air Specification G-7.1)

GRADES	D	E	F	G	H	I
Hydrocarbons	5 mg/m^3	5 mg/m^3				
CO	20	10	5	5	5	1
CO$_2$	1000	500	500	0.5		
Hydrocarbons, gaseous (as methane)			25	15	10	0.5
Nitrogen Dioxide				2.5	0.5	0.1
Nitrous Oxide						0.1
Sulfur Dioxide				2.5	1	0.1
Halogenated Solvents			10	1	0.1	
Acetylene						0.05

typical high-efficiency dust, fume, and mist filter consists of a flat sheet of material that is pleated and placed in the filter cartridge. The pleating provides a large filtering area which both improves the particle loading capacity, and reduces the breathing resistance. Some filters for protection against fumes of various metals, used on the so-called fume respirators, look similar. The difference is that the fume filter is less efficient (90-99%) against 0.6 m particles and is approved only for contaminants whose permissible exposure limit is 0.05 mg/m^3 or more. To aid in identification of filter use, they are color coded.

Single Use

The single-use, or disposable, dust respirator is very controversial. Because of problems with fitting and employee training, many institutions prohibit single-use respirators. Single-use respirator filters are an integral part of the facepiece, or the facepiece itself. When the filtering surface is permanently attached to the facepiece, the material often is a resin-impregnated natural wool fiber or synthetic fiber felt. In some currently approved devices, the entire facepiece is a fabric filtering medium. Single-use, disposable respirators are approved for pneumoconiosis and fibrosis-producing dusts, toxic dusts, dusts and mists, or dust, fumes, and mists.

Vapor- and Gas-Removing Respirators

Respirators that remove gases and vapors function by the interaction of contaminant molecules with a granular, porous material, commonly called the sorbent (sorption). Adsorption, absorption, and chemisorption are three sorptive mechanisms used in vapor- and gas-removing respirators. Adsorption, retains or holds the contaminant on the surface of the sorbent by physical or chemical attraction. In physical attraction, the molecules are weakly held. If heated, these bonds may be broken and the gas and vapor molecules released. If chemical forces are involved, the adsorption process is called chemisorption. The bonding in chemisorption is much stronger, and thus the media cannot easily be reactivated by heating.

A characteristic common to all adsorbents is a large surface area. Activated charcoal is the most common adsorbent. It is used primarily to remove organic vapors although it does have some capacity for adsorbing acid gases. Activated charcoal can also be impregnated or coated with other substances to increase its selectivity.

Absorbents differ from adsorbents in that, although they are porous, they do not have as large a surface area. Absorption is also different because the gas or vapor molecules penetrate deeply into the molecular spaces throughout the sorbent, where they are held chemically. Adsorption occurs instantaneously, whereas absorption is a slower process. Most absorbents are used for protection against acid gases. They include mixtures of sodium or potassium hydroxide with lime and/or caustic silicates.

A catalyst is a material that speeds a chemical reaction between substances without being consumed. One catalyst used in respirator cartridges and canisters is hopcalite, a mixture of porous granules of manganese and copper oxides. This catalyzes the reaction between the gases carbon monoxide (toxic) and oxygen, to form nontoxic carbon dioxide.

Limitations

Water vapor reduces the effectiveness of some sorbents and increases that of others. Vapor- and gas-removing cartridges must generally be protected from the atmosphere while in storage. The effect of humidity and temperature can be significant.

If a vapor or gas lacks adequate warning properties (odor, taste, irritation) in a concentration above the established time-weighted average concentration, a vapor- and gas-removing, air-purifying respirator should not be used. This is because there is no adequate way to gauge the continued effectiveness of the PPE.

Cartridges and canisters have a finite capacity. These respirators either remove vapors and gases from air or catalyze a reaction that converts toxic vapors or gases to nontoxic products. If an odor or taste of gas is detected in the inspired air, or eye or throat irritation is felt, the wearer should leave the hazardous area immediately and go to a safe area containing breathable air. Because of the limited useful service time of canisters and cartridges, they should be replaced periodically. This may be daily, after each use, or even more often, if the wearer detects or experiences an odor, taste, or an irritation.

A respirator wearer may detect an odor, or experience a taste or irritation for a very short time and then the sensation disappears. This is not necessarily because the penetration of an air contaminant into the respiratory or inlet covering has ceased. Rather, the nerve endings that cause a sensation of odor, taste, or irritation become fatigued, or their response has been dulled by low concentrations of substances. Thus, one may fail to detect low concentrations of some substances in air. This phenomena frequently occurs when the concentration increases very slowly.

Labeling

Vapor- and gas-removing cartridges and canisters are designed for protection against specific contaminants, or classes thereof. An American National Standard, ANSI K.13.1, established a color code for the various types of sorbent cartridges and canisters. (Table 13-3) The code identifies the contaminants the cartridges are designed to protect against. The printed approval label also clearly lists these contaminants. Whether the color code is memorized or not, the wearer should always check by reading the label! This is the only foolproof way of ensuring use of the correct cartridge or canister. ANSI K.13.1 has been included verbatim in the OSHA regulations, 29 CFR 1910.134(g).

Other types of canisters are designed for protection against multiple vapors or gases. In them, the sorbents are either arranged in layers or intermixed. In certain

Acid gases	White
Hydrocyanic acid gas	White with ½ inch green stripe completely around the canister near the bottom
Chlorine gas	White with ½ inch yellow stripe completely around the canister near the bottom
Organic vapors	Black
Ammonia gas	Green
Acid gases and Ammonia gas	Green with ½ inch white stripe completely around the canister near the bottom
Carbon monoxide	Blue
Acid gases and Organic vapor	Yellow
Hydrocyanic acid gas and Chloropicrin vapor	Yellow with ½ inch blue stripe completely around the canister near the bottom
Acid gases, Organic vapors, and Ammonia gases	Brown
Radioactive materials (except Tritium and noble gases)	Purple (Magenta)
Particulates (dusts, fumes, mists, fogs, of smokes) in combination with any of the above gases or vapors	Canister color for contaminant as designated above with ½ inch gray stripe completely around the canister near the top
All of the above canister near the top contaminants	Red with ½ inch gray stripe completely around the atmospheric
Gases not included in this table.	Orange; refer to label to determine protection

Table 13-3

Color code for vapor- and gas-removing canisters

instances, one type of construction has an advantage over the other, but mostly it is a matter of manufacturing convenience, with sorbent layering being most common.

Powered Air-Purifying Respirators

The powered air-purifying respirator uses a blower to draw contaminated air through an element that removes the contaminants. The purified air is, subsequently, supplied to the respiratory-inlet covering. The purifying element may be a filter to remove particulates, a cartridge to remove vapors and gases, or a combination filter and cartridge. The covering may be a facepiece, helmet, or hood.

Typically, the powered air-purifying respirator consists of an air-purifying element attached to the housing of a small battery powered blower. The blower is connected by flexible tubing to the respiratory inlet on the covering. The wearer carries the entire assembly on his person. A battery-powered, air-purifying respirator should supply air for at least 4 hours without recharging the battery.

The greatest advantage of the powered air-purifying respirator is that it supplies air at positive pressure. Thus, any leakage is outward from the facepiece. Theoretically, if the fit is poor, contaminated air cannot enter. The type and degree of protection depend on the air-purifying element. The protection level and useful service time depend on the size, shape, and composition of the filter and the nature and concentration of the contaminant. Unfortunately, at high work rates, it is possible, through rapid breathing, to create a negative pressure in the facepiece, thereby increasing leakage into the respirator around the facepiece.

Some of the advantages of air-purifying devices are that they are

- small
- relatively inexpensive
- easily maintained
- least restrictive to the worker's movement

Some of the disadvantages of air-purifying respirators are that they cannot be used

- in IDLH atmospheres
- when the contaminant has poor warning properties (except for escape)
- in oxygen-deficient atmospheres

Atmosphere-Supplying Respirators

As the name implies, in atmosphere-supplying respirators, air is supplied to the worker from some external source. The different types of atmosphere-supplying respirators are categorized by air supply method and regulation.

Self-Contained

The distinguishing feature of self-contained breathing apparatus (SCBA) is that the wearer is not connected to a stationary air source. The air source is carried with him or her. SCBAs can be classified as closed-circuit or open-circuit.

Closed-Circuit SCBA

Another name for closed-circuit SCBA is *rebreathing device*, which is indicative of the mode of operation. The air is rebreathed after the exhaled carbon dioxide has been removed. The oxygen content is restored from a compressed or liquid oxygen source or an oxygen-generating solid. Rebreathers are designed primarily for a 1 to 4 hour use in oxygen-deficient atmospheres as might be encountered during mine rescues. Rebreathers are still similar in design to those that were in use since the early 1900s, when the Gibbs and McCaa devices were first developed. It would be uncommon to find one of these in a regular industrial facility today.

Open-Circuit SCBA

In a typical open-circuit SCBA, a tank of high pressure compressed air, carried on the back, supplies air to a two-stage regulator that reduces the pressure for delivery to the facepiece. This regulator also serves as a flow regulator by passing air to the facepiece only on demand. The device must deliver 6 cubic feet per minute (cfm) to a helmet or hood. The flow rate must not exceed 15 cfm.

A flexible corrugated hose connects the regulator to the respirator-inlet covering, or facepiece. Because it must provide the total breathing requirements (not just the oxygen requirements, as in the closed-circuit SCBA), the service life of the open-circuit SCBA is shorter. Most open-circuit devices have a service life of 30 minutes.

In a demand-type regulator, air at approximately 2,000 psi is supplied to the regulator through the main valve. A bypass valve passes air to the facepiece in case of regulator failure. Downstream from the main valve, a two-stage regulator reduces the pressure to approximately 50–100 psi at the admission valve.

The admission valve is actuated by movement of a diaphragm and its associated levers. The admission valve stays closed as long as positive pressure in the facepiece (during exhalation) presses the diaphragm away from the valve assembly. Inhalation creates negative pressure in the face piece, and the diaphragm contracts, opening the admission valve and allowing air into the facepiece. In other words, air flows into the facepiece only on demand by the wearer, hence the name.

The pressure-demand regulator is very similar to the demand type, except that there is usually a spring between the diaphragm and inside the case of the regulator. This spring tends to hold the admission valve slightly open, theoretically allowing continual air flow into the facepiece. Because of the positive pressure, any leakage is outward. Therefore, a pressure-demand SCBA provides excellent protection.

Air-Line Respirators

Air-line respirators are supplied from an external source via an air line. They are available in demand, pressure demand, and continuous flow configurations. The respiratory-inlet covering may be a facepiece, helmet, hood, or complete suit, although there are presently no approval tests for suits. A full facepiece, helmet, or hood provides special protection against impact and abrasion from rebounding abrasive material.

Continuous flow air-line respirators maintain air flow at all times, rather than only on demand. In place of a demand or pressure demand regulator, an air flow control valve or orifice partially controls the air flow.

A flow of at least 4 cfm, to a tight fitting respiratory-inlet covering, and 6 cfm, to a loose-fitting one, must be maintained at the lowest air pressure and longest hose length specified. This means that, by design, the control valve cannot be closed completely, or a continually open bypass is provided to allow air to flow around the valve and maintain the required minimum rates.

Never replace an air flow control valve with another type of valve, even one from another type of respirator, or one from another manufacturer's air-line respirator. Besides possibly creating a hazard owing to improper air flow rates, substitution of another component negates NIOSH and MSHA approval of the device.

Air-line respirators provide a high degree of protection, but their use is limited to atmospheres not immediately hazardous to life. The reasoning is that the wearer is totally dependent upon the integrity of the air supply hose.

To be usable in an atmosphere immediately hazardous to life, an air-line respirator must have an auxiliary air supply to protect against potential failure of the primary supply. This is provided by adding a self-contained tank of high-pressure compressed air to a Type C or CE air-line respirator.

Because of the brief service time of the self-contained breathing air supply, combination units generally are used for emergency entry into and escape from atmospheres immediately hazardous to life. The self-contained air supply is used only when the air line fails and the wearer must escape, or when it is necessary to disconnect the air line, e.g., while changing locations. A combination air line and

SCBA may be used for emergency entry into a hazardous atmosphere (to connect the air line), if the SCBA part is classified for 15 minutes or longer service.

Written Respirator Program

Under a revised respiratory protection standard published January 8, 1998, by the Occupational Safety and Health Administration, employers will have to develop a written respiratory protection program (63 FR 1152). This final rule replaces the respiratory protection standards that were adopted by OSHA in 1971. The new respiratory protection final rule is effective as of April 8, 1998. A written respirator program must be developed by all facilities anytime a respirator is used. However, a written respiratory protection program is not required if a worker voluntarily wears a "filtering facepiece (dust mask)" in a situation in which a respirator is not required. The program should contain all the information needed to maintain an effective respirator program to meet the user's individual requirements. Standard Operating Procedures (SOP) should be written so as to be useful to those directly involved in the respirator program. This includes the program administrator, those fitting the respirators and training the workers, respirator maintenance workers, and the supervisors responsible for overseeing respirator use on the job.

The program should contain all information needed to ensure proper respiratory protection. It should be targeted towards a specific group of workers and against a specific hazard or several particular hazards. The hazard(s) must have been assessed thoroughly; otherwise the written procedures will have only limited validity.

Respirator Program Effectiveness

The program must outline how the company will achieve the following:

1. Determining what guidance, or hazard evaluation, was used to select the approved respirator(s) for protection against particular hazard(s).
2. Providing detailed instructions for training workers in proper use of the respirators including respirator fitting, fit testing.
3. Ensuring detailed maintenance procedures for these tasks:
   - cleaning, disinfecting, drying, inspection
   - repair or replacement of worn or defective components
   - storage
4. Developing administrative procedures for these tasks:
   - purchase of NIOSH approved respirator(s)
   - control of spare parts inventory
   - obtaining new respirators
   - getting respirators ready for reissue after maintenance
5. Developing procedures for issuance of respirators to ensure use of the proper one for a given hazard.
6. Ensuring the guidance of supervisory personnel in the continued surveillance of respirator use and determination of worker exposure to respiratory hazards.
7. Providing instructions for respirator use during emergencies.
8. Ensuring that all employees who utilize respirators are enrolled in a medical surveillance program. The program must state what is required and how employees are enrolled. The components vary depending upon the hazard being considered and the respirators selected.

9. Ensuring the respirator program's effectiveness. Employees must be trained sufficiently to be able to demonstrate a knowledge of why the respirator is necessary; how improper fit, usage, or maintenance can compromise the protective effect of the respirator; the limitations and capabilities of the selected respirator; how to deal with emergency situations involving the use of respirators or with respirator malfunction; how to inspect, don and remove, and check the seals of the respirator; procedures for maintenance and storage of the respirator; the medical symptoms and signs that may limit or prevent effective use of respirators; and the general requirements of the Respirator Standard.

10. Establishing specific provisions for annual training and retraining, as is required when changes in the workplace or in the type of respirator used render previous training obsolete.

The exact format of a written program may vary. The company with many workers wearing respirators and, perhaps, several respiratory hazards to consider, may formulate separate procedures for the selection and use of respirators for each hazard. The company with only a few workers to protect from one or two hazards may use a much simpler document. In any case, it is imperative that the program be complete for any and all contingencies.

Minimally Acceptable Respirator Program

ANSI Standard Z88 lists the following points to be included in a minimally acceptable respirator program:

1. Instruction in the nature of the hazard, whether acute, chronic, or both, and an honest appraisal of what may happen if the respirator is not used.

2. Explanation of why a more positive control is not immediately feasible. This should include a recognition that every reasonable effort is being made to reduce or eliminate the need for respirators.

3. Discussion of why this is the proper type of respirator for a particular purpose.

4. Discussion of the respirator's capabilities and limitations.

5. Instruction and training in actual use of the respirator and close frequent supervision to ensure that it continues to be used properly. This is especially essential for emergency use of respirators.

6. Classroom and field training in recognizing and coping with emergencies.

As stated above in Item 9 under "Respirator Program Effectiveness," the Respirator Standard now requires that the participant be able to demonstrate knowledge and understanding of these points.

Supervision Program

Supervisors, or those who oversee the daily activities of one or more workers who wear respirators frequently, should have a comprehensive knowledge of respirators and respiratory protection practices. Their training should include, but not necessarily be limited to, knowledge of the following:

1. Basic respiratory protection practices.

2. Selection and use of respirators to protect workers against every respiratory hazard to which they may be exposed.

3. The nature and extent of the respiratory hazards to which the workers may be exposed.

4. Structure and operation of the entire respirator program.

5. Understand of the supervisor's responsibility to facilitate functioning of the program, including these tasks:
   - maintenance that workers may be expected to do themselves
   - issuance of respirators
   - control of respirator use
   - evaluation of the program's effectiveness

6. Legal requirements pertinent to the use of respirators in his or her situation.

7. Training and education in proper use of respirators. For safe use of any respirator, it is essential that the user be properly instructed in its selection, use, and maintenance. Both supervisor and workers should be properly instructed by competent persons.

Training is designed to provide workers with an opportunity to handle the respirator, have it fitted properly, test its facepiece-to-face seal, wear it in normal air for a long familiarity period, and wear it in a test atmosphere.

Fit Testing

Every respirator wearer should receive fitting instructions, including demonstrations and practice in how the respirator should be worn, how to adjust it, and how to determine if it fits properly. Respirators should not be worn when conditions prevent a good face seal. Such conditions may be a growth of beard, sideburns, a skull cap that projects under the facepiece, dentures, or temple pieces on glasses. The presence of one of these conditions can seriously affect the fit of a facepiece. The supervisor should periodically evaluate the worker's diligence in observing these factors. To assure proper protection, the facepiece fit should be checked by the wearer each time the respirator is donned.

Air sampling data are important in the selection of the proper respirator. A thorough assessment includes the following information:
- identification of the contaminant
- nature of the hazard
- concentration at the breathing zone

The data are also essential in estimating the levels of exposure that may occur during respirator use.

Protection Factors

Air purifying respirators are typically rated with a **protection factor.** This is a combination of an empirical, measured value, and a **safety factor**. The protection factor is a ratio and relates to the rated efficiency of the respirator. For example, a protection factor of 10 implies that, at worst, the concentration inside of the respirator will be $\frac{1}{10}$ the concentration outside.

A half mask respirator with an approved cartridge may be used up to ten times the permissible exposure limit ($10 \times PEL$). Therefore, if the TLV for a material is 200 ppm, and the level in the environment is 1,000 ppm, a respirator with a protection factor of 10 would present the wearer, maximally, with a concentration of 100 ppm, so:

$\frac{1}{10}$ of 1000 ppm = 100 ppm

Some might say that it would therefore be safe to use this type of protection in this concentration, but this is not necessarily the case. Additional concerns must be evaluated.

For many contaminants such as asbestos, ACGIH and NIOSH have developed guidelines indicating that the concentration inside the worker's respirator should not exceed a level that would be acceptable to the unprotected individual. On asbestos worksites, the rule of thumb is that the concentration inside the worker's mask must never exceed the level for clearance of the space. This is the level at which unprotected workers, occupants, and visitors would be allowed to re-enter the space.

Therefore, the protection factor formula may have many applications. Protection factors can be used to determine the maximum outside concentration that would be desirable with a specific type of respirator. They can be used to predict the probable inside concentration given a type of respirator and outside concentration. Given the desired maximum inside concentration and the anticipated outside concentration, we can determine what protection factor would be necessary. This information can be used to aid in respirator selection.

In all of these cases, an air monitoring program should be in place to conduct sampling and evaluate the effectiveness of the instituted actions.

Advantages and Disadvantages

The respirator does not protect the skin or body against radiation from airborne concentrations of gaseous radionuclides. These include the radon and other relatively biologically inert gases. However, in these cases, the proximity of the gases to the tissue in the lung, and its possible circulation through the tissue by the blood stream, are the primary hazards.

The extent of the hazard in space and time and its physical location should be considered during respirator selection. Factors of import include these:

1. Length of time for which protection will be needed

2. Entry and exit times

3. Accessibility of a fresh air supply

4. Ability to use air lines or move about freely while wearing the respirator

Special consideration should be given to the location of the contaminated area with respect to a possible source of breathable air. In using a hose mask, supplied-air respirator, or abrasive-blasting respirator, the distance that the wearer can go into a contaminated atmosphere is limited by the length of hose. Furthermore, the presence of the hose requires the wearer to enter and leave the area by the same route. The suit should be equipped with an auxiliary air cylinder, chemical canister or cartridge, or particulate filter appropriate for use in withdrawal. While wearing a SCBA or a gas mask, a person may leave the contaminated area by an exit. However, the wearer should take into account possible delays to make certain that the device will afford protection until breathable air is reached.

Determining the length of time for which respiratory protection is needed includes the time needed to enter and leave a contaminated area and the work time. The work area to be covered, work rate, and mobility required of the wearer in carrying out work should be considered in respirator selection. The SCBA, gas mask, or chemical-cartridge respirator provide respiratory protection for relatively short periods. These respirators are best for short excursions.

The hose mask with blower, air-line respirator, and abrasive-blasting respirator provide protection for as long as the facepiece is supplied with adequate breathable air. If the atmospheric particulate loading is low, particulate-filter respirators can provide protection for extended periods, without the need for filter replacement. In such cases, for protracted missions, the Type B respirator, hose mask with blower, and air-line respirators are preferable.

Some respirators have a means for indicating the remaining service life. All SCBA have some type of warning. Some gas masks are similarly equipped. The warning device may be a pressure gauge, timer, audible or physical alarm, or window indicator in the canister. The user should become familiar with, and understand the operation and limitations of, each type of warning device. Most gas masks and chemical-cartridge respirators do not have an indicator of the remaining service life. Prevention is the only solution. Keep a log of canister and cartridge use, and change canisters and cartridges according to the manufacturer's directions.

Air-purifying respirators present minimal interference with the wearer's movement. Supplied-air respirators with trailing hoses severely restrict site access coverage. Hoses present a potential hazard, since the trailing hose can come in contact with machinery, objects, and obstructions. If the hose becomes damaged, the wearer is left in an awkward and hazardous situation. The SCBA presents a size and weight penalty, which may restrict climbing and movement in tight places.

Respiratory Minute Volume (RMV) is determined by the wearer's work rate, maximum inspiratory flow rate, and inhalation breathing resistance. The RMV is of great significance in self-contained and air-line respirators operated from cylinders. This is because the RMV determines the operating life of the respirator. Useful life under moderate work conditions may be only one-third of that under rest conditions.

Peak airflow rate is important in the use of constant flow air-line equipment. To maintain the respiratory enclosure under positive pressure, the air-supply rate should always be greater than the peak inspiratory flow rate. High breathing resistance, in air-purifying respirators, under conditions of heavy work, can result in negative pressure. Such a situation may permit the ingress and breathing of contaminated air.

At best, communication is difficult when encumbered with PPE. Speech is difficult when the mouth is covered with a respirator. Hearing is frequently restricted by encapsulating suits. Effective speech communication may be required in jobs for which the respirator is being selected. Alternatives are equipment incorporating two-way radios and hand signals. It is imperative that communication strategies be worked out in advance of any mission.

The ability to cope with stress is necessary when using PPE. Several varieties of **stress** need to be considered. First, due to the hazardous nature of work requiring PPE, there will be some **anxiety**. Second, loss of mobility and restricted movement due to the equipment is inevitable. Thus there is **added work** required to perform job functions while using PPE. Third, the added work and the extra effort generates **heat**. Because of the confinement of the equipment, normal heat loss is reduced. The result is an increase in core body temperature. Cognizance of performance degradation caused by temperature extremes is especially important in emergency situations.

Eye protection and eyepieces tend to obstruct vision. All facepieces will somewhat restrict the wearer's vision, increasing the potential for accidents. Other problems include the wearing of prescription glasses and fogging of the respirator lens. Respirators with a specially ground lens for vision correction are available. A hood allows the use of eyeglasses under low level protection situations. **Eyeglasses should never be worn under high level protection conditions** since, when in an encapsulating suit, they cannot be accessed and therefore may create a hazard if dislodged or fogged. **The use of contacts with PPE is prohibited**. When required, full facepiece respirator eyepieces and eye protection worn with half mask facepieces shall meet the pertinent requirements found in the American National Standard Institute (ANSI) Practice for Occupational and Educational Eye and Face Protection.

Worker Instruction and Training

The extent and frequency of the worker training depends primarily on the nature and extent of the hazard. If the hazard is a nuisance particulate, for example, the danger from misuse of the respirator is minimal. However, against highly toxic particulates, a single misuse may have serious consequences. The same holds true, of course, for gases and vapors. If the respirator is to be used in an emergency, training in its use should be comprehensive to the point of becoming reflexive. In any case, the worker requires basic instruction and training in respiratory protection practices.

Minimally, workers and supervisors should be trained in basic respiratory protection practices. Also, each should be trained in the use of the respirator selected for his or her particular situation. Because proper respirator use depends upon the wearer's motivation, it is important that the need for the appropriate respirator be fully explained. It is particularly important to discuss issues which directly affect the wearer since these can be used as motivational incentives.

OSHA regulations require that workers must be allowed to test the facepiece-to-face seal of the respirator and to wear it in a test atmosphere. During any fitting test, the respiratory head straps must be as comfortable as possible. Loosening or tightening the straps will sometimes reduce facepiece leakage. It is imperative that the wearer be able to tolerate the presence of the respirator for an extended length of time, or at least, for the duration of the mission. Because respirators from different manufacturers have different configurations, it is good to have an inventory of different sizes and styles for individuals to choose from to ascertain the best fit.

Negative Pressure Check

The wearer should perform the negative pressure check in the field before each use. It consists of closing off the inlet of the canister, cartridge(s), or filter(s) by covering with the palm(s), replacing the seal(s), or squeezing the breathing tube so that it does not pass air. Next, the wearer should inhale gently so that the facepiece collapses slightly and be able to hold his or her breath for 10 seconds. If the facepiece remains slightly collapsed and no inward leakage is detected, the respirator is probably tight enough. This test, of course, can be used only on respirators with tight fitting facepieces.

Because this check is simple, it has severe drawbacks. For instance, the wearer must handle the respirator after it has been positioned on his face. This handling can modify the facepiece seal. It is strongly recommended that the negative pressure check be used only as a gross determination of fit. While passage does not ensure a perfect fit, failure is a definite indication of a problem. The wearer can easily use this test as a last check before entering any hazardous environment.

Positive Pressure Check

This test is like the negative pressure test, and it has the same advantages and limitations. It is conducted by closing off the exhalation valve and exhaling gently into the facepiece. The fit is considered satisfactory if slight, positive pressure can be built up inside the facepiece without any evidence of outward leakage. For some respirators, this method requires that the wearer remove the exhalation valve cover and carefully replace it after the test. This is often a most difficult task.

Removing and replacing the exhalation valve cover often disturbs the respirator fit even more than the negative pressure test itself. Therefore, this test should be used sparingly when it requires removing and replacing a valve cover. The test is easy for respirators whose valve cover has a single small port that can be closed by the palm or finger. The wearer should perform this test just before entering any hazardous atmosphere.

Qualitative Fit Tests

Isoamyl Acetate Vapor

Isoamyl acetate has a pleasant, easily detectable odor. Therefore, it is used widely in qualitatively checking respirator fit. This test provides the wearer with an OSHA required opportunity to wear the respirator in a test atmosphere. Generally, it consists of creating a banana oil-permeated atmosphere around the wearer. It may be used with an atmosphere-supplying or air-purifying respirator with an organic vapor-removing cartridge(s) or canister. If the hazard is particulate matter or a nonorganic vapor or gas, the organic vapor cartridge(s) or canister must be replaced with a particulate filter(s) or proper cartridge(s) or canister after the test.

The isoamyl acetate fitting test is performed as follows:

1. The wearer puts on the respirator in a normal manner. If it is an air-purifying device, it must be equipped with a cartridge(s) or canister specifically designed for protection against organic vapors.

2. The wearer enters the test enclosure, or the saturated cloth or stencil brush is passed close to the respirator sealing surfaces.

3. If the wearer smells banana oil, he or she returns to clean air and
   • readjusts the facepiece, and/or
   • adjusts the head straps without unduly tightening them

The wearer repeats the second step. If the scent of banana oil is not detected, the wearer should assume a satisfactory fit has been achieved. With any detection of the banana oil scent, an attempt should be made to find the leakage point. If the leak cannot be located, another respirator of the same type and brand should be tried. If this leaks, another brand of respirator with a facepiece of the same type should be tried.

4. After a proper fit is obtained, if the respirator is used as a air-purifying device, it must be equipped with the correct filter(s), cartridge(s), or canister for the anticipated hazard.

5. During the test, the subject should make movements that approximate a normal working situation. These may include, but not necessarily be limited to, the following:
   • normal breathing
   • deep breathing, as during heavy exertion
     (This should not be done long enough to cause hyperventilation.)
   • side-to-side and up-and-down head movements
     (These movements may be exaggerated, but should approximate those that take place the job.)

- talking
  (This is most easily accomplished by reading a prepared text loudly enough to be understood by someone standing nearby.)
- other exercises, depending upon the situation.
  (For example, if the wearer is going to spend a significant part of his time bent over at some task, it may be desirable to include an exercise approximating this bending.)

If the test is used in training, the worker should select the respirator that fits him or her best and perform the complete set of exercises. The number of exercises may be reduced when the test is used as a quick field check before routine entry into a contaminated atmosphere.

There are several disadvantages to the isoamyl acetate test. The major drawback is that the odor threshold varies widely among individuals. Furthermore, the sense of smell is easily dulled and may deteriorate during the test so that the wearer can detect only high vapor concentrations. Another disadvantage is that isoamyl acetate smells pleasant, even in high concentrations. Therefore, a wearer may say that the respirator fits although it has a leak. This is usually because the particular respirator feels comfortable or because the wearer is following the lead of someone else and selecting the same respirator. Conversely, a wearer may claim that a particular respirator leaks if it is uncomfortable. Therefore, unless the worker is highly motivated toward wearing respirators, the results of this test may be suspect.

Irritant Smoke Test

The qualitative irritant smoke test is similar in concept to the isoamyl test. It involves exposing the respirator wearer to an irritating aerosol produced from commercially available smoke tubes. Normally, these smoke tubes are used to check the quality of ventilation systems. They are sealed glass tubes, approximately 12 cm long by 1 cm in diameter, filled with pumice impregnated with stannic chloride. When the ends of the tube are broken and air is passed through them, the material inside reacts with the moisture in the air to produce a dense, highly irritating smoke consisting of hydrochloric acid adsorbed on small solid particles.

As a qualitative means of determining respirator fit, this test has a distinct advantage in that the wearer usually reacts involuntarily to leakage by coughing and sneezing. The likelihood of the wearer giving a false indication of proper fit is reduced. On the other hand, the aerosol is very irritating and must be used carefully to avoid injury. It is advisable to have exhaust ventilation behind the subject. This ventilation serves to protect the person taking the test.

This test can be used for both air-purifying and atmosphere-supplying respirators, but an air-purifying respirator must have a high-efficiency filter(s). After the test, it may be necessary to replace the high-efficiency filter(s) on the air-purifying respirator with another type of air-purifying element(s). Ultimately, filter selection depends upon the hazard to which the respirator wearer is to be exposed. This test can be used for both worker training or respirator selection.

The irritant smoke test must be performed with proper safeguards since the aerosol is highly irritating. A suggested procedure is as follows:

1. Wearers put on the respirator normally, taking care not to tighten the head straps uncomfortably. They stand with their backs to a source of exhaust ventilation, such as a chemical fume hood.
2. The tester tells the wearers to close their eyes, even if they are wearing a full facepiece respirator and to keep them closed until told to open them.

3. The tester lightly puffs smoke over the respirator, holding the smoke tube at least 2 ft. from it. At this time, the tester should keep the amount of smoke minimal and pause between puffs to note the wearer's reaction.

4. If wearers detects no leakage, the tester may increase the smoke density and move the smoke tube progressively closer to the subjects, while still remaining alert to their reactions.

5. When the smoke tube has been brought to within about 6 inches of the respirator, with no leakage detected, the tester may start to direct smoke specifically at the potential sources of leakage, around the sealing surface and exhalation valve, while subjects hold their heads still.

6. At this point, if no leakage has been detected, the wearers may cautiously begin the head movements mentioned in the isoamyl acetate test. The tester should remain alert and be prepared to stop producing smoke if an emergency arises.

7. If leakage is detected at any time, the tester should stop the smoke and let the wearers readjust the facepieces or head strap tension. The tester should then start the test at the second step.

In all fairness, this test is not as time-consuming as it sounds. And if wearers keep their eyes closed and the smoke is increased gradually, there is little danger or discomfort.

Quantitative Fit Testing

For a number of reasons Quantitative fit testing is conducted in operational situations. The main advantage of the quantitative fit testing over the qualitative testing outlined above, is that a number is obtained. This result, calculated for each person, is called an individual protection factor, commonly referred to as a fit factor. The test is conducted in much the same way as the qualitative test, except that the individual is placed into a sealed chamber and the facepiece has been probed to allow for simultaneous measuring of the concentration inside and outside the facepiece, thereby testing the inside the chamber as well. This provides us with a much better understanding of exactly how well the facepiece fits the individual wearer. Therefore, we are not relying upon the general protection factors (which would apply to the average population).

General Maintenance

A program for the maintenance and care of respirators should be adjusted to the type of plant, working conditions, and hazards involved. A maintenance program should contain the following five elements:

1. inspection for defects (including a leak check)
2. cleaning and disinfecting
3. repair
4. storage
5. inspection

All respirators should be routinely inspected before and after each use. A respirator that is not routinely used, but is kept ready for emergency use, should be inspected after each use, and at least monthly when not in use. This is required to ensure that it is in satisfactory working condition. A SCBA should be inspected monthly. Air and oxygen cylinders should be fully charged according to the manufacturer's instructions. The regulator and warning devices should be checked to ensure proper operation.

Respirator inspection should include a check of the tightness of connections, the conditions of the facepiece, headbands, valves, connecting tube, and canisters. Rubber or elastomer parts should be inspected for pliability and signs of deterioration. A record should be kept of inspection dates and findings, especially for respirators maintained for emergency use.

Cleaning and Disinfection

Routinely used respirators should be collected, cleaned, and disinfected periodically. This should be as frequently as necessary to insure that proper protection is provided for the wearer, and definitely before it is worn by someone else. Each worker should be briefed on the cleaning procedure and be assured that he or she will always receive a clean, disinfected respirator. Such assurances are of greatest significance when respirators are not individually assigned to workers. Respirators maintained for emergency use should be cleaned and disinfected after each use.

The following procedure is recommended for cleaning and disinfecting respirators:

1. Remove any filters, cartridges, or canisters.
2. Wash the facepiece and breathing tube in cleaner-disinfectant or detergent solution. (See following paragraph) Use a soft-bristled handbrush to facilitate removal of dirt.
3. Rinse completely in clean, warm water.
4. Air dry in a clean area.
5. Clean other respirator parts as recommended by manufacturer.
6. Inspect valves, head straps, and other parts for defect, wear, or other damage.
7. Replace defective parts with new parts.
8. Insert new filters, cartridges, or canisters. Make sure seal is tight.

Cleaner-disinfectant solutions are available that effectively clean the respirator and contain a bactericidal agent. The bactericidal agent is generally a quaternary ammonium compound. The respirator should be immersed in the solution, rinsed in clean, warm water, and air dried.

Alternatively, respirators may be washed in a liquid detergent solution, then immersed in one of the following:

- hypochlorite solution (50 ppm of chlorine) for 2 minutes
- aqueous iodine solution (50 ppm of iodine for 2 minutes
- quaternary ammonium solution (200 ppm of quaternary ammonium compounds in water with less than 500 ppm total hardness.

Surveillance of the Work Area Conditions and Worker Exposure

For permanent installation, conditions should be altered such that PPE use can be avoided. Changes in operation or process, implementation of engineering controls, temperature, and air movement can affect the concentration of the substance(s) which originally required the use of respirators or other PPE. To determine the continued necessity of respiratory protection, or need for additional protection, measurements of the contaminant concentration should be made after changes are made or observed. These measurements should be documented and records kept on file.

Program Evaluation

The health and safety program and the respirator program should be evaluated at least annually. Program adjustments should be made as appropriate to reflect the evaluation results. Common steps in a program review include the following:

1. Assessing and inspecting each site to ensure compliance with regulations applicable to that specific site.
2. Assessing the number of person-hours that workers wear various protective ensembles.
3. Assessing levels of exposure.
4. Analyzing adequacy of equipment selection.
5. Analyzing adequacy of the operational guidelines.
6. Analyzing adequacy of decontamination, cleaning, inspection, maintenance, and storage programs.
7. Assessing and analyzing adequacy and effectiveness of training and fitting programs.
8. Assessing and analyzing accident and illness records.
9. Analyzing overall safety and health program element coordination.
10. Assessing program element fulfillment rating
11. Analyzing adequacy of program records.
12. Assessing and analyzing program costs.
13. Developing recommendations for program modifications and improvements.

Program Administration

It is appropriate to regularly evaluate the following:

- Is the program responsibility vested in an individual who is knowledgeable and who can coordinate all aspects of the program?
- What is the present status of the implementation of engineering controls, if feasible, to alleviate the need for respirators?
- Are there written procedures/statements covering the various aspects of the respirator program, including:
    designation of administrator
    respirator selection
    purchase of approved equipment
    medical aspects of respirator usage
    issuance of equipment
    fitting
    maintenance, storage, repair
    inspection
    use under special conditions

Personal Protective Clothing

Protective clothing is designed to protect the wearer from chemicals, heat, or cold. Protective clothing can be divided into three groups:

1. **Chemical protective clothing**
   Suits specifically designed to protect the wearer's skin from direct chemical contact. These suits may be either encapsulating or nonencapsulating.

2. **Insulative protective clothing**
   Clothing designed to protect the wearer from short-term temperature extremes.

3. **Fire fighters' protective clothing**
   Suits designed to protect the wearer from extremes of temperature, steam, hot water, hot particles, and the ordinary hazards of structural fire fighting. Please note that this clothing is NOT designed to prevent chemical skin contact.

Chemical Protective Clothing

Chemical protective clothing is made from materials that are compatible with specific chemicals and groups of chemicals. Most chemical protective clothing does not provide thermal protection and is not rated for exposure to fire or intense heat. In fact, because of the impervious nature of most chemical protective suits, ventilation is a definite problem with the subsequent elevation of body core temperatures. A suit's durability varies, as it may be designed for single use (disposable) or multiple use, (reusable).

Chemical protective clothing is designed to resist permeation, degradation, and penetration. **Permeation** refers to the diffusion of chemicals through the materials of the suit. **Degradation** refers to the breakdown of the suit material. **Penetration** refers to the entrance of a substances through pores in the suit material. The ease of decontamination depends upon if and how a chemical reacts with a suit. Chemicals that permeate or degrade the suit materials are not easily removed.

Chemical suit materials must be selected for a particular chemical resistance. No single material of construction will provide protection in all situations. Mixtures of chemicals present special protection problems. This is because it may not be possible to obtain a material impervious to all of the chemicals in the mixture. An additional problem is flexibility. Materials of construction must be sufficiently flexible to allow movement. Care must be taken that the suit material will not react with a chemical and stiffen.

Fully encapsulated suits are one-piece, completely enclosed garments. They provide chemical protection for the entire body. Boots and gloves may be an integral part of the unit. For maximum protection, an encapsulated wearer, with positive pressure SCBA, is enclosed within a fully encapsulating suit.

Nonencapsulating clothing consists of a variety of gear designed to protect the skin and the eyes. Hard hats, goggles, face shields, splash hoods, aprons, gloves, sleeves, leggings, and boots are common items. A number of coveralls made of Tyvek, or other materials, protect against particulate and some chemical hazards. With the use of duct tape, nonencapsulating clothing can be configured into an almost encapsulating suit.

An ensemble is the combination of individual components into a fully protective PPE suite. Ensembles are created in response to specific needs. Compatibilities of materials must be considered when creating an ensemble.

Thermal Protective Clothing

Thermal protective clothing is designed to protect the wearer for short-term high temperature exposure. Frequently, construction is of heat reflective fabric and insulated linings. Such a suit may provide short duration and proximity protection to radiant heat temperatures as high as 1100°C. These suits are designated as proximity or fire entry, depending on whether flame entry is permitted. Respiratory protection must be provided.

Firefighters' Protective Clothing

Firefighters require protection against extremes of temperature, steam, hot water, hot particles, and a variety of other hazards while fighting structural fires. Structural fire protective clothing is commercially available. These suits are not designed for chemical exposure and may provide little to no protection. Respiratory protection must be provided for fire entry and the prevention of smoke inhalation.

Miscellaneous Protective Clothing

There are numerous circumstances where specialized protective clothing is required. Blast and fragmentation suits are available for work in proximity to explosives. Bulletproof vests or fragmentation vests of Kevlar are standard issue to emergency workers in many cities.

Cooling garments are advantageous for those working in a fully encapsulated ensemble. Cooling core body temperature reduces stress, increases the duration and speed of work, and improves worker comfort. Design differs depending upon the particular need and the required level of protection. With an external air source, cooling can be accomplished by air circulation. Cold vests can remove excess heat for limited periods of time. Liquid coolant suits are available. Any cooling garment requiring an external connection has the same disadvantages as an external air-supplied respirator.

Radiation protective suits are required for proximity work with hot, radioactive materials. Generally, these are lead-lined and thus very heavy.

Study Exercises

1. Describe the conditions under which PPE is required.

2. List the criteria used to assess level of hazard.

3. Describe the four EPA levels of protection (Levels A–D) with respect to respiratory and skin protection.

4. Select the appropriate personal protective equipment level to reduce the potential for adverse health effects.

5. List four potential hazards in wearing and using PPE.

6. Describe what is permitted under the OSHA standard for workers who wear prescription glasses or contact lenses.

7. Explain the benefit in maintaining a positive pressure inside the facepiece.

8. List four limitations in using an air-purifying respirator.

9. Compare the advantages and disadvantages of a Self-Contained Breathing Apparatus (SCBA) with those of a supplied-air respirator.

10. Describe and list the principal components of a respiratory protection program under OSHA standard 29 CFR 1910.134.

11. Describe how the Protection Factor is determined and how it is used to determine the Maximum Use Limitation (MUL) for a given airborne contaminant.

12. Describe the three principal factors that affect chemical breakthrough time of Chemical Protective Clothing (CPC).

13. List five factors, other than those having to do with chemical resistance, that should be considered when selecting CPC.

14. Describe and/or demonstrate the proper maintenance of various types of PPE.

Demonstrations/Exercises

These optional in-class demonstrations and exercises may be conducted by the instructor with class participation:

- donning and doffing of personal protective clothing
- fit-testing using various qualitative measures

During these demonstrations, each student should make up five questions with answers in a Jeopardy format to be used during an in-class exercise. In addition to these, each student will be assigned a category in which to write five questions from the text dealing with respirators types, dealing with personal protective clothing, a written program, respirator selection, or protection factors. The five questions should be ranked by level of difficulty.

In class, groups of 2 or 3 individuals will be formed, and each group will edit the questions written by the group members during the above demonstrations. Fifteen minutes will be allowed for this purpose. All students will then turn in their questions, and these, along with questions compiled by the instructor, will be used for PPE Jeopardy.

The board will be set up by the instructor, with assistance.

The class will be divided into 3 to 5 groups to play. Each group must select a spokesperson. A timekeeper will be nominated. The first group will select the category and level. The answer will be read and the first group will have one minute to discuss the answer. Then the spokesperson must answer at that time. If the answer is incorrect, the group loses the value of the question and the question goes to group two. The two highest ranking groups will go into Second Level, Double PPE Jeopardy.

14 Exposure Control Methods

Arleen Goldberg, CIH
Contributing Author

Overview

This chapter presents the principal control technologies used in the industrial setting. The function of these controls is to reduce worker exposure to hazardous materials. We will introduce the different types of engineering controls, the principles and types of ventilation control, and examples of administrative (work practice) controls.

Classifications

Airborne hazards are grouped by their physical forms: **solids**, **liquids**, **gases**, and **vapors**. Hazards of these types may be further subdivided based upon a variety of physical and chemical characteristics.

Particulates

Particulates are solid or liquid particles suspended in the air. The size of the particles can range from individual molecules to approximately 100 microns in diameter. (One micron is one one-millionth of a meter.) The smallest visible particle is approximately 50 microns. Airborne particulates can be subdivided into four categories: dusts, fibers, fumes, and mists.

Dusts

Airborne dusts are solid particles. They are released into the air by mechanically breaking up hard solids. Sanding, grinding, and crushing are operations that generate airborne dust. Airborne dusts are also generated by dispersing powdered material, or by dislodging settled dust. This is often noticed near unpaved highways or on construction sites. Windstorms in dry areas , for example the Owens lake bed east of the Sierras in California, are another example of airborne dusts. Airborne dust particles tend towards irregular shapes.

Large dust particles tend to fall out of the air faster than small ones. This is due to their relative size and density as compared to air. Small dust particles flow with the air and can remain airborne for weeks increasing the chance of their being inhaled. Small airborne particles, usually 5 microns or smaller, can penetrate deep into the respiratory system. Particles larger than 1 micron in diameter are normally caught in the mucous membranes of the nose and throat, but smaller particles find their way beyond these defense mechanisms and into the lungs.

Examples of airborne dusts that are regulated by OSHA and CAL-OSHA are silica and carbon black. (See General Industry Standard, 29 CFR Section 1910.1000.)

Fibers

Fibers are solid particles which are several times longer than they are wide. Based on their elongated shape and aerodynamic properties, fibers longer than 10 microns can occasionally be inhaled. Most often, particles of this size are caught in the nasal passages.

Examples of regulated airborne fibers are asbestos and fibrous talc.

Fumes

Extremely small particles—0.001 to 1 μ in diameter—introduced into the air from condensation of metal vapors, are referred to as fumes. The fume particles remain airborne and **agglomerate** (stick together) to form larger particles. Settled fume particles, usually of the larger particle size, can be reintroduced into the air by air currents or vibrations. Examples of fume production processes include welding, smelting, and metalworking activities.

Fumes can occasionally be formed from materials other than metals, for example, polytetrafluoroethylene (Teflon).

Mists

Mists are liquid droplets of a substance that become airborne. These droplets may result from boiling, bubbling, splashing, or spraying of a liquid or from condensation of liquid vapors. Individual mist droplets vary widely in size, from below 0.01 microns to greater than 10 microns in diameter. Mist droplets or particles are usually spherical in shape.

Examples of mists regulated by 29 CFR Section 1910.1000 are oil mist, acid mist, and certain metal mists.

Gases

Gases are molecules of a substance which are a gas at room temperature. Gases have varying densities. A gas may be escaping or leaking from a tank, vessel, or pipe, and will mix with air according to the specific gravity and ambient temperature of the gas. If a gas is cold, it may settle and then spread as it warms. Also, if a gas is hot, it may rise rapidly and then spread or diffuse. If a gas is light (like helium) it may rise. If a gas is heavier than air, like carbon dioxide, it may sink to the lowest level and accumulate until disturbed by air movement.

Examples of gases regulated by 29 CFR Section 1910.1000 are sulfur dioxide, ozone, phosgene, chlorine, and hydrogen sulfide.

Vapors

Vapors are the gaseous form of materials which are liquid or solid at room temperature. Vapors behave much as gases do, but condense to form mists or fogs as they cool or as pressure increases. Vapors evolved from volatile liquids and solids may increase as temperature increases or as pressure decreases.

Examples of vapors regulated by 29 CFR Section 1910.1000 are benzene, methyl alcohol, methyl ethyl ketone (MEK), styrene, toluene, trichloroethylene, mercury, and vinyl chloride

Self-Test Exercise: Name the types of airborne hazards and how they are formed, giving an example of each.

Hazard Control

The health hazard associated with most airborne substances depends upon the following factors:

1. toxicity
2. chemical affinity of the substance
3. rate of emission
4. concentration of the substance in the air
5. size of the airborne particles (dust, fibers, mists, and fumes)
6. length of exposure

Control of these hazards depends upon minimization or elimination of these factors. These factors are evaluated to develop a Degree of Hazard.

Hazard Control Strategies

Control strategies include **engineering control**, **work practice control**, **administrative control**, and as **interim measures**, PPE.

Engineering Controls

Engineering controls deal with substances, processes, and equipment. The control of airborne hazards associated with a process or piece of equipment should be an integral part of the preliminary engineering design.

Design

The first step in engineering control is good design. New construction, remodeling, and retrofitting should all be designed to minimize hazard exposure and to ensure current and future compliance with health and safety standards. Such up-front design is the most cost effective methodology for dealing with worker and workplace safety.

Many industrial facilities were designed with little or no attention to exposure hazards. Therefore, it becomes necessary to install retrofit control measures. Retrofitting can be awkward, inefficient, and expensive. This is due to existing limitations, such as space, process design, and other structures.

Occasionally, no good design data exists regarding airborne hazard control for a specific situation. Substances may be involved for which no standards are in place. This requires an engineer to predict what controls might be necessary. It also means proceeding without knowing whether, ultimately, a more stringent or less stringent standard may be promulgated. A more stringent standard may result in the need to retrofit additional controls.

Substitution

A hazard may be reduced by simply substituting a less toxic chemical. An example would be the substitution of carbon dioxide for vinyl chloride as a propellant in aerosol cans.

Improvement may also be gained by substituting equipment or a different process which does not require the use of, or does not emit, the hazardous substances (or requires or emits less of it.)

An example would be substitution of electric forklifts or front-end loaders, which do not emit carbon monoxide, for gasoline or propane fueled ones.

Isolation

Simply isolating or enclosing the source of the contaminant can control the hazard in many situations. Examples of isolation are
- storing dry sand in an enclosure such as a tank or small building, rather than in a pile which is disturbed by traffic or wind; and
- limiting escape surface by putting a lid on a tank; and,
- using antifoam products which are inert.

Process Controls

Processes operate within a range of acceptable operating parameters called the operating envelope or simply, the envelope. By changing the operating condition within the envelope by varying operating parameters such as temperature or pressure, you may reduce exposure. These changes would minimally affect output while greatly reducing the evolution of vapors from the process.

Ventilation

Ventilation control is the exhausting of the contaminant from the workplace or dilution of the contaminant in the workplace. Ventilation is by far the most popular method of engineering control for airborne contaminants. (Ventilation is discussed in more detail later in this chapter.)

It is important to note that general dilution ventilation only spreads the hazard over a larger area, but does not eliminate the hazard. In the 70s the slogan "the solution to pollution is dilution" was widely discussed. However, we now know that the solution to pollution is **not** dilution but rather prevention.

Review

Give an example of each of the following:
- Forms of engineering control
- Design
- Substitution
- Isolation (enclosure)
- Process Control
- Ventilation

Give an example of **Pollution Prevention (P2)** for each form of engineering control.

Work Practice Controls

The employee is an important factor in reducing exposure to airborne hazards. Good work practices and work practice controls result from an employee's appreciation of the hazard, well-developed work procedures, as well as good equipment arrangement and design. For example, a maintenance operation to clean up settled dust might be conducted by vacuuming instead of sweeping; vacuuming creates less airborne dust.

Waiting until the pressure of a reaction vessel is equilibrated with the atmosphere before opening the top is a simple work practice that can reduce the emission of vapors from a solvent system or dusts from a dry powder system. Increasing drag out time with a liquid process allows more draining from the processed part and can generate significant cost savings.

Administrative Controls

The first level of administrative control is employee training. General, industrial, and workplace hazards, as well as specific, individual process and equipment hazard identification should be part of the employer's business plan. Familiarity with a business's emergency response plan and pollution prevention plans should be incorporated into any employee training program.

OSHA has made the training and education of employees mandatory. Employees must be educated not only to recognize hazards in their work situation, but also to understand why they are hazards and how to deal with them. It is necessary to promote safety in the workplace from a dual standpoints of safety and health.

Administrative controls for airborne hazards reduce the duration of individual exposures while often increasing the number of exposed employees. The magnitude of one employee's exposure is reduced by shortening the length of time he or she remains in a particular environment.

Personnel are rotated from areas of high exposure to those of low or no exposure. This reduces the time-weighted average exposure of each individual. Such administrative controls are effective, but advisable in only a few circumstances. They could be a suitable alternative if body functions eliminate the contaminant in a short period of time. The rotation should be based on sound medical recommendations and testing to ensure that there are no cumulative effects. Moreover, the size of the available workforce may be a limiting factor.

Proper equipment maintenance is vital for a variety of reasons. Proper maintenance ensures the life of the equipment and thus contributes to its cost effectiveness. Properly performing equipment minimizes hazards and pollution. Therefore, it is beneficial to the employee's safety.

The level of employee exposures may be reduced by performing contaminant generating operations during off-shifts, e.g., sweeping during early morning hours when no workshift is scheduled..

Exposures may also be reduced by adjusting production scheduling and sequencing contaminant producing operations, rather than performing them simultaneously. These process production controls are effective and often inexpensive.

Monitoring of contaminants is key to scheduling and sequencing. Proper recordkeeping is a necessary adjunct to scheduling and sequencing. Not only does recordkeeping benefit the employee by ensuring compliance with safety standards, it also protects the employer by documenting compliance for liability and insurance concerns.

Personal Protective Equipment

PPE is a last line of defense. It should only be used temporarily as a stopgap measure while engineering controls are being designed, evaluated, and installed. Another use is when all feasible engineering, work practice, and administrative controls are being utilized and operating properly, but the contaminant concentration is still above permissible exposure levels. Such equipment is also appropriate in emergency situations.

Total employee exposure may include contributions from inhalation, absorption, and ingestion. Inhalation in many industrial settings is often considered to be the primary route of entry. In some situations, minor absorption and ingestion may become critically important when coupled with inhalation. In these situations, personal protective clothing and equipment, as well as good sanitation practices, i.e., personal hygiene, are necessary to provide adequate protection.

Ventilation

Ventilation deals with air movement. In order to understand ventilation we must have a basic understanding of the properties of air.

Properties of Air

Air is a mixture of many gases. These include oxygen at about 20%, nitrogen at about 80%, and trace amounts of carbon dioxide, carbon monoxide, argon, ozone, and an assortment of many other gases and vapors. Gases expand when heated and they contract upon cooling. They expand as pressure decreases and contract when pressure increases. Because of these physical properties, air can move in a manner similar to liquids. (For ease of analysis, we will assume that air moves in an incompressible way, resembling the movement of liquids.)

A compressor creates an area of higher pressure air. The high pressure air can be released to a lower pressured area (such as a work room) by flowing through a conductor, such as a tube, and terminating at a vent or nozzle.

Wind blowing over a building creates a difference in pressure. With the windows open, air flows from the area of higher pressure to that of lower pressure, creating ventilation or air movement. We cannot always depend upon atmospheric conditions to create air movement or to ventilate. A fan can be used to create a difference of pressure and move the air through ductwork. A fan, turning in a duct, creates a region of negative pressure before it and a region of positive pressure after it. These pressure differentials move air from one point in the duct to another.

Types of Ventilation

1. General ventilation
2. Dilution ventilation
3. Local exhaust ventilation
4. Makeup air ventilation
5. Makeup local supply ventilation

General Ventilation

General ventilation is the most common type of ventilation. We also know it as comfort ventilation. This type of ventilation is found in classrooms, office areas, warehouses, and small manufacturing operations where air contaminants are absent. Components of this system include a ventilation inlet register and a control system which regulates the ventilation system. It may or may not include heating and cooling elements, ductwork, and a way to move the air, such as a fan or blower. The movement of air is called a **flow**, while the mover, the fan, or the impeder, friction, is called a **force**. Mathematical models may be constructed which relate the flows and forces to the elements of a system. These equations are utilized to design systems meeting the particular need or situation.

Dilution Ventilation

A dilution ventilation system requires, at a minimum, an exhaust fan or an exhaust system. Also required is some means for providing makeup air to replace that air which has been exhausted. Typically a makeup air unit which heats or cools entering air is used to complement the exhaust system.

This is one of the systems used for contaminant control. If, for example, an operation in an enclosed room is emitting a contaminant, and we measured the amount of the contaminant, it would be possible to dilute the contaminant below the permissible exposure limit by simply adding a known volume of air.

Dilution ventilation is appropriate in only limited situations. Dilution ventilation is normally effective for lower levels of gases or vapors. It is a greatly misused method of control!

Dilution ventilation is not effective or economical if a highly toxic material is present. Highly toxic substances have low permissible exposure limits. Dilution ventilation is also inefficient where large quantities of contaminants are evolved or where the contaminant is emitted nonuniformly. Dilution ventilation is practical only when makeup air is readily available and where the heating and tempering of makeup air is feasible and economical. Dilution ventilation can provide sufficient contaminant control to protect employees in many situations, such as the soldering of electronic parts using simple lead-tin solder. A uniform evolution of contaminants, coupled with small quantities being evolved, makes dilution ventilation a workable solution.

For example, some welding operations may use dilution ventilation. However, there are many limiting factors in its use, including the room volume afforded per welder, the ceiling height, and the limiting of the operation to work with mild steel, iron, or aluminum. Welding of other metals such as nickel-chromium steel and zinc coated steel, which produce hazardous fumes, requires other types of ventilation.

Fabrication or welding of metalwork involving many diverse operations can be controlled with dilution ventilation. In some situations, gases emitted by forklift trucks can be controlled adequately with dilution ventilation.

Design characteristics should be identified by anyone evaluating a dilution ventilation system. The structural characteristics of the building must be suitable for dilution ventilation. For example, an industrial building with a high and pointed roof, which may contain exhaust louvers or exhaust fans, is not appropriate for this method.

Design characteristics should include placement of exhaust openings near the source of contaminants. Dilution air should pass through the zone of contamination in order to dilute the contaminant. The contaminant source is located between the operator and the exhaust intake. The exhaust intake should never be placed near the outlet for the makeup air unit. This is because we would like the two air flow streams to behave independently. If the two outlets are adjacent, the flows mingle and the equations used to predict each system couple. This situation requires the use more complex equations and modeling. It is also critical that the exhaust outlet should not be located near the makeup air unit intake. This is because the tainted air from one system would enter and contaminate the second system.

Local Exhaust Ventilation

Local exhaust ventilation is the method most often used by the industrial hygienist to control employee exposure. The contaminant is collected as it is evolved. Local exhaust systems require less air movement and thus fewer makeup air requirements. Point-of-operation exhaust uses smaller motors and saves energy.

A local exhaust system has four basic parts:

1. hood or collection device
2. ductwork
3. contaminant separator
4. motor-fan

Terms associated with local exhaust ventilation include the following:

Capture velocity	The air velocity at the contaminant source required to overcome the contaminant's own velocity, room air currents, operator generated air movements, and bring the contaminant into the local exhaust system.
Face velocity	The velocity of air measured at the true face of the hood.
Static pressure	That pressure which presses against the sides of a duct. Static pressure is the potential energy required to overcome the friction and other losses within the system. Static pressure is abbreviated as SP.
Total Pressure	The sum of the static pressure and the pressure exerted by molecules of air trying to enter the end of a measuring device such as a U-tube manometer which is juxtaposed between the system and the outside atmosphere. Total pressure is abbreviated as TP.
Velocity Pressure	The difference between total and static pressure. This difference is abbreviated as VP. The velocity pressure is that pressure due to the velocity of the air itself. If there is no movement of air in the duct, there cannot be any velocity pressure.

An algebraic way of relating total pressure to static pressure and velocity pressure is given by the formula

Total pressure = static pressure + velocity pressure, or

TP = SP + VP

In local exhaust ventilation systems, the velocity pressure is always positive. The static pressure and the total pressure are negative values on the upstream side of the fan. On the downstream side of the fan, the total pressure, static pressure, and the velocity pressure are all positive.

Effective local exhaust systems have certain characteristics which can be identified without any special equipment. An effective system encloses the operation as much as possible. A good way to think of it is to draw a box around the operation. Then open the box where mandatory to move materials in and out and to allow the operator to have access to the operation as minimally required. The more closed the system, the more effective it will be. A good system does not require the operator to come between the source and the exhaust hood.

A well-designed system uses the natural velocity of the contaminant to assist in its capture. Particulate materials require air moving at a minimum velocity, **transport velocity**, in order to be captured and carried along with the flow.

A properly designed system captures this evolving material at or near the source. One common misconception is that industrial contaminants that are heavier than air will settle to the floor, and thus the hood should be located on the floor. It is actually better to catch them where they are emitted before they can diffuse or spread.

Hoods

Local exhaust hoods are instruments used to capture the contaminant where it evolves. The two basic types of hoods are exhaust hoods and receiving or canopy hoods.

Exhaust hoods come in a great variety of shapes and sizes depending on the needs of the exhaust system and the characteristics of the air contaminant. Slot hoods are commonly used in ventilation of open surface tanks. Another type of hood is the booth or lab hood. In a downdraft, by-products are sucked through the grating and pulled away from workers.

Push-pull systems are rather unique but difficult to maintain and must be carefully designed in order to work properly. The air flow from the fan or duct behind the worker pushes clean air over the contaminated area. In addition, the suction hood on the work side of the worker pulls the contaminated air from the workplace. Design of these systems requires the coupling of two sets of equations for system flows and forces. Thus they are more complicated to design and operate.

Portable hoods can be built and maintained which effectively control situations such as welding and soldering. Flexible or jointed pipes with flared ends are moved to points of operation. The biggest problem is actually repositioning as the job progresses or as the work changes.

The canopy hood or receiving hood is generally reserved for hot processes. The canopy hood will overflow if it receives hot air at a greater rate than its exhausting rate. Hot processes create hot air. Hot air is less dense than cold air and is propelled upward by the resulting buoyant force. As hot air rises from the source, it entrains other air and contaminants with it.

Ducts

Ductwork moves air and contaminant flows to the fan or separator. Ducts can come in round, square, and rectangular shapes. Round ducts are most efficient for moving air, but occupy the most space in a building. A large main duct is called a **plenum**. Ductwork can be made of various grades and gages of materials. PVC and fiberglass are used occasionally where corrosive contaminants are being removed from the work area.

Ductwork in a well-designed local exhaust system will have smooth entries and large radius elbows. In the entries and elbows, energy losses occur due to friction, and this affects the overall performance of the system and raises the energy requirements necessary to move the air through the system.

Fans

There are two basic types of fan used in industrial exhaust ventilation: the axial fan and the centrifugal fan.

Axial fans are used where the fan must be located in the duct. Common applications include paint booths, exhaust ducts in windows, and in-moving non-dusty materials. With an axial fan, flow continues in a straight line.

Centrifugal or squirrel-caged fans include three basic types: the forward curve blade, the straight blade, and the backward curve blade. The centrifugal fan is used in many local exhaust systems. Typical applications include lab hoods, exhaust systems for vapors or gases, and dust systems (where the straight-bladed centrifugal fan is most commonly found.) These fans typically change the direction of air flow by ninety degrees, a right angle.

Separators

There are many different types of separator. Which of these types is appropriate depends upon the particular material being separated from the air stream.

The electrostatic precipitator is known in the industry by ESP or Cottrell. It is a highly efficient particulate material separator. Its separating action is produced by electrically charging the particulate material and collecting the charged particles on an oppositely charged plate. These plates are periodically cleaned and the collected material is captured below and taken away for disposal.

A baghouse filter system can have a very high efficiency. It consists of a metal shell, or box, with ductwork for air entry and exit. At the bottom of the baghouse is a funnel-shaped hopper which receives collected dust. Air containing particulates is channeled into bag filters inside the baghouse. As the air flows through the material of the bag, the particles are trapped on the inside of the bag. Clean air passes out of the bag and is directed towards the discharge of the baghouse.

The particles trapped inside the bag form a filter coating, or cake, on the inside of the bag. This particulate cake actually improves collection efficiency. However, when the cake becomes too thick, the resistance to air flow through the bag greatly increases and the bags must be cleaned by either shaking or by introducing quick blasts of reverse airflow. The dislodged material from the dust cakes and falls to the hopper at the bottom of the baghouse. From the hopper, the dust may be recycled into the process or sent to a disposal site.

A common device in many local exhaust ventilation systems is an expansion area where the flow rate drops below the required transport velocity and the large particles are allowed to settle out. These may be used in conjunction with baghouses or electrostatic precipitators to remove large particles, preventing overload of the high-efficiency, small-particle collectors.

Wet Collectors (Scrubbers)

Wet collectors are used in many applications involving the collection of combustion products, gases, and vapors. Wet collectors can remove gases, vapors, and particulate contaminants from an effluent air stream.

Particle removal is achieved by impinging the airborne particles on discrete droplets or sheets of liquid. This is accomplished by mixing the particulate-laden air with a liquid spray or the liquid itself. After this air/liquid interaction, the liquid containing the wetted particles stays in the scrubber and the clean air emerges.

Scrubbers remove gases and vapors from an air stream in much the same way, except that the mixing of air and liquid results in absorption of the gas or vapor into the liquid.

Cyclones

Cyclones use centrifugal force to separate particulate materials. Air entering a cyclone collector is caused to flow in a circular path. Particles which cannot stay in the air stream due to their mass and momentum impinge on the sides of the cyclone collector and fall to the hopper at the bottom. Cyclones are not suitable for particles which are smaller than 5 micrometers in diameter as they tend to follow airflow stream lines. Therefore, they do not impinge on the walls of the collector.

Charcoal or Catalytic Conversion

Charcoal or catalytic conversion systems are used in a variety of situations. The air or exhaust is directed to move in and about the material in a packed column or tower. Charcoal has the capacity to adsorb many gases and vapors. The gas and vapor molecules are adsorbed onto the surface of the tiny pores in the charcoal. Contaminated charcoal may be heated or flushed with a solvent to remove the contaminants collected, or in some situations, it may require disposal when its capacity to remove contaminants has been exhausted.

Catalytic conversion involves passing the exhaust or contaminated air containing an undesirable gas through a bed containing a material which will catalyze its chemical conversion to a less toxic or nontoxic form. The catalytic converters on automobiles cause carbon monoxide to be chemically reacted with oxygen to form carbon dioxide.

Makeup Air Systems

OSHA regulations require that makeup air be heated and cleaned if it is to be recirculated. Makeup air should be introduced so that it is dispersed properly throughout the contaminant evolution area. Makeup air must pass through the zone of contamination.

More commonly, doors and windows are opened in order to introduce air into the room. This works well in temperate climates, but closing doors and windows will adversely affect a local exhaust or dilution system in the wintertime if it has not been designed to operate in this manner.

One indication of the lack of makeup air is when doors slam shut or are opened with difficulty. This condition reveals a difference of pressure between the room air and the outside air, thus indicating that makeup air is not being introduced properly into the room.

Local Ventilation

There are three basic types of local exhaust ventilation systems. These are the plenum system, the blastgate system and the pre-designed system, which has no blastgates.

Plenum

The plenum system uses a large duct, or plenum, from which small ducts emerge. This system is adaptable to changes in intake parts and will maintain constant static pressure on the duct. It is commonly found in such applications as motor vehicle garage ventilation. Its main disadvantage is that it does not maintain a transport velocity and is therefore not suitable for dusty operations.

Blastgate

The blastgate system involves designing ductwork with blastgates or damper controls for air flow. A blastgate may be a flat piece of metal which is inserted into a duct at a right angle, or perpendicular, to the airflow. The blastgate reduces the cross-sectional area available for airflow. A butterfly damper inside the duct itself may perform a similar function. Most blastgates can be locked into position by bolts or pegs.

The advantage of the blastgate system is that it can be adjusted periodically and closed off for operations that are not running. Its greatest disadvantage is that it is difficult to balance. It is also prone to misadjustment by employees, throwing the system off balance. Blastgates also have a tendency to stick, clog, or collect dust.

Predesigned

The predesigned system is a system where the ductwork is initially designed to its proper size. There are no blastgates or other controls that can be adjusted. Once the system has been installed, it is not easily adapted to different operations. Its main advantage is that it cannot be tampered with, and that once designed, installed, and properly operating, it should continue to provide the same efficient operation for years of service.

System Evaluation

There are certain design characteristics to be evaluated that the industrial hygienist and technician can understand and look for during an evaluation of an industrial atmosphere.

The hood should capture all, or the significant part of, the contaminant. This can usually be seen by the unaided eye. The use of a smoke tube (to be discussed later) is also useful in determining the capture characteristics of a hood.

A local exhaust system may be adversely affected by air currents found in the room. You should be sharply aware of open windows, open doors, and fans which may affect the system. These situations arise normally if an employee is dissatisfied with the general, or comfort, ventilation in the room and attempts to alter the heating or cooling systems by the use of open windows, doors, and fans. It is frequently difficult to obtain consensus from employees on the optimal comfort levels and conditions.

A well-designed local exhaust system is not easily tampered with and functions over a variety of local variations in conditions. A well-designed local exhaust system does not interfere with the worker's duties. It does not make the work more difficult and it does not tempt the employee to tamper with the system. In addition, a well-designed system allows for easy maintenance and access to the duct system. Provisions for cleanout should be made for the ductwork, the hood area, and the fans.

A **pitot tube** is an instrument, typically located within a ventilation system, used to measure the air velocity by sensing pressure. Normally the technician or industrial hygienist would not have to access the pitot tube. Many plants have records that indicate the velocity in the duct. The face velocity of a hood can be measured using any of a variety of techniques or instruments. This is a critical measurement.

One simple and inexpensive piece of equipment which can be used to evaluate local exhaust systems is the smoke tube. The smoke tube uses a material which emits smoke when air is moved across it. A rough idea of the face velocity can be obtained by squeezing a small amount of smoke out directly in front of the capture hood. The time for the smoke to move one or two feet can be measured roughly on a watch. If the smoke takes one second to move one foot, then it can be seen that the face velocity is roughly 60 feet per minute. If the smoke moves two feet in one second, then the face velocity is roughly 120 feet per minute. Many ventilation parameters are still given in feet. Fan capacity is commonly expressed in **cubic feet per minute (cfm)**.

The smoke can be used to check the air currents in and out of the hood around the edges. Smoke should be emitted from a smoke tube tester around the entire edge of the hood to see that all edges are operating properly.

More expensive and elaborate equipment is available. Such equipment includes the thermal anemometer, the vane anemometer, and other velocimeters.

On/Off Switch

Occasionally a system deficiency can be traced to the on/off switch of the equipment. Many times the local exhaust systems are not turned on for one reason or another. Often this obvious problem is simply overlooked.

Flow Obstructions

Plugged ducts and ducts within which dust has accumulated will significantly alter the capture and airflow characteristics of a local exhaust system.

Fan Rotation

The fans and motors on all equipment should be checked. Broken fan belts, slipping fan belts, and missing fan belts indicate less-than-perfect operation. The fan housing should be checked to determine whether the fan rotation is occurring in the proper direction. Many centrifugal fans will exhaust air when turning backward, but at a lower rate. If possible, the original design should be compared with the existing system.

Motors and fans are sized to meet specific requirements. Many times motors burn out and are replaced by smaller or larger motors which do not meet the requirements of the system. Most industrial fans have a life of about ten years. When the fans are replaced, occasionally they are replaced with inappropriate fans. These should be checked against the original design specifications.

As fan blades wear out, holes in the blades may appear or the edges may be worn away. The effectiveness of the fan is greatly affected and then the system will not operate properly. Most fan housings can be opened to check the fan blades. When fan blades become coated with dust or other accumulated materials, their airflow characteristics change.

Questions

1. What are the physical properties of air?
2. What are the three basic types of industrial ventilation?
3. What are the four basic components of a local exhaust system?
4. Why must makeup air be provided whenever dilution or local exhaust ventilation is used as a control technique?

Additional Sources of Information

The Industrial Environment: Its Evaluation and Control, National Institute for Occupational Safety and Health, U.S. Department of Health, Education, and Welfare, U.S. Government Printing Office, Washington, D.C., 1973.

Industrial Ventilation: A Manual of Recommended Practice, 14th Ed., Committee on Industrial Ventilation, The American Conference of Governmental Industrial Hygienists, Lansing, Michigan, 1976.

Introduction to Industrial Hygiene for Safety Officers, U.S. Department of Labor, U. S. Government Printing Office, Washington, D. C.

Code of Federal Regulations, Part 1910, Section 1000.

State Codes of Regulations

Study Exercises

1. Distinguish between a gas, a vapor, a mist, and a fume.

2. Rank the methods for controlling exposure to hazardous materials.

3. Identify basic methods of controlling airborne hazards.

4. Compare the different types of engineering controls:
   A. Change in plant design and equipment selection
   B. Substitution of materials
   C. Isolation (equipment enclosure)
   D. Process controls
   E. Ventilation

5. Explain what is meant by *work practices control.*

6. Describe the types of administrative controls to be considered:
   A. Employee training
   B. Work scheduling
   C. Equipment maintenance
   D. Employee selection and placement
   E. Monitoring of contaminants
   F. Recordkeeping

7. When is personal protective equipment considered to be an acceptable exposure control measure.

8. Differentiate between the major types of ventilation:
   A. General ventilation
   B. Dilution ventilation
   C. Local-exhaust ventilation
   D. Makeup air ventilation
   E. Makeup local supply ventilation

9. Describe the basic principles of operation and the advantages and disadvantages of general (dilution) ventilation.

10. Describe a local exhaust system including:
    A. Principles of operation
    B. Components: hood, ductwork, contaminant separator, and motor-fan
    C. Design factors
    D. Advantages and disadvantages

11. Describe the differences between a centrifugal fan system and an axial flow (tube) fan system, including the typical uses for each type.

12. Compare the various types of air-cleaning devices:
    A. Fabric filters
    B. Electrostatic precipitators
    C. Air scrubbers
    D. Carbon adsorbers
    E. Catalytic

13. Describe characteristics of good and bad ventilation systems.

14. Give examples of good and bad ventilation systems.

16. Define the following terms:
    • fibrosis
    • edema
    • ulceration
    • defatting agent
    • necrosis
    • narcosis

17. Identify the principal types of airborne hazardous substances.

18. Explain why particulates between 0.5 and 5.0 microns in diameter are of primary concern.

15 Medical Monitoring, Treatment, and Management

Overview

This chapter introduces the principles, elements, and requirements of a medical program. A medical surveillance program and the use of the Biological Exposure Indices (BEI) complement other exposure control and assessment methods. A number of medical screening tests and biological monitoring tests are presented. Examples of toxic substances, that can be assessed by each type of test, are provided. The basic principles of medical diagnosis are reviewed and suggested initial responses to toxicity are discussed.

Medical Program

A medical program is essential to assess a worker's health and fitness. It is desirable to evaluate a worker's health and fitness prior to employment and critical to continue health monitoring upon employment, during the course of work, and at termination. It is imperative to provide emergency and other treatment as needed and to keep accurate records for future reference. OSHA recommends a medical evaluation for employees required to wear a respirator (29 CFR 1910134b), and certain OSHA standards include specific medical requirements (29 CFR 1910.95 and 29 CFR 1910.1001 through 1045). Information from a site medical program may be used to conduct future epidemiological studies, to adjudicate claims, to provide evidence in litigation, and to report a worker's medical condition to federal, state, and local agencies as required by law.

Recommendations regarding medical surveillance programs should be used as an adjunct and complement to other controls. It is assumed and implied that workers will first have adequate protection from exposures through administrative and engineering controls. Secondly, appropriate personal protective equipment and decontamination procedures are mandatory and paramount. Although a medical surveillance program is of primary importance, its need should be a secondary, not a primary, preventive measure.

A medical program should be developed for each site based on the specific needs, location, and potential exposures of employees at the site. The effectiveness of a medical program depends on active worker involvement. Management should have a firm commitment to worker health and safety. Management commitment is expressed through the medical surveillance and treatment program and through encouragement of employees to maintain good health through exercise, proper diet, and avoidance of tobacco, alcohol, and drug abuse.

Program Components

The four basic components of a site medical program are these:

1. Surveillance
2. Treatment
3. Recordkeeping
4. Program review

Surveillance

Surveillance includes a pre-employment screening, periodic medical examinations (and follow-up examination when appropriate), and a termination examination. Pre-employment screening should consist of a number of elements. These include a medical history, an occupational history, a physical examination, a determination of fitness to work wearing protective equipment, and baseline monitoring for specific exposures.

Periodic medical examinations should include a variety of elements. These include a yearly update of medical and occupation history and a yearly physical examination, with testing based upon these considerations:

- examination results
- exposures
- job class and task

(More frequent testing may be required based upon specific exposure data.)

Treatment

Treatment plans include both emergency and nonemergency (case-by-case) treatment. An emergency treatment plan has several elements. It should provide for emergency first aid onsite, develop a liaison with local hospital and medical specialists, arrange for decontamination of victims, arrange in advance for transport of victims, transfer medical records, and communicate details of incident and medical history to the next care provider.

Recordkeeping

OSHA requires all employers with ten or more employees to keep records of workplace injuries. An employer must maintain a log and summary of all occupational injuries or illnesses that cause a fatality or loss of a workday or that require medical treatment. Within six working days, the employer must complete a supplementary record for the injury or illness. The summary of injuries and illnesses must be posted annually, and the records must be retained for five years and be available to employees (29 CFR 1904.2-5). The records may be inspected by OSHA, the Department of Health and Human Resources, union representatives, and the injured employee (29 CFR 1904.7). If an occurrence at work results in a fatality or the hospitalization of five or more employees, the incident must be reported to the nearest office of the area director of OSHA within 48 hours (29 CFR 1904.8).

The OSHA 200 log must contain detailed information concerning the injury, including the date of the injury, the employee's name, regular occupation, department, a description of the injury or illness, the number of workdays during which the employee's ability to perform regular work was restricted, the type of occupational illness, and, if the employee dies due to a work injury or illness, the date of death.

The OSHA 101 supplemental report requires additional information, including the employee's address, the accident location, a description of the accident or injurious exposure and how it occurred, identification of the object that caused the injury, and identification of the treating physician and hospital.

OSHA regulations mandate that, unless a specific occupational safety and health standard provides a different time period, the employer must maintain and preserve medical records on exposed workers for 30 years after they leave employment (29 CFR 1910.20). OSHA has assessed fines of $10,000 for each nonrecorded or misrecorded injury or illness. (See *Recordkeeping Guidelines for Occupational Injuries and Illnesses,* O.M.B. No. 1220-0029.)

Treatment for injuries that merely require first aid need not be recorded if the injury does not involve death, loss of consciousness, restriction of work or motion, or transfer to another job, such as the following:

- application of antiseptics during first visit
- treatment of first-degree burns
- application of bandages
- use of elastic bandages
- removal of foreign bodies not embedded in eye, if only irrigation is required
- removal of foreign bodies from wound by tweezers or other simple objects, if procedure is uncomplicated
- use of nonprescription medications and administration of a single dose of prescription medication on first visit
- soaking therapy on initial visit or removal of bandages by soaking
- application of hot or cold compresses during first visit
- applications of ointments to abrasions to prevent drying or cracking
- applications of heat therapy during first visit
- use of whirlpool bath therapy during first visit
- negative x-ray diagnosis
- observation of injury during visit
- administration of tetanus shots or boosters

A restricted work day is defined to include any day (including the day of injury) during which the worker cannot perform all the regular duties of his or her job because of a work injury or illness. This time includes both those days on which the worker is officially assigned to light duty or has work restrictions from a doctor and any time he or she is not able to perform regular job functions, even if the performance of those job functions is not specifically required on those particular days. A missed work day does not include the day of the injury.

It is suggested that the employer provide all workers the following information in a written company document:

The company is required to keep records on every workplace injury and to record any time a worker is unable to perform his full duties because of a work-related injury. The company safety or personnel department maintains these records. Thus, any time a worker incurs a workplace injury, an accident form must be sent to the company safety or personnel department no matter how minor the injury. Similarly, no worker should be excused from performing any of his regular work duties, except by a work restriction form. Thus, if a worker is unable to perform all his regular work duties, the foreman must send him to the safety or personnel department for issuance of a work restriction form. If a worker's work restriction form expires and the worker is still unable to perform his regular duties, the worker must be sent back to the safety or personnel department for re-issuance of a work restriction form. The foreman may not accommodate an employee without a work restriction form.

Evaluation

Regular evaluation of the medical surveillance program is important to ensure its effectiveness. Program review includes several elements. It should provide regular reviews of the site safety plan to determine if additional testing is needed. It should review the program periodically, focusing on current site hazards, exposures, and industrial hygiene standards.

Maintenance and review of medical records and test results aid medical practitioners, public health officers, and employers in assessing the effectiveness of the health and safety program. The site safety officer, medical consultant, and management representative should do the following at least annually:

1. Ascertain that each accident or illness was promptly investigated to determine the cause and to make any necessary changes in health and safety procedures.
2. Evaluate the efficacy of specific medical testing in the context of potential site exposures.
3. Add or delete medical tests as suggested by current industrial hygiene and environmental data.
4. Review potential exposures and site safety plans at all sites to determine if additional testing is required.
5. Review emergency treatment procedures and update lists of emergency contacts.

Biological Exposure Indices (BEIs)

Biological monitoring provides occupational health personnel with an additional tool for assessing a worker's exposure to chemicals. Biological monitoring is complementary to air monitoring. It may offer an advantage when supplementing air monitoring and can be used to substantiate air monitoring and to test the efficacy of Personal Protective Equipment, PPE. Biological monitoring may be used to determine the potential for absorption via the skin and the gastrointestinal tract or to detect nonoccupational exposure.

Biological Exposure Indices (BEIs) are reference values intended as guidelines for the evaluation of potential health hazards. These are particularly important in the practice of industrial hygiene. A BEI represents the level of a **determinant**, i.e., a chemical which is most likely to be observed in a specimen taken from a worker. The existence of BEI does not indicate the need to conduct biological monitoring. It is assumed that the specimen is collected from a healthy worker who was exposed to a chemical. It is further assumed that this worker is exposed to the same extent as a worker with inhalation exposures to the TLV.

BEIs apply to a classic **workweek** of eight hours of exposure per day for a five-day workweek. Other working schedules require an extrapolation of BEIs based on the principles of pharmacokinetics and pharmacodynamics. In these cases, BEIs should not be applied either directly or through a conversion factor per se. BEIs are **not** meant to be used to provide determination of **safe levels** for nonoccupational exposures to air, water, or food pollutants or contaminants. Nor are BEIs intended for use as a measure of adverse effects or for diagnosis of occupational illness.

The data bases for BEI recommendations are human-exposure based. They consist of information from a variety of available sources. Correlated information on the absorption, elimination, and metabolism of chemicals is used; the correlations are between exposure intensity and biological effects in workers. Some BEIs are based on the relationship between intensity of exposure and biological levels of the determinant. Others are based on the relationship between biological levels and health effects. These relationships are derived from human data obtained from controlled and field studies. Animal studies have not been able to provide data suitable for the establishment of BEIs. However, BEIs are related to exposures to TLV, and therefore, the BEI is indirectly based on the dose-response animal studies used in establishing the TLV.

Intra-individual and inter-individual differences in tissue levels of determinants for essentially the same exposure level occur. This is because the total body burden is partitioned throughout the body's compartments (see Chapter 3) and typically is in flux due to activity. This variability must be considered in the interpretation of the data. Such differences arise on account of variations in pulmonary ventilation, hemodynamics, body composition, efficacy of excretory organs, and the activity of enzyme systems that mediate metabolism of the chemicals. The major sources of inconsistency are outlined below:

1. Physiological and health status of the worker, such as
   - Body build
   - Diet (water and fat intake)
   - Enzymatic activity
   - Body fluid composition
   - Age
   - Sex
   - Pregnancy
   - Medication
   - Disease state

2. Exposure source, such as
   - Intensity of the physical work load
   - Fluctuation of exposure intensity
   - Skin exposure
   - Temperature
   - Humidity
   - Coexposure to other chemicals

3. Environmental sources, such as
   - Community air pollution
   - Home air pollution
   - Water contamination
   - Food contamination

4. Individual life style sources, such as
   - After-work activities
   - Personal hygiene
   - Working habits
   - Eating habits
   - Smoking
   - Alcohol intake
   - Drug intake
   - Exposure to household products
   - Exposure to chemicals from another workplace
   - Exposure to chemicals from hobbies

5. Methodological sources, such as
   - Specimen contamination during collection
   - Specimen deterioration during storage
   - Poor analytical sample preparation
   - Bias of analytical methodologies used

Timing in collecting the sample with respect to the exposure is critical. In many instances, the level of the determinant changes rapidly due to rapid clearance or, conversely, may accumulate due to very poor clearance rates. The sampling time is specified according to the differences in the uptake and elimination rates of the chemicals and their metabolites. The three general sample collection protocol are listed below:

1. Assume complete clearance overnight. Determinants with prior to shift, during shift, and end of shift samplings are eliminated rapidly (half-life of less than five hours).

2. Assume relatively complete clearance over the weekend. Determinants with timing beginning of the workweek and end of the workweek tend to accumulate in the body during the workweek. For chemicals with multiphase elimination, the timing may be both shift and workweek exposure.

3. Assume months to years for clearance. Determinants with timing that is not critical or discretionary have very long elimination half-times and accumulate in the body over a period of years or for a lifetime. (Hair and nail analysis can provide information on chronic exposures for substances such as lead.)

Three principal types of biological specimens are commonly collected. These are urine, exhaled air, and blood. Each type of specimen has distinctive causes of variability which affect the level of the determinant in the specimen.

For urine analysis, variation in urine volume is the most significant factor. Quantitative collection of urine output over extended periods of time tends to be problematic. Therefore, qualitative methods work best with urine sampling.

Exhaled air analysis may be end-exhaled air (alveolar air) or mixed-exhaled air. Concentrations tend to change rapidly during the expiration phase. Exhaled air specimens collected from workers with altered pulmonary function may not be suitable for exposure monitoring.

Blood analysis determinants may be partitioned among the various components of blood, such as plasma binding. Generally, the analysis of whole blood, plasma, serum, or erythrocytes is specified, as required. The concentration difference between arterial and venous blood, induced by pulmonary uptake or clearance, should be considered when measuring volatile chemicals. (BEIs are based upon venous blood and not capillary, or arterial, blood.)

Typically, multiple sampling is necessary to reduce the effects of variable factors. Monitoring protocols should be developed based on health and safety requirements. TLVs may serve as a reference value. Measurements are made of the appropriate determinants in biological specimens collected from the worker. These specimens are collected at specific times. The protocol should be designed to definitively explain any differences in TLV and BEIs.

BEIs do not indicate a sharp distinction between hazardous and nonhazardous exposures. Also, due to biological variability, it is possible for an individual's measurements to exceed the BEI without incurring an increased health risk. However, the cause of excessive values must be investigated. This may be from specimens obtained from a single worker on different occasions, or from the majority of measurements in specimens obtained from a group of workers, in the same workplace. When measurements persistently exceed the BEIs, proper action should be taken to reduce the exposure.

Ability to Work in Protective Equipment

A significant industrial hazard that is often overlooked or ignored is heat stress. Because workers in many industries have endured heat stress for decades and have survived with minimal ill effects, the problem is all too frequently ignored. Therefore, managers and administrators do not consider the problem. One reason for evaluating heat stress factors stems from the increasing need for, and use of, ensembles to protect against exposures to hazardous materials and chemical contaminants. PPE, and in particular respirators, is frequently a critical and integral element of worker health and safety plans. The use of a respirator increases worker stress. The level of heat stress increases when protective clothing hampers the body's ability to cool itself through sweat evaporation. This loss of cooling capacity must be balanced by lower physical work demands or by lower environmental heat. It is essential that each worker be initially and periodically evaluated for the use of PPE. Medical screening should follow the following protocol:

- Disqualify individuals who are clearly unable to perform based upon the medical history and physical exam (e.g., those with severe lung disease, heart disease, or back or orthopedic problems).

- Note limitations concerning the worker's ability to use protective equipment (e.g., individuals who must wear contact lenses cannot wear full facepiece respirators).

- Provide additional testing (e.g., chest X ray, pulmonary function testing, electrocardiogram) for ability to wear protective equipment, where necessary.

- Base the determination on the individual worker's profile (e.g., medical history, physical exam, age, previous exposures, and testing).

- Make a written assessment of the worker's capacity to perform while wearing a respirator, if wearing a respirator is a job requirement. Note that the OSHA respirator standard (29 CFR 1910.134) states that no employee should be assigned to a task that requires the use of a respirator unless it has been determined that the person is physically able to perform under such conditions.

The greater the level of protection, the greater the level of stress experienced by the user. Basically, if a job requires clothing heavier than the typical work uniform's 6 oz/yd cotton, there is an added heat stress burden, and the effects on productivity can be pronounced. Simple coveralls effectively increase the ambient temperature a couple of degree Celsius while vapor barrier ensembles increase the temperature 8 to 10 degrees. It is suggested that a worker in full protection (i.e., impermeable ensemble) has only one-third of normal capacity for a particular temperature. Work in a confined space at an elevated temperature radically decreases operational capacity. Table 15-1 provides data on suggested capacity while wearing PPE. Of special importance in situation assessment and predicting performance evaluation is realizing that these values are for a fit and acclimatized worker. Something as innocent as a weekend of rest and recreation can significantly degrade fitness and acclimatization.

Table 15-1

Suggested work duration for a fit and acclimatized worker

Adjusted Temperature °C	°F	Normal Work Ensemble Minutes of Work	Impermeable Ensemble Minutes of work
32.2	90 and above	45	15
30.8 – 32.2	87.5 – 90	60	30
26.1 – 30.8	82.5 – 87.5	90	60
25.3 – 26.1	77.5 – 82.5	120	90
22.5 – 25.3	72.5 – 77.5	150	120

Investigation and Evaluation

To find evidence of heat stress, examine how work place temperatures vary from cool to hot. Identify average temperatures for shifts over the course of the day and the year—day versus night shifts and summer versus winter. The preferred method of temperature measurement is with a wet bulb thermometer; it automatically factors in the humidity factor. Watch for changes in the frequency of accidents and injuries, in the quality of product, and in employee relations such as morale, absenteeism, and turnover. If there is a worsening of these indicators during periods of increased average temperature, heat stress is a reasonable candidate. Most work conditions that are already recognized as being hot will likely be found to be above the TLVs for heat stress.

Control Methodology

Control methods were presented in Chapter 14. They include general controls, such as training, heat stress hygiene, and medical surveillance. These are the first line of action to reduce the risk of heat related disorders and are the minimum for good practice. Training should include causes and effects of heat stress and when they may occur on the job; signs, symptoms, and first aid for heat-related disorders, and procedures for reporting them; heat-stress hygiene practices; and site-specific considerations. Heat stress hygiene includes these topics:

- **Fluid replacement**
  Employees should drink small amounts of water every 15 to 20 minutes. The water should be readily available and cool (about 50 °F).

- **Self-determination**
  Employees need reasonable discretion in regulating their work under heat-stress conditions. They should report extreme discomfort or early symptoms of heat-related disorders to their supervisors and should seek relief from the heat.

- **Health status**

  Employees should be aware that illness affects the body's tolerance of heat stress. They should tell their physicians that their occupation involves heat stress and should inform their supervisors of any chronic or acute illness and medication they are taking. The medical department may choose to reduce such employee's exposure to heat stress.

- **Lifestyle and diet**

  Employees are expected and encouraged to maintain good health. They should exercise regularly, avoid recreational drugs, limit alcohol consumption, get adequate sleep and limit off-the-job heat exposure. A well-balanced diet is important. As in cases of illness and medication, those on weight-loss diets or restricted or low-salt diets should inform the medical department and mention it during the interview portion of their periodic physical exams.

- **Acclimatization**

  Under normal circumstances the body adapts to heat stress during the first week or so of exposure. An acclimatized worker can tolerate heat or other stress better than unacclimatized workers. Therefore, reduced productivity may be expected during the acclimatization process.

Job-specific controls include both engineering and administrative controls. They are typically designed and selected for specific jobs or tasks within economic and technical constraints. Engineering controls lower the metabolic demands of the task or reduce the effects of the environment. Metabolic requirements are lowered by reducing the external work demands. It is important to match clothing requirements to both chemical and heat-stress hazards. Vapor-transmitting fabrics provide excellent particle protection, with workers wearing such fabrics being 20% more productive than those wearing spun-bond polyethylene. Vapor-transmitting fabrics also provide adequate protection from some liquids and may provide a more comfortable substitute to vapor-barrier clothing.

Chemical Hazards	
Behavioral changes	Irritability
Breathing difficulties	Irritation of eyes, nose and throat
Changes in complexion or skin color	Irritation of respiratory tract
Coordination difficulties	Irritation of skin
Coughing	Light-headedness
Diarrhea	Nausea
Dizziness	Sneezing
Drooling	Sweating
Fatigue and/or weakness	Tearing
	Tightness in the chest

Heat Exhaustion (Warning Condition)	
Clammy skin	Heat rash
Confusion	Light-headedness
Dizziness	Nausea
Fainting	Profuse sweating
Fatigue	Slurred speech
	Weak pulse

Heat Stroke (Medical Emergency)	
Confusion	Incoherent speech
Convulsions	Staggering gait
Hot skin	Sweating stops
High temperature (chilled feeling)	Unconsciousness

Table 15-2

Signs and symptoms— emergency response and hazardous conditions

Work practices can be regulated to reduce the risk of heat stress. Adequate rest periods, monitoring of body temperature, heart rate and sweat loss, and redistribution of task assignments can be used to minimize heat stress. Personal cooling units, such as air circulating or cold vests, may be very effective at reducing heat stress.

Toxic Substances, Testing, and Screening

Preliminary medical screening and assessment can be initiated for some general chemical groups. A few of the more common follow. A screening protocol should be developed in response to anticipated job and workplace hazards.

Aromatic Hydrocarbons

Examples	Benzene, ethyl benzene, toluene, xylene
Sources	Commercial solvents and intermediates for synthesis in the chemical and pharmaceutical industries
Affected Areas	Bone marrow, blood, CNS, eyes, respiratory system, skin, liver, and kidney
Effects	All cause CNS depression, decreased alertness, loss of consciousness, and defatting dermatitis
	Benzene suppresses bone marrow function, causing blood changes. Chronic exposure can cause leukemia.
Note	Because other aromatic hydrocarbons may be contaminated with benzene during distillation, benzene-related health effects should be considered when exposure to any of these agents is suspected.
Screening	Occupational and general medical history, emphasizing prior exposure to these or other toxic agents. Medical examination with focus on liver, kidney, nervous system, and skin.
	Laboratory testing **Complete Blood Count (CBC)** (See Frequently performed health tests), platelet count, and measurement of kidney and liver function.

Asbestos or Asbestiform Particles

Examples and Sources	A variety of industrial uses including building, construction, cement work, insulation, fireproofing, pipes and ducts for water, air, chemicals, and automobile brake pads and linings.
Affected Areas	Lungs and gastrointestinal system
Chronic Effects	Lung cancer, mesothelioma, asbestosis, and gastrointestinal malignancies.
	Asbestos exposure coupled with cigarette smoking has been shown to have a synergistic effect in the development of lung cancer.
Screening	History and physical examination should focus on the lungs and gastrointestinal system. Laboratory tests should include a stool test for occult blood evaluation as a check for possible hidden gastrointestinal malignancy.

A high quality chest x-ray and pulmonary function test may help to identify long-term changes associated with asbestos disease. However, early identification of low-dose exposure is unlikely.

Asbestosis bodies are golden-yellow bodies of various shapes formed by the deposition of calcium, iron salts, and proteins on a spicule of asbestos occurring in the **sputum**, lung secretion, and feces of patients with asbestosis.

Halogenated Aliphatic Hydrocarbons

Examples   Carbon tetrachloride, chloroform, ethyl bromide, ethyl chloride, ethylene dibromide, ethylene dichloride, methyl chloride, methyl chloroform, methylene chloride, tetrachloroethane, tetrachloroethylene, trichloroethylene, and vinyl chloride

Sources   Commercial solvents and intermediates in organic synthesis

Affected Areas   CNS, kidney, liver, and skin

Effects   All cause CNS depression: decreased alertness, headaches, sleepiness, and loss of consciousness. Kidney changes: decreased urine flow, swelling (especially around eyes), and anemia. Liver changes: fatigue, malaise, dark urine, liver enlargement, and jaundice.

Vinyl chloride is a known carcinogen. Several others in this group are potential carcinogens.

Screening   Occupational and general medical history emphasizing prior exposure to these or other toxic agents. A medical examination should focus on liver, kidney, nervous system, and skin. Laboratory testing for liver and kidney function, and carboxyhemoglobin where relevant.

Heavy Metals

Examples   Arsenic, beryllium, cadmium, chromium, lead, and mercury

Sources   Wide variety of industrial and commercial uses

Affected Areas   Multiple organs and systems, including blood, cardiopulmonary, gastrointestinal, kidney, liver, lung, CNS, and skin

Effects   All are toxic to the kidneys. Each heavy metal has its own characteristic symptom cluster. Lead causes decrease in mental ability, weakness (especially hands), headache, abdominal cramps, diarrhea, and anemia. Lead can also affect the blood-forming mechanisms, kidneys, and the peripheral nervous system.

Long-term effects also vary. Lead toxicity can cause permanent kidney and brain damage. Cadmium can cause kidney or lung disease. Chromium, beryllium, arsenic, and cadmium have been implicated as human carcinogens.

Screening   History taking and physical exams, and search for symptom clusters associated with specific metal exposure—i.e., for lead, look for neurological deficit, anemia, and gastrointestinal symptoms.

Laboratory testing	Measurements of metallic content in blood, urine, and tissue (i.e., blood lead levels; urine screen for arsenic, mercury, chromium, and cadmium), and CBC. Measurement of kidney function and liver function where relevant.
	Chest X ray or pulmonary function testing where relevant.

Herbicides

Examples	Chlorophenoxy compounds, 2,4-dichlorophenoxyacetic acid, 2,4-D and 2,4,5-trichlorophenoxyacetic acid, 2,4,5 T. Dioxin (tetrachlorodibenzo-p-dioxin TCDD—which occurs as a trace contaminant in these compounds) poses the most serious health risk.
Sources	Vegetation control
Affected Areas	Kidney, liver, CNS, and skin
Effects	Chlorophenoxy compounds can cause chloracne, weakness or numbness of the arms and legs, and may result in long-term nerve damage. Dioxin causes chloracne and may aggravate pre-existing liver and kidney diseases.
Screening	History and physical exam should focus on the skin and nervous system.
	Laboratory tests include measurement of liver and kidney function, where relevant, and urinalysis.

Organochlorine Insecticides

Examples	Chlorinated ethanes, DDT, cyclodienes, aldrin, chlordane, dieldrin, endrin, and chlorocyclohexanes, lindane
Sources	Pest control
Affected Areas	Kidney, liver, and CNS
Effects	All cause acute symptoms of apprehension, irritability, dizziness, disturbed equilibrium, tremor, and convulsions.
	Cyclodienes may cause convulsions without any other initial symptoms.
	Chlorocyclohexanes can cause anemia. Cyclodienes and chlorocyclohexanes cause liver toxicity and can cause permanent kidney damage.
Screening	History and physical exam should focus on the nervous system.
	Laboratory tests include measurement of kidney and liver function, and CBC for exposure to chlorocyclohexanes.

Organophosphates and Carbamate Insecticides

Examples	Organophosphate—diazinon, dichloravos, dimethoate, trichlorfon, malathion, methyl parathion, and parathion
	Carbamate—aldicarb, baygon, and zectran

Sources	Pest control
Affected Areas	CNS, liver, and kidney
Effects	All cause a chain of internal reactions leading to neuromuscular blockage. Depending on the extent of poisoning, acute symptoms range from headaches, fatigue, dizziness, increased salivation and crying, profuse sweating, nausea, vomiting, cramps, and diarrhea to tightness in the chest, muscle twitching, and slowing of the heartbeat. Severe cases may result in rapid onset of unconsciousness and seizures. A delayed effect may be weakness and numbness in the feet and hands. Long-term, permanent nerve damage is also possible.
Screening	Physical exam should focus on the nervous system.
	Laboratory tests should include RBC, cholinesterase levels for recent exposure (plasma cholinesterase for acute exposures). Measurement of delayed neurotoxicity and other effects.

Polychlorinated Biphenyls (PCBs)

Sources	Wide variety of industrial uses
Affected Areas	Liver, CNS, respiratory system, and skin
Effects	Various skin ailments, including chloracne, may cause liver toxicity; carcinogenic to animals.
Screening	Physical exam should focus on the skin and liver.
	Laboratory tests should include serum PCB levels, triglycerides and cholesterol. Measurement of liver functions.

Frequently Performed Occupational Health Tests

A number of standard occupational health tests are used for assessing worker health and safety. The liver, kidney, and blood are frequently evaluated to assess worker status.

Liver

General Blood tests	Total protein, albumin, globulin, total bilirubin.
Obstruction Enzyme test	Alkaline phosphatase
Cell injury Enzyme tests	Gamma glutamyl transpeptidase (GGTP) Lactic dehydrogenase (LDH) Serum glutamic-oxioacetic transaminase (SGOT) Serum glutamic-pyruvic transaminase (SGPT)

Kidney

| General
Blood tests | Blood urea nitrogen (BUN), creatinine, uric acid |

Multiple Systems and Organs

Urinalysis   Including color, appearance, specific gravity, pH, qualitative glucose, protein, bile, and acetone; occult blood; microscopic examination of centrifuged sediment.

Blood-Forming Function

Complete blood count (CBC) with differential and platelet evaluation, including white cell count (WBC), red blood count (RBC), hemoglobin (HBG), hematocrit or packed cell volume (HCT), and desired erythrocyte indices. Reticulocyte count may be appropriate if there is a likelihood of exposure to hemolytic chemicals.

First Aid and Specific Response

Three common worker inflictions are reaction to chemical hazards, heat stroke, and heat exhaustion. Workers in enclosed impermeable ensembles are at risk to all of these. Therefore, it is advantageous to differentially diagnose the most likely candidate for when to initiate treatment.

Carbon monoxide is an exhaust gas from combustion processes. Poisoning symptoms are those of oxygen deficiency. (See Table 15-3) The carboxyhemoglobin (caused by inhaling carbon monoxide) as a percentage of hemoglobin in heavy smokers is close to, or can exceed, the 8% reached by exposure to a concentration of 50 ppm, the TLV of carbon monoxide.

Treatment should be initiated as soon as possible. Since there are typically multiple symptoms, treatment order must be given consideration. Order is typically determined by the degree of risk. A standard treatment order protocol is shown below:

Treatment Order, General

1. Cessation of Breathing/Cardiac Arrest
2. Eye Injury
3. Skin Contact
4. Shock

Chemical Inhalation

1. Remove from contaminated area
2. Lay supine with legs raised
3. Loosen collar and belt
4. Cover with blanket; keep warm
5. Calm and reassure patient

Chemical Ingestion

1. Remove from contaminated area
2. Rinse mouth with cold water
3. Loosen collar and belt
4. Lay supine with legs raised
5. Cover with blanket; keep warm
6. Minimize moving and speaking

Signs and symptoms	Percentage of Blood Saturation
No symptoms	0 – 10
Tightness across forehead, possible slight headache, dilation of cutaneous blood vessels	10 – 20
Headache, throbbing in temples	20 – 30
Severe headache, weakness, dizziness, dimness of vision, nausea and vomiting, collapse	30 – 40
Same as previous item with greater possibility of collapse or syncope; increased respiration and pulse	40 – 50
Syncope, increased respiration and pulse; coma with intermittent convulsions; Cheyne-Stokes respiration	50 – 60
Coma with intermittent convulsions, depressed heart action and respiration, possible death	60 – 70
Weak pulse and slowed respiration; respiratory failure and death	70 – 80

Table 15-3

Signs and symptoms of carbon monoxide poisoning

(Adapted from Sayers, P.R. and S. J. Davenport, *Public Health Bulletin No. 195*, U. S. Government Printing Office, Washington, DC, 1930)

Chemical Contact on Skin

1. Take to nearest shower
2. Remove clothing from affected areas
3. Gently wipe off any excess chemical
4. Wash affected area under shower with soap
5. Rinse affected area with lukewarm water
6. Dry skin gently with soft towel

Chemical Contact with the Eyes

1. Take to nearest eyewash or shower
2. Gently wipe off any excess chemical
3. Wash with eyelids held open
4. Irrigate with slowly running water for at least 15 minutes

References

Marc J. Lefevre, "First Aid Manual for Chemical Accidents," ISBN 0-442-20490-6 Van Nostrand Reinhold Co; New York, 1980.

"Occupational Safety and Health Guidance Manual for Hazardous Waste Site Activities," DHHS (NIOSH) Pub. No. 85-115.

B. Plog, "Fundamentals of Industrial Hygiene," National Safety Council, 3rd Ed. 1988.

Study Exercises

1. Describe which workers are required by the OSHA standard to be in a medical surveillance program.

2. Explain the value of medical surveillance.

3. List the factors that could affect an individual's particular response to an exposure, as well as cause variabilities in test results.

4. Define total body burden.

5. Describe the components of a medical surveillance program.

6. Describe the rights of the worker in obtaining medical records.

7. Describe three categories of biological exposure indices.

8. Identify an example of a common workplace substance being monitored for each of the following tests:
   - urine tests
   - blood analysis
   - breath analysis
   - hair analysis
   - liver enzyme tests
   - kidney function tests

9. Describe the signs and symptoms of chemical exposure.

10. Describe the order for treating symptoms.

11. Describe the treatment for chemical inhalation.

12. Describe the treatment for chemical ingestion.

13. Describe the treatment for chemical contact on the skin.

14. Describe the treatment for chemical contact in the eyes.

16 Risk Assessment and Epidemiology

Overview

This chapter introduces and formalizes the complex processes of risk assessment, perception and management, and epidemiology. The four components of risk assessment are presented and explained. The various factors that complicate their use and cause uncertainties in their application are reviewed. The Reference Dose (RfD) for noncarcinogenic (threshold) effects and the Cancer Potency Factor(q_1*) for carcinogenic (nonthreshold) effects are developed and related to risk assessment. The factors affecting the public's perception of a risk as being acceptable or not acceptable are examined and the theoretical basis for action levels are derived. The relationships between risk and epidemiology are explored and the need and purposes for epidemiologic studies is developed. The types, benefits, and limits of epidemiologic studies are explained.

Risk

Risk is the probability of injury, disease, or death under specific circumstances. It may be expressed in quantitative terms, through probability theory, taking values from zero (certainty that harm will not occur) to one (certainty that it will). In many cases risk can only be described qualitatively as high, low, or trivial.

Inherent Risk

All human activities carry some degree of risk. Many risks are known with a relatively high degree of accuracy because epidemiologists have collected data on their historical occurrence, one of the methods of epidemiology. For example, if deaths from motor vehicles in the United States in a particular year are known to be 46,000, then the individual risk for you, as a driver, is 2.2×10^{-4} or 1 in 4,500. Furthermore, assuming the annual risk remains constant, this would imply a lifetime (70 years) risk of 1 in 65. This type of manipulation of data is known as **statistics**.

The risks associated with many activities cannot be readily assessed and quantified. This is especially true of the exposure to various chemical substances. There are considerable historical data on the risks of some types of chemical exposure, e.g., the annual risk of death from intentional overdoses or accidental exposure to drugs, pesticides, and industrial chemicals. Unfortunately, such data are generally restricted to those situations leaving little doubt about causation. Generally, these are cases in which a single, very high exposure resulted in an immediately observable form of injury.

Assessment of the risks of lower levels of chemical exposure is far more problematic. These situations do not cause immediately observable forms of injury or disease, or cause only minor forms, such as transient eye or skin irritation. In such cases, the exposure may have been brief, extended but intermittent, or extended and continuous. The assessment of these risks is complex, complicated by the lack of a direct cause-effect relationship, and confused by the presence of contributory or extraneous sources.

Risk Assessment

Risk assessment is a determination of the probability that an adverse effect will be produced. The National Academy of Sciences defines risk assessment as the scientific activity of evaluating the toxic properties of a chemical, and the conditions of human exposure to it, in order to both ascertain the likelihood that exposed humans will be adversely affected and to characterize the nature of the effects they may experience. (*Risk Assessment in the Federal Government: Managing the Process,* National Academy Press, Washington, D.C., 1983). (See Figure 16-1)

Toxicologically, risk assessment is a methodological approach in which the toxicities of a chemical are identified, characterized, and analyzed for dose-response relationships. The validity of the risk estimate is related to the validity of the data and the assumptions inherent to the process. Uncertainty is central to the process. Therefore, there is controversy over the usefulness and validity of quantizing risk and using assessments of these values as decision-making tools.

Frequently, a mathematical model is applied to the data. The model is used to generate a numerical estimate in a process that may act as a guideline to decisions concerning allowable exposures. Laboratory animal data is commonly a data source. Utilization of this data requires extrapolation from the very high doses administered to the animals to the much lower human doses. Extrapolations are often made over 4 to 6 orders of magnitude; dose differences of 10,000 to 1,000,000 are typical.

Quantitative Risk Assessment

Quantitative risk assessment generates a numerical measure of the risk or safety of a chemical exposure. (Safety is the inverse of risk, i.e., the probability that an adverse effect will not occur.) Qualitative risk assessment relates the hazards of one chemical or substance to the hazards of another chemical or substance (e.g., as dangerous as) or groups the chemicals into categories (e.g., mutagen or carcinogen) which connotes certain risks.

Perceived Risk

Perceived risk is a very personal matter and thus a very complex one. The closer the public's perception of a specific risk agrees with the risk determined by an

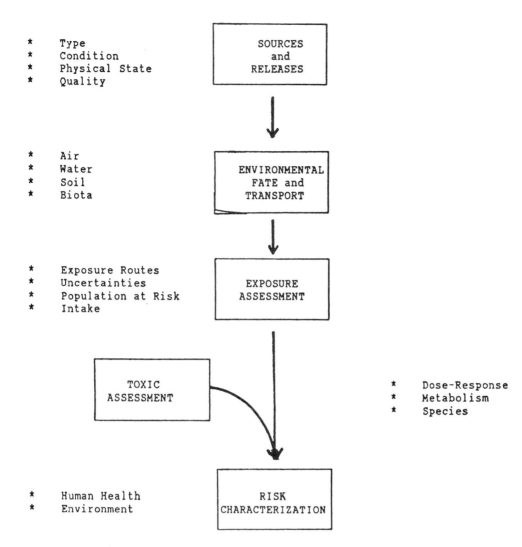

* Type
* Condition
* Physical State
* Quality

SOURCES
and
RELEASES

* Air
* Water
* Soil
* Biota

ENVIRONMENTAL
FATE and
TRANSPORT

* Exposure Routes
* Uncertainties
* Population at Risk
* Intake

EXPOSURE
ASSESSMENT

TOXIC
ASSESSMENT

* Dose-Response
* Metabolism
* Species

* Human Health
* Environment

RISK
CHARACTERIZATION

Figure 16-1

Health risk assessment and characterization. We are assuming a goal of characterizing the risks to both humans and the environment due to any source, spill, or release. The first step is to examine the environmental fate of the materials and substances and to determine its movement. Given this information, it is possible to assess the exposure potential and determine which populations are at risk. This information, combined with information about toxicity, is then used to characterize the risk.

objective scientific and statistical risk assessment process, the greater the chances that regulatory decisions will be based on knowledge rather than on emotions, and that good and lasting benefit will accrue to society from the controls imposed. Education is an obvious and critical factor in risk perception. People who have more education perceive risks differently than do people who have lesser education. Other factors include social status, economic status, age, sex, sexual orientation, national origin, place of residence, racial and cultural background, religious orientation, and the opinions of friends and relatives.

Acceptable (*De Minimis*) Risk

Acceptable risk may be viewed from two levels, societal and personal. On a personal level, the determination as to whether a risk is acceptable or not depends greatly on the perceived risk. On a daily basis, most of us analyze risk and make personal risk benefit decisions. *Should I drive a car* or *should I climb a ladder* are risk assessments and analyses. These analytical activities are associated with risk benefit decisions. Because of the inherent risks involved in these activities, subpopulations (e.g., children) are discouraged from both activities, either by parental or public policy.

The EPA defines negligible risk for adults as a one-in-a-million chance of getting cancer from a particular residue in a lifetime. One interpretation of this is that in California, with a population of 30 million, it would be acceptable for 30 people to develop cancer due to exposure to a particular residue. (This erroneously implies that an extra case of cancer in a population of a million people could actually be measured and the cause identified.) Whether such a risk is acceptable to an individual, however, is a different story. A person may discover that he or she, or a loved one, has become that statistical one-in-a-million, to someone else's benefit. This is a frightening concept!

Components of Risk Assessment

Every risk assessment has four components:

1. **Hazard Identification**
   This involves gathering and evaluating data on the types of health injury or disease that may be produced by a chemical and on the conditions of exposure under which injury or disease is produced. Hazard identification may also involve characterization of the behavior of a chemical within our bodies, as well as the interactions it undergoes with organs, cells, or even parts of cells. Data on the latter types may be of value in answering the ultimate question of whether the forms of toxicity known to be produced by a substance in one population group or in experimental settings are also likely to be produced in humans. Hazard identification is not risk assessment. It simply determines whether it is scientifically correct to infer that toxic effects observed in one setting will occur in other settings. For example, are substances found to be carcinogenic or teratogenic in experimental animals likely to have the same result in humans?

2. **Dose-Response Evaluation**
   This involves describing the quantitative relationship between the amount of exposure to a substance and the extent of toxic injury or disease. Data may be derived from animal studies, or less frequently, from studies in exposed human populations. There may be many different dose-response relationships for a substance if it produces different toxic effects under different conditions of exposure. The risks of a substance cannot be ascertained with any degree of confidence unless dose-response relations are quantified, even if the substance is known to be toxic.

3. **Human Exposure Evaluation**
   This involves describing the nature and size of the population exposed to a substance and the magnitude and duration of their exposure. The evaluation could concern past or current exposures, or exposures anticipated in the future.

   Some exposures are additive. The human total dose is equal to the sum of doses from five routes. Principal exposure routes occur through drinking, inhalation, ingestion, and skin absorption from water and contaminated soil. For example, consider the contamination of water. Assume an adult consumes two liters of water every day with a concentration of contaminant at just 5 mg per liter. That is almost 4 kilograms or 8 pounds of contaminants per year.

4. **Risk Characterization**
   This generally involves the integration of the data and analysis of the first three components to determine the likelihood that humans will experience any of the various forms of toxicity associated with a substance. In cases where exposure data are not available, hypothetical risk can be character-

ized by the integration of hazard identification and dose-response evaluation data alone.

Risk is generally characterized as follows:

A. For noncarcinogens, and for the noncarcinogenic effects of carcinogens, the **margin-of-safety (MOS)** is estimated by dividing the experimental **No Observable Effects Level (NOEL)** by the estimated daily human dose.

B. For carcinogens, risk is estimated for a human dose by multiplying the actual human dose by the risk per unit-of-dose projected from the dose-response modeling. A range of risks may be obtained using different models and assumptions about dose-response curves and the relative susceptibilities of humans and animals.

These steps are far more complex than detailed above. There are numerous problems associated with the timing and duration of exposures. However, the MOS and the carcinogenic risk are the ultimate measures available to us on the likelihood of human injury or disease from a given exposure or range of exposures.

The **Acceptable Daily Intakes (ADIs)** are not a measure of risk; rather, they are considered a safe value. As will be discussed later, they are derived by imposing a specified **safety factor** to a dose level which causes no observable effect. The purpose is not to specify an ADI, but to ascertain risk. (This does not apply to noncarcinogens.) The MOS is used as a surrogate for risk; as the MOS becomes larger, the risk becomes smaller. At some point, the MOS is so large that human health is ensured to be safe, and most certainly, as can best be determined, not jeopardized. The magnitude of the MOS needed to achieve this condition varies between substances. Its selection is based upon the same, or similar, factors as those used to select safety factors for the establishment of ADIs.

Risk Communication

Risk communication is the interactive process of exchange of information and opinion among individuals, groups, and institutions. It involves multiple messages about the nature of risk and other messages, not strictly about risk, that express concerns, opinions, or reactions to risk messages or to legal and institutional arrangements for risk management.

Risk communication embraces the very delicate task of explaining all aspects of risk to the public. Average citizens are capable of understanding complex subjects that are of interest and importance to them. Government, industry, and academic scientists are not always capable of making complex subjects understandable. Scientists who attempt to explain science to the public objectively and in nontechnical terms face the risk of being labeled by one side or the other, or both, of an issue. Journalists are often blamed for the public's misunderstanding or lack of appreciation of the safety or danger of some technologies. The popular press is an important avenue for risk communication. Citizen reaction to risk messages is a fascinating field coming under increasing study by communication, psychology, and social science experts. These expert findings provide insights that help formulate reports so they increase public understanding.

Outrage Factor

Risk communication experts point to a factor they call **outrage**. Outrage refers to the level of public anger and fear about an environmental risk issue. Outrage has a much greater influence on a citizens' reactions to a hazard than the scientifically calculated risk. When people become outraged, they may overreact. Conversely, if people are not outraged, they may underreact. People become outraged—fearful, angry, frustrated—if risk is perceived to be:

Involuntary
People do not like to be forced to face a risk—like trace chemicals in tap water. (But they will voluntarily assume risks like drinking diet soda.)

Uncontrollable
When preventing risk is in someone else's hands (government or industry), citizens feel helpless to change the situation. If the citizen can prevent or reduce the risk (using household chemicals properly), the risk is more acceptable.

Immoral
Pollution is viewed as an evil. Therefore, people consider it unethical for governments and industries to claim that a risk is acceptable based on cost-benefit analysis or because there is "only" a low incidence of harm.

Unfamiliar
People become uneasy when scientists are not certain about the risk posed by a hazard—its exact effect, severity, or prevalence.

Catastrophic
A risk resulting in a large-scale disastrous event (plane crash, nuclear reactor meltdown), is more dreaded than a risk affecting individuals singly (auto accidents, radon).

Memorable
A potential risk similar to a remarkable event imbedded in the memory, like Bhopal or Three Mile Island, is viewed as much more dangerous than the risk of some unheard-of or little-known disease.

Unfair
People become outraged if they feel they are being wrongfully exposed. For example:

- exposure to a risk that people in a neighboring community or a different economic bracket are not being exposed to.

- exposure to a risk with no benefit, e.g., living next to a nuclear waste dump, but receiving no benefit from nuclear power generation. In contrast, people will assume the risk of exposure to something like medical X rays because they perceive a benefit that equals or outweighs the risk.

From an untrustworthy source
People become outraged if they have no confidence in the source of the risk, such as industry or government. In contrast they will accept risks from what they view as a reliable risk source, such as a doctor.

Clearly, emotions play a large role in public perception of risk. No explanation of scientific findings makes much impression if people are either hysterical or agitated. This emotional response may be viewed as irrational and, therefore, ignored or condemned. In fact, individual emotional responses are based on psychologically valid factors and are perfectly rational. When people become aware of a threat, they are naturally inclined to react in certain ways:

- Fear the unknown
- Want to maintain control
- Protect home and family

- Be alienated by dependence on others (government, industry officials)
- Protect their belief in a just world.

By contrast, technically trained officials tend to trust scientific analyses, accept the effectiveness of engineering solutions and contingency plans, and believe that experts know best. Thus, much of the conflict surrounding risk issues is a result of groups with vastly different values becoming pitted against one another. Communications experts urge those involved in communicating risk—officials and reporters—to accept the reality and validity of the public's emotions and seek ways of communicating that take these emotions into account.

Risk Communication Guidelines

There are some ways that officials and reporters can address the psychological factors influencing citizen response to hazards. The point is not to diminish legitimate concerns or heighten illegitimate ones but to encourage constructive action and provide a sense of empowerment. Some ways to address these factors are as follows:

- Describe what individuals can do to reduce their exposure.
- Describe what industry and government are or are not doing to reduce the risk.
- Describe the benefits as well as the risks to the specific audience (not just society in general) of the substance or process of concern.
- Describe the alternatives and their risks.
- Describe what people can do to get involved in the decision-making process.
- Provide information that will help the audience to evaluate the risk.

Helping the Audience Evaluate Risk

Ultimately, citizens judge how dangerous a risk is and whether they should take action to reduce it. Officials and reporters can play a key role in encouraging sound decisions by providing information that will help their audience evaluate the risk. Some fundamental information needed by citizens is as follows:

- How much of the substance is the audience actually being exposed to?
- What is the likelihood of accidental exposure? What safety/backup measures are in place?
- What is the legal standard for the substance? Is the standard controversial or widely accepted as sound?
- What health or environmental problems is the standard based on? Are there other problems that should be considered?
- Is the source of the risk information reputable? Who funded the work? What do other sources say?
- Were the studies done on a population similar to this audience?
- What are the benefits of the substance/facility? What are the trade-offs?

There are many references, both articles and books, produced by environmental, professional, and consumer groups and government agencies describing aspects of the subject of risks presented by natural and synthetic chemicals and substances. The majority of people initially have open minds and are willing to listen. As usual, the first impression is the most enduring and lasting impression and may never be reversible.

Risk Management

Risk management is the process of evaluating the assessed risk by value judgments and social bias so as to modify the risk assessment according to need, cost, or technical feasibility (Figure 16-2). Inherent to this process are risk-benefit and cost-benefit analyses. Risk-benefit analysis assesses the risk involved in some action or event against the benefits derived. Cost-Benefit analysis examines the cost of some action or event, such as the cost of risk reduction in relation to the quantitative risk reduction.

Ideally, risk management is the control of risk by eliminating or modifying the conditions that produce the risk. However, few actions are without repercussions. For example, water supplies are chlorinated and fluorinated for public health reasons. Unfortunately, these actions carry their own risks: carcinogenesis and osteoporosis, respectively. In these cases, the risk and cost benefit analysis resulted in the general implementation of these practices. The government practices risk management by legislating and regulating specific procedures for controlling and mitigating risks.

Figure 16-2

The interrelationship between risk assessment and risk management. The four components of risk assessment are utilized as input to the management process with the objective of reaching regulatory decisions suitable for mitigating and reducing risk.

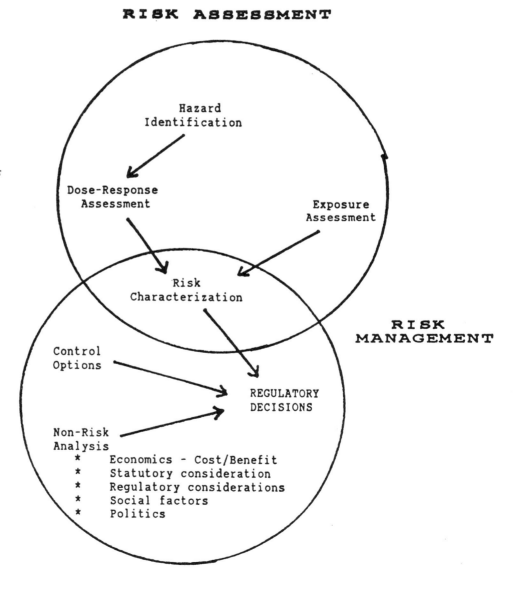

Maximum Tolerated Dose (MTD) versus Acceptable Daily Intakes (ADI) or Reference Dose (RfD)

As discussed in the dose-effects Chapter, the threshold dose is approximated by the No-Observable-Effect Level (NOEL). If insufficient data is available, then a **Suggested No-Adverse-Response Level (SNARL)** is developed. **Maximum Tolerated Dose (MTD)** is the maximum dose that an animal species can tolerate for a major portion of its lifetime without significant impairment of growth or observable toxic effect other than carcinogenicity. Concerns about the use of MTDs are these:

1. The underlying biological mechanisms that lead to the production of cancer may change as the dose of the carcinogen changes.

2. Current methods for estimating an MTD for use in an experiment do not usually take these mechanisms into account.

3. The biological mechanisms at work under conditions of actual human exposure may be quite different from those at work at or near the MTD.

4. Observations at or near a MTD, as determined by current methods, may not be qualitatively relevant to conditions of actual human exposure.

5. How can it be known that the most serious effects of a substance have been identified?

The **Acceptable Daily Intake (ADI)** and the **Reference Dose (RfD)** are essentially the same. MTD and threshold are used to determine NOELs. However, there are numerous difficulties in actually determining NOELs. For these reasons, standard-setting and public health agencies protect populations from substances displaying threshold effects by dividing experimental NOELs by large **safety factors (SFs)**. The magnitude of safety factors varies according to the nature and quality of the data from which the NOEL is derived; the seriousness of the toxic effects; the type of protection being sought (e.g., are we protecting against acute, subchronic, or chronic exposures?); and the nature of the population to be protected (e.g., the general population, or populations such as workers expected to exhibit a narrower range of susceptibilities). Safety factors of 10, 100, 1,000, and 10,000 have been used.

Safety factors suggested by the National Academy of Science's Safe Drinking Water Committee depend upon the uncertainty factors used in combination with NOELs. A safety factor of 10 is suggested when valid human data based on chronic exposure are available. A safety factor of 100 is suggested when human data are inconclusive (i.e., limited to acute exposure histories) or absent, but long-term reliable animal data are available for several species. A safety factor of 1,000 should be utilized when no long-term or acute human data are available and the experimental animal data are scanty. In addition, if the lowest-observable-effect dose is used rather than a true threshold dose, then an additional factor of 10 should be added to the SF.

The ADI for humans for chemical exposure is derived by dividing the experimental NOEL into mg/kg/day, with the toxic effect appearing at lowest dose by one of the safety factors. There is no way to determine that exposures at ADIs estimated in this fashion are without risk. The ADI represents an acceptable, low level of risk, but not a guarantee of safety. Conversely, there may be a range of exposures well above the ADI, perhaps including the experimental NOEL itself, that bear no risk to humans.

Risk Sensation

This is an area of much uncertainty and carries a high potential for misunderstanding. In 1989, an environmental group identified the plant growth regulator Alar (daminozide) as a potent carcinogen. The Natural Resources Defense Council (NRDC) predicted that exposure to Alar "during the first six years of life alone is estimated to result in a cancer risk of approximately one case for every 4,200 preschoolers exposed." These results were dramatically announced on the popular CBS program *60 Minutes*. The result was national consumer panic and a loss in sales in excess of $100 million for the U.S. apple industry.

Risk Comparisons

Comparing a new, unfamiliar risk with an old, familiar one is appealing because the comparison provides a concrete way to express a numerical concept (such as one death in a million). Risk comparisons appear to establish a scale of severity by which people can judge whether the new risk is something to be concerned about. However, risk comparisons must be used with great care. Often, an involuntary risk is compared with a voluntary one, e.g., the risk form nearby chemical plant emissions is compared with smoking, dietary habits, or some other lifestyle choice. Such comparing of an involuntary exposure to risk with a voluntary exposure tends not to influence people's perception. If such a comparison is done in the spirit of minimizing the importance of the involuntary risk, it will generate anger.

The value of risk comparisons is also limited by the fact that risks tend to accumulate in people's minds. No matter how small the new risk, people are inclined to see it as simply one more unwelcome vexation to add to their already heavy burden of coping with modern-day problems. Several types of risk comparisons are generally more useful than comparing involuntary risks with voluntary ones:

- **Comparisons of similar risks**
  How do synthetic pesticide levels in a food compare with the levels of natural pesticides found in many foods?

- **Comparisons of risks with benefits**
  The risk to human health of using chlorine to disinfect drinking water versus chlorine's role in protecting human life from infectious disease.

- **Comparisons of alternative substances/methods**
  Which actually causes the most pollution of the environment—incinerating a waste or landfilling it?

- **Comparisons to natural background levels**
  How does the level of a substance in a suspected contaminated area compare with natural background levels, such as the level of lead in someone's backyard compared with the average natural lead levels in soils in the United States?

- **Comparisons with a regulatory standard**
  How does the comparison of arsenic in a city's drinking water compare with the standard set by the Environmental Protection Agency?

Epidemiology

Epidemiology is the most direct method of assessing risk to humans, but like any scientific method, it has its own limitations and problems of interpretation. Knowing the principles and pitfalls of epidemiology will help in interpreting epidemiological studies. Epidemiology is the study of patterns of disease in human populations. Because epidemiological studies look directly at humans rather than extrapolate from animals, they provide the most compelling evidence for measuring environmental risks to humans. Most studies in recent decades that have linked environmental factors to human diseases were designed using the principles of epidemiology.

Examples of epidemiological studies that have provided critical linkage evidence are these:

- Toxic shock syndrome to tampon use
- Leukemia to on-the-job exposures to benzene
- Heart attacks to cholesterol
- Lung cancer, heart attacks, and low birth-weight to cigarette smoking
- Legionnaires' disease to a building's contaminated cooling units

Epidemiological studies provide evidence, not proof. Remember the basic tenant of the scientific method, *You cannot prove anything!*, and its corollary, *All you can do is disprove!* Epidemiologists cannot be absolutely sure that the effect they see corresponds to the suspected cause. There is always a level of uncertainty. Uncertainty is inherent in the tools used, and while it may be very small, it can never be zero.

Epidemiologists compare two or more groups of people to determine what characteristics distinguish groups of individuals who become diseased from groups of individuals who do not. These distinguishing characteristics are then examined to elucidate the how and why they are associated with disease. Some of the characteristics sought are these:

- Consumption of certain foods
- Contact with bacteria, chemicals, or viruses
- Gender, race, or socioeconomic status
- Daily activities and behaviors
- Genetic background
- Metabolic characteristics, such as cholesterol level and blood pressure

Risk Factors and Exposures

Epidemiologists prefer not to use the word *cause* when looking for clues to disease because many characteristics associated with disease are not true causes. For example, cigarette smoking is associated with heart attacks because chemicals in the smoke trigger the attacks. Race, gender, and socioeconomic status also are strongly associated with heart attacks, not because they directly cause the attacks, but because they are proxies for many hard-to-define behaviors, environmental factors, and genetic factors that increase the risk of heart disease.

So epidemiologists use the term *risk-factor* to describe anything that increases the risk of disease. Cigarettes, race, and socioeconomic status are all risk factors for heart disease. Risk factors may also be called exposures; a person with a risk factor is said to be exposed; a person without that particular risk factor is unexposed. (However, it is not usual to describe risk factors that are inherent characteristics of an individual, such as sex and race, as exposures.)

Types of Epidemiological Studies

Epidemiologist favor two types of studies for searching out risk factors for disease: case-control studies and cohort studies. In **case-control** studies a group of people with disease (cases) and a group without disease (controls) are surveyed about their histories. The survey may involve direct questioning or examination of medical or other records. The important question is *What differs in the histories of these two groups that could explain why one is diseased and the other is not?*

As an example, in the spring of 1980, U.S. doctors diagnosed hundreds of cases of **toxic shock syndrome (TSS)**, a potentially fatal, previously rare disease. Most cases occurred in young women during their menstrual periods. Investigators at the Center for Disease Control questioned 50 women with toxic shock syndrome (cases) about their use of sanitary products in the month before they got sick. Then they asked each woman for the names of three friends who did not have TSS (controls), and the investigators asked the friends the same questions. Women with TSS were more likely than their friends to have used tampons. In particular, they were almost 8 times as likely to have used one brand— Rely. This brand was withdrawn from the market in September 1980, and the incidence of TSS decreased dramatically.

A **cohort** (follow-up) study begins with a group of people who do not have the disease being studied. Group members differ on one or more characteristics suspected of causing the disease (for example, some may smoke while others do not). The group is followed over time to see if members with the suspect characteristic are more likely to develop the disease. The important question here is *Are the people with the suspect characteristic at greater risk of getting disease?*

An example was a cohort study used to evaluate the effect of environmental lead exposure on children's IQs. Researchers followed 516 children in the lead-smelting town of Port Pirie, Australia, from birth to age seven, periodically taking blood samples to measure lead levels. At age seven, children with highest blood lead levels over the years had the lowest IQs.

Case-control studies are more common than cohort studies because they are faster and cheaper. Also, for relatively uncommon diseases like childhood leukemia, they are often the only practical way to look for causes of disease. Cohort studies are more convincing for two reasons:

1. They provide a much better opportunity to establish a cause-effect relationship because they begin with the exposure (cause) and move forward in time to the disease (effect). In contrast, case-control studies begin with the disease (effect) and look back to the exposure (cause). It is not always clear that the identified cause actually did come first.

2. Case-control studies are more prone to certain study design problems, such as bias or chance.

Cohort studies have their own drawbacks:

1. They are very expensive.

2. They take a long time (because they start with well people and wait for them to become sick).

3. They are difficult to conduct properly because study subjects tend to drop out of the study over time.

Two other types of epidemiological studies are **cross-sectional** and **clinical trials**. The cross-sectional study identifies a population of interest (people in a particular neighborhood, people coming to a clinic) and asks its members about current diseases and current exposures. Cross-sectional studies offer epidemiologist a quick way to determine whether a problem exists that warrants further study. For example, do workers in a particular industry have an unusually high rate of disease? However, this kind of study is not useful for establishing cause and effect because it is difficult to determine whether the exposures actually caused the disease.

To illustrate, let us examine a misinterpretation from a cross-sectional study. It is well known that cigarette smoking increases the risk of a heart attack. But if researchers did not know this and surveyed a city's residents to determine who had heart disease and who smoked, they might find that healthy people smoke more than people with heart disease. The real reason for this result is that people tend to quit smoking after they are diagnosed with heart disease. (In effect, the outcome is influencing the cause.) However, to the researchers it might appear that cigarette smoking protects against heart disease. Many cross-sectional studies suffer from this cause-and-effect problem.

A clinical trial is a study done to test the effectiveness of a drug or other treatment. Patients with a particular disease are randomly assigned to receive either the treatment under study or an inactive placebo (or the standard treatment, if one exists). Depending upon who knows whom is receiving which treatment, the study is labeled blind or double-blind. Patients are then followed for a specified period to determine whether patients receiving the new treatment do better than those getting the standard treatment or the placebo. Clinical trials are the best of epidemiological studies in terms of the quality of the information they provide. Unfortunately, they cannot be used to explore causes of disease because it is unethical to assign people to be exposed to suspected toxins. Such trials may be very useful for studying preventive measures, such as vaccines.

Estimating Risk

At the end of a study, researchers calculate the **risk ratio** or relative risk, by comparing the occurrence of disease in the two groups—the one with a suspect characteristic and the other without. This becomes the basis of statements such as "people who smoke are 10 times as likely to get lung cancer as people who do not." When analyzing relative risk from a study we are examining the data, or results, obtained from a study of, typically, two groups. In this case we are comparing two groups and looking at the incidence of cancer. If ten persons per thousand in the smoking group developed cancer and one person per thousand of the control group developed cancer, the incidence ratio is 10:1. How can we interpret the significance of these ratios?

- Risk ratios close to 1 suggest the characteristic has no effect on disease.
- Risk ratios greater than 1 suggest the characteristic increases risk of disease.
- Risk ratios less than 1 suggest the characteristic protects against disease.

Scientists followed 34,445 British male physicians from 1951 to 1961 to see if those who smoked had a higher rate of lung cancer. At the end of 10 years, the cancer mortality rate statistics were as follows:

- Among nonsmokers—7 per 100,000
- Smokers of up to half a pack daily were 8 times as likely ($54/7 = 7.7$)
- Smokers of up to a pack a day were 20 times as likely ($139/7 = 19.9$)
- Smokers of more than a pack a day were 32 times as likely ($227/7 = 32.4$)

This data tends to support a link between smoking and increased mortality rate from lung cancer, assuming all other factors are held constant. This latter assumption is very critical for determining or promoting study validity and is of utmost concern when designing the original study. These assumptions are also the first ones examined by those analyzing the validity of the study or seeking to invalidate the study. Furthermore, the study appears to connect increased smoking to increased lung cancer mortality rates.

How, then, would this relate to an individual interviewed who attributes her longevity of 100+ years to smoking a pack a cigarettes a day. Not at all! While the basic tenant of the scientific method is that you can only disprove, and, yes, it is true that a single instance, example, can invalidate the hypothesis, care must be taken to explicitly link the hypothesis to the argument. Here the hypothesis is not that *individuals who smoke will not live a long time*; rather it is that *individuals who smoke have statistically a significantly greater risk, or chance, of dying from lung cancer, a smoking-related ailment, than nonsmokers*. Nowhere is it being said that you cannot live a long time if you smoke; it is just that the odds are against you in doing so!

If an association between an exposure and a disease has been observed and **bias**, or **confounders**, and other possible errors have been reasonably accounted for, then researchers can address the question of whether the association is likely to reflect a true cause-effect relationship. These are some commonly used criteria:

- **Strength of association**
  The exposure is associated with a large increase (or decrease) in the risk of disease. (The stronger the association, the less likely it is to be due to bias or an unknown confounder.)

- **Dose-response relationship**
  Higher doses of the exposure are associated with higher rates of disease.

- **Biologic credibility**
  A plausible biologic mechanism is available to explain how the exposure causes disease.

- **Consistency**
  Other studies done in different ways and in different populations have found the same association.

- **Time sequence**
  The exposure can be shown to occur before the disease.

- **Specificity**
  The exposure is associated with a specific disease.

These criteria are guidelines, not rules. Some toxicants that clearly cause disease do not meet all these criteria. For example, cigarette smoking does not meet the specificity criterion, for it is associated with many diseases.

The perfect study has never been done. Each study is performed at a distinct time and place with a unique group of subjects by investigators who utilize a particular study design. Diverse methods are used to execute the study and to analyze its results. Chance always plays a part; this is why different studies sometimes produce contradictory results. In laboratory studies the most common cause of distortion is chance, including the effects of sample size. For epidemiological studies, bias and confounding, as well as chance, are important factors to consider.

Bias is an imbalance in the way researchers choose people for a study or systematic mistakes in the way they classified people as sick or well, exposed or unexposed, and can produce a false relationship between an exposure and a disease. Inevitably, some people in a study will be put in the wrong category of exposure or disease because of clerical errors, mistakes in the design or execution of the study, or an imperfect test for disease or exposure. It is well known that health status distorts memory of past events; people who are healthy are more likely to forget past exposures than those who are unhealthy. If the study is not a blind study, problems may arise from investigators' and interviewers' unconscious desires to see what they want or expect to see. A selection bias may occur whenever the method of choosing participants results in a study group that differs from the parent population in ways that affect the study's conclusion. Bias can be limited by using the following strategies:

- Use memory aids to avoid recall bias.

- Lessen the chance of interviewers influencing the responses by not revealing to interviewers the study's exact purpose or whether the people they are interviewing are cases or controls.

- Minimize dropouts by recruiting participants for cohort studies from groups that are likely to be cooperative and easy to trace. Doctors and nurses are favored subjects for follow-up studies because they tend to be interested in research and are easy to trace through professional associations.

- Choose more than one control group for a case-control study. For example, if the study gets the same results with a control group selected from the cases' friends and a control group selected from the cases' work associates, it provides some assurance that the findings are real and not just artifacts of the control selection processes.

All studies have some bias. What is important is whether investigators took care to reduce bias as much as possible and whether the findings could be easily explained by bias.

Confounding occurs when a characteristic not considered by the researchers is in fact associated with both the disease and the suspected disease-causing agent. In the 1970s, a group opposed to fluoridation of water in the United States reported a dramatic increase in cancer death rates in 10 U.S. cities that had switched to fluoridated water. But other investigators noted that the populations of these cities had changed dramatically in the same period, with growing proportions of elderly and black people, both groups that experience higher risk of cancer. After taking into account the confounding effects of age and race, the investigators found that these cities' cancer death rates actually had dropped since the introduction of fluoridated water. Age, sex, race, income, and cigarette smoking are among the most common confounders, because they affect the risk of many diseases and also are closely linked with other exposures that might be investigated as causes of disease. If these factors are not taken into account when looking at the causes of disease, the study is suspect.

Acknowledgment

My thanks to the Foundation for American Communications (FACS), facs@facsnet.org;http://www.facsnt.org; for their significant contributions in the preparations and material on Epidemiology. A further discussion may be found in their *Reporting on Risk, a Journalist's Handbook on Environmental Risk Assessment*.

Study Exercises

1. Describe five applications of the Risk Assessment methodology.

2. Define the following:
   - Risk
   - *de minimis* risk
   - Safe
   - Maximum Exposed Individual (MEI)

3. Identify and describe the four components of the risk assessment process.

4. Identify the five principal exposure pathways.

5. Compute the following:
   - The average daily intake of a substance for an adult given a concentration of 1 mg/ml of a substance in drinking water.
   - Normalize this average daily intake for body weight.

6. Explain under what conditions it is appropriate to sum the doses for each different exposure route and when it is not appropriate.

7. Define the following:
   - SNARL
   - Reference Dose (RfD)
   - Cancer Potency Factor(q_1*)
   - Action level

8. Compute the following (Values to be provided by instructor):
   - The Individual Lifetime Cancer Risk (ILCR) given a substance's average lifetime exposure and its cancer potency factor.
   - The Aggregate Lifetime Cancer Risk (ALCR) given the size of a population and given a substance's average lifetime exposure and its cancer potency factor.
   - The annual incidence of cancer in a population as a result of the exposure, given the size of a population, the lifetime exposure, and its cancer potency factor.
   - The risk of harm from a noncarcinogenic (threshold) effect given the RfD and the estimated daily human dose.
   - The carcinogenic (nonthreshold) risk given the estimated daily dose and the cancer potency factor.

9. Distinguish between risk assessment and risk management.

10. Identify and explain five factors that affect the public's perception of a risk.

11. Compute the Acceptable Daily Intake (ADI) given a substance's NOEL of 150 and a safety factor of 100.

Hazardous Materials Classes

The hazard class of a hazardous material is indicated either by its class (or division) number or its class name. For a placard corresponding to the primary hazard class of a material, the hazard class or division number must be displayed in the lower corner of the placard. However, no hazard class or division number may be displayed on a placard representing the subsidiary hazard of the material. For other than Class 7 or the OXYGEN placard, text indicating a hazard (e.g., CORROSIVE) is not required. The class or division number must appear on the shipping paper after each shipping name.

Material Classification

Class 1   Explosives

Division 1.1	Explosive with a mass explosion hazard. Explosive A
Division 1.2	Explosive with a projection hazard. Explosive A or B
Division 1.3	Explosive with predominately a fire hazard. Explosive B
Division 1.4	Explosive with no significant blast hazard. Explosive C
Division 1.5	Very insensitive explosive; blasting agents.
Division 1.6	Extremely insensitive detonating substances.

Class 2   Gases

Division 2.1	Flammable gas
Division 2.2	Nonflammable, nonpoisonous compressed gas
Division 2.3	Gas poisonous by inhalation. Poison A
Division 2.4	Corrosive gas (Canadian).

Class 3   Flammable Liquid and Combustible Liquid

Division 3.1   Low flash point group (below 0° F)

Division 3.2   Intermediate flash point (0 to 73° F)

Division 3.3   High flash point (greater than 73° F)

Class 4   Flammable Solid, Spontaneously Combustible Material, and Dangerous-When-Wet Material

Division 4.1   Flammable solid

Division 4.2   Spontaneously combustible material

Division 4.3   Dangerous-when-wet material

Class 5   Oxidizers and Organic Peroxides

Division 5.1   Oxidizer

Division 5.2   Organic peroxide

Class 6   Poisonous Material and Infectious Substance

Division 6.1   Poisonous materials

Division 6.2   Infectious substance

Class 7   Radioactive Material

Class 8   Corrosive Material

Class 9   Miscellaneous Dangerous Substances and Other Regulated Materials

Color Code Classification

The Department of Transportation (DOT), the National Fire Protection Agency (NFPA), and others use the following color code classification:

- **Blue**   health hazard potential
- **Red**   fire hazard potential
- **Orange**   explosive
- **Yellow**   reactivity hazard potential
- **White**   special hazard potential
- **Green**   nonflammable gas

Numerical Code Classification

Agents are given four-digit identification numbers. These numbers are available from various books and tables. For example, see the *Emergency Response Guidebook*. Either the ID number or the name of the agent may be included on placards and labels. Additionally, many various symbols are used on placards and labels. The symbols represent a fourth way of conveying information with placards and labels.

Conversion Factors and Measurement Units

Dusts and particulates are often measured in metric units of milligram/cubic meter (mg/m³).

Gas and vapor concentration is expressed as the volume of toxic gas or toxic vapor in a certain volume of air. This is a small number. It is multiplied by one million volumes of toxic gas or vapor volume of contaminated air. This is called parts per million, or ppm. In the metric system, the units have prefixes which divide the basic unit into smaller units. The units of area and volume are units of length that are squared and cubed. The following table lists the various units and their relationships.

Conversion Table

1.0 meter (m)	=	100 centimeters (cm), or
	=	1000 millimeters (mm), or
	=	1,000,000 microns (μ)
1 m	=	39.37 inches
1 inch	=	2.54 cm
1 mm	=	$^{39}/_{1000}$ inch
1 μ	=	.039 thousandth of an inch
1.0 gram (g)	=	.001 kilogram (kg)
1.0 milligram (mg)	=	0.0353 ounce (oz)
1 mg	=	0.0022 pound (lb)
1 kg	=	2.2 lbs
1.0 liter(L)	=	.001 cubic meter (m³)
	=	1000 ml
	=	1000 cm³
	=	1000 cc
1.0 quart (qt)	=	61.0250 cubic inches (in³)
1 kl	=	1000 liters (L)
1000 liters (L)	=	1 cubic meter (m³)

Notes: A cubic centimeter is approximately the size of a sugar cube

The approximate size of a cubic meter is the inside volume of a standard refrigerator.

Consider a quantity of dust or paper one-eighth the size of a postage stamp and weighing about one milligram. If this paper were ground up into fine pieces and dispersed throughout an area about the size of the inside of a typical refrigerator, or 1 m^3, then the dust concentration would be 1 mg/m^3.

Glossary

abiotic — nonliving; especially the nonliving elements in ecological systems

abortifacient — an agent that produces an abortion

abortus — the aborted products of conception

abscissa — the horizontal axis of a graph

absorbent material — material used to soak up liquid (hazardous) material

absorbed dose — the amount of a chemical that enters the body of an exposed organism

absorption — the movement of a chemical from the site of initial contact with the biologic system, across a biologic barrier, and into either the bloodstream or the lymphatic system

absorption factor — the fraction of a chemical making contact with an organism that is absorbed by the organism

acaricide — a pesticide used to control spiders, ticks, and mites

Acceptable Daily Intake (ADI) — the daily intake of a chemical substance that appears to be without appreciable risk during a lifetime based up all known facts; often expressed as milligrams per kilogram of body weight (mg/kg)

acclimation — the adaptation over several time periods to a marked change in the environment

accumulation — the buildup of a material in on organism or in the environment due to repeated exposure and absorption

accumulative effect of a chemical — the effect of a chemical on a biologic system when the absorption exceeds eliminations and, thus, the total body burden is increasing

acetic acid — major component of vinegar, glacial acetic acid is the pure compound

acetylation — a Phase II metabolic reaction involved in either activating or deactivating aromatic amines

acetylcholine — acetic acid ester of choline which serves as a neurotransmitter and neuromuscular transmitter at the synaptic junctions

acetylcholinesterase — an enzyme present in nerve and muscle tissues that hydrolysizes acetylcholine to choline and acetic acid

ACGIH — American Conference of Governmental Industrial Hygienists, a professional organization which recommends exposure limits (TLVs and BEIs) for toxic substances

acid — a substance which dissolves in water and releases hydrogen ions (H+); a proton donor or an electron acceptor; acids cause irritation, burns, or more serious damage to tissue, depending on the strength of the acid, which is measured by pH, 1 strongest to 6 weakest

acid cleaning — using any acid for the purpose of cleaning materials; methods of acid cleaning are pickling and oxidizing

acidity — the quantitative capacity of aqueous solutions to react with hydroxyl ions

acidosis — a condition of decreased pH in the body which may be due to decreased respiratory rate or increased acid consumptions; a pathologic condition resulting from accumulation of acid in, or loss of base from, the body

action potential — an electrical potential that is formed as a result of the movement of sodium ion movement into a cell through sodium channels followed by the movement of potassium ions movement out of the cell through potassium channels; used to propagate or transmit neural impulses

active ingredient — the chemical that has action (e.g., pesticide); active ingredients are listed in order on a label as percentage by weight or as pounds per gallon of concentrate

active transport — an energy-expending mechanism by which a cell moves a chemical across the cell membrane from a point of lower concentration to a point of higher concentration, against the diffusion gradient

acute — actions occurring within a short period after initiation

acute effect — an adverse effect on a human or animal body that takes place soon after exposure

acute poisoning — poisoning by a single exposure to a toxic chemical

acute toxicity — any poisonous effect produced by a short-term exposure; the LD_{50} of a substance is typically used as a measure of its acute toxicity

additive effect — a biological response to an exposure to multiple chemicals which is equal to the sum of the effects of the individual agents

Adenosine TriPhosphate (ATP) — the "energy coin" of the cell; the primary source of energy required for cellular processes; comprised of a nucleic acid base, adenosine and the sugar deoxyribose coupled to three phosphate complexes

administrative control — a method of controlling employee exposures to contaminants by job rotations and variations in work assignments

adrenal glands — a pair of endocrine glands superior to the kidneys that produce steroids, hormones related to water balance, and norepinephren

adsorption — the attachment of the molecules of a liquid or gaseous substance to the surface of a solid

adulterant — chemical impurities or substances that, by law, do not belong in a food, plant, animal, or pesticide formulation

aerobic — a life process that depends on the presence of molecular oxygen

aerosol — a suspension of a liquid or solid particles in a gas

afferent nerves — nerves that carry stimuli from the periphery nervous system (PNS) to the central nervous system (CNS)

affinity — the strength with which a chemical messenger binds to its receptor

agglomerate — to come together into an amophous mass as when small particles stick together to form a larger particle

agonist — a chemical messenger that binds to a receptor and triggers the cell's response; often refers to a drug that mimics a normal messenger's action

agranular leucocyte — white blood cell (WBC) with little to no visible granules in the cytoplasm; lymphocytes and monocytes

AIDS — Acquired Immunodeficiency Syndrome; epidemic disease caused by a virus that attacks the human immune system

airborne — referring to that which can be carried by or in the air; e.g., dusts and mists

air pollution — the presence of contaminant substances in the air that interfere with human health or air quality; a level of pollutants that exceed air qualify standards, as prescribed by law, and may not be exceeded over a specified period of time in a defined area

air reactive — materials that will ignite at normal temperatures when exposed to air

albinism — a recessive genetic condition characterized by the inability to produce melanin pigments. Expresses itself in white or translucent skin tone, white or colorless hair, and eyes with pink or blue iris, and deep-red pupils. (Trait exhibited by white lab mice and rats)

albumin — a blood protein that binds to and transports various types of materials, including toxic substances, through the blood

aldehyde — a group of various reactive compounds (e.g., acetaldehyde) characterized by the presence of a CHO group

aldrin — an organochlorine insecticide

alkali — same as a base (See *base*)

allergen — a substance that causes an allergy

allergy — same as hypersensitivity (Also see *sensitivity*) An allergy is a reaction to a substance that occurs through a change in the immune system caused by the production of antibodies, and is usually experienced by only a small number of people exposed to a substance. Allergic reactions in the workplace tend to affect the skin (see *dermatitis*) and lung (see *asthma*)

alopecia — a loss of hair, baldness

alpha particle — generally, low energy ionizing radiation comprised of a helium nucleus

alveolitis — an inflammation of the alveoli

alveolus — microscopic air sacs that form the terminal ends of the air passages of the lungs

Alzheimer's Disease — an illness of the brain associated with degeneration of mental capacities, loss of orientation, and forgetfulness

ambient — environmental or surrounding conditions

amelia — the congenital absence of a limb or limbs

Amyotrophic Lateral Sclerosis (ALS) — a syndrome characterized by muscular weakness and atrophy due to degeneration of myelin and motor neurons in the spinal cord/CNS

analgesia — a loss of pain sensation

anaphylactic — pertaining to an extreme allergic reaction

anaplasia — a loss of cell differentiation and function; characteristic of most malignancies

anemia — any condition that results in less than normal hemoglobin in the red blood corpuscle (RBC) or lower than normal numbers of erythrocytes in the blood

anencephaly — an abnormality caused by lack of development of the brain and skull bones resulting in a condition that is incompatible with life

anesthetic — a chemical that causes a total or partial loss of sensation. Overexposure can cause impaired judgment, dizziness, drowsiness, headache, unconsciousness, and even death.

angiotensin — a protein synthesized in the liver that interacts with renin resulting in vasoconstriction leading to increased cardiac output and blood pressure

anhydrous — containing no water

animal — one of the major kingdoms of living organisms; multicellular eukaryotes capable of locomotion

animal studies — investigations using animals as surrogates for humans, on the expectation that results in animals are pertinent to humans

anions — negatively charged ions in a solution; e.g., hydroxyl ions, OH

anode — the positively charged electrode in an electrochemical cell

anomaly — a thing or organism that deviates from normal

anoxia — a complete reduction in the oxygen concentration supplied to cells or tissues

ANS — *See* Autonomic nervous system

ANSI — American National Standards Institute, a private organization that recommends safe work practices and engineering designs

antagonism — situations whereby two chemicals interfere with each other's actions such that the net impact is less than the action of either chemical or substance individually; a molecule that competes for a receptor with a chemical messenger normally present in the body. The antagonist binds to the receptor but does not trigger the cell's response

anterior chamber — space, chamber, behind the cornea and in front of the lens filled with an aqueous fluid, that is drained by the Canals of Schlemm; loss of drainage leads to an increase of intraocular pressure, a condition termed Glaucoma

anthrax — an acute, infectious disease caused by an increase in the concentration in the bacterium *Bacillus anthracis*, usually found in cattle, sheep, horses, and goats. It may attack the lungs producing respiratory distress, cyanosis, shock, and coma; death may occur if it is left untreated. The disease in contracted by man as a result of handling of contaminated products from these animals

antibody — a protein (immunoglobulin) molecule that has a specific amino acid sequence, by virtue of which, it interacts only with the antigen that induced its synthesis, or with antigens closely related to it

anticoagulant — a chemical that prevents clotting of the blood

antigen — a substance that, when introduced into the body, is capable of inducing the formation of antibodies and, subsequently, of inducing a specific immune response

antipyretic — an agent that relieves or reduces fever

aplasia — lack of development of an organ or tissue, or of the cellular products of an organ or tissue

apnea — cessation of breathing; asphyxia

aqueous humor — a transparent fluid filtered from the blood plasma found in the anterior chamber of the eye

aquifer — an underground river or lake contained in a bed, or layer, of earth, gravel, or porous storage that contains water

aromatic hydrocarbons — refers to a class of chemicals consisting of a resonant electron structure, i.e., the benzene ring

arrhythmia — any irregular beating of the heart; atrial and ventricular fibrillations; irregularities in the EKG

arteriole — smaller branch of the arterial blood supple that regulates blood flow within the body

asbestos — a widely used mineral, often of a fibrous nature, which can contaminate either air or water and promote cancer after inhalation or ingestion

aspartame — an artificial sweetener formed by combining the two amino acids aspartic acid and phenylalanine

asphyxia — a condition in which decrease in the amount of oxygen in the body is accompanied by an increase of carbon dioxide and leads to loss of consciousness or death. Assume relatively complete clearance over the weekend.

asphyxiant — a vapor or gas that can cause loss of consciousness and death due to lack, or reduction, of breathable oxygen

aspirate — to inhale liquid into the lungs; to suck-up liquids

assay — a test for a particular chemical or effect

ASTDR (Agency for Toxic Substances and Disease Registry) — established in 1980 to produce health-based information in support of chemical waste disposal sites

asthma — constriction of the airways (bronchial tubes) to the lungs, producing symptoms of cough and shortness of breath. It may be an allergic or an emotional response.

astrocytes — large-footed star shaped neuroglia cells which form and maintain the blood-brain barrier

ataxia — failure of muscular coordination; irregularity of muscular action

atmosphere — the layer of gas over the surface of a planet; on earth, a mixture of gases, typically 20 percent oxygen, 80 percent nitrogen, and assorted trace gases; suitable for breathing and supporting life

atmospheric pressure — weight of the air above the earth's surface, typically 14.7 pounds of pressure per square inch (760 mm Hg)

atom — the smallest particle of an element that can exist while maintaining all of the properties of that element

atomic number — number of protons in the nucleus of an atom

ATP (Adenosine TriPhosphate) — energy coin of the cell, produced by the mitochondria

atria — the chambers of the heart that receive blood from veins and pump blood into the ventricles prior to ventricular contraction

atrial fibrillation — characterized by rapid, irregular unsynchronized contractions of the atria, which produces irregular filling of the ventricles and decreases cardiac output

atriaoventricular (AV) node — a group of myocardial cells located in the right atrial wall which serve to transmit the action potential into the ventricular region of the heart

atrioventricular bundle (Bundle of His) — track of Purkinje fibers which conducts the action potentials from the AV node along the interventricular septum throughout the walls of the ventricles

atrophy — to decrease in size or waste away; opposite to hypertrophy

attributable risk — the proportion of exposed cases that would not have gotten the disease if they had not been exposed; the number of individuals who actually develop a particular illness or disease associated with an exposure

autoimmune response — disorder resulting from the production of antibodies that attack normal body tissue; the immune system erroneously identifies the cells of the body as not-self agents that must be eliminated

autonomic nervous system (ANS) — portion of the nervous system responsible for automatic activities, e.g., heartbeat, respiration, digestion, blood pressure

autopsy — the examination of a dead body. Most autopsies involve an examination of the internal organs and some microscopic examination of tissues.

axon — the single process in every neuron responsible for conducting information away from the cell body

Bacillus anthracis — a bacterium, capable of living for years in soils or on animal remnants, that is responsible for causing the disease anthrax in cattle, goats, sheep and horses

background level — normal ambient environmental concentration of a chemical

bacteria — typically prokaryotic microorganisms forming one of the major kingdoms of living organisms

baghouse — an air pollution abatement device used to trap particulates by filtering gas streams through large fabric bags; similar to a vacuum cleaner bag

BAL (2,3-Dimercaptopropanol) — chelating agent capable of binding a variety of toxic metals such as mercury, antimony, bismuth, cadmium, chromium, cobalt, and nickel

baroreceptor — groups of sensory neurons sensitive to stretching caused by the increase in pressure

basal cell carcinoma — a slow-growing, locally invasive, neoplasm of the skin derived from the basal cell layer of the epidermis

base — a substance that dissolves in water and releases a hydroxyl ion (OH^-); an electron donor or proton acceptor. It has the ability to neutralize an acid and form a salt. Strong alkalis, pH 14, are irritating and may damage tissue. (Also see *caustic*)

base excision repair — a DNA molecular repair mechanism in which a damaged nucleotide is removed and replaced through enzymatic activity

basophil — a WBC which manufactures chemicals such as hisamine and heparin

BEI — Biological Exposure Index, the maximum recommended value of a substance in blood, urine, or exhaled air, recommended by the ACGIH

benign — harmless; a noncancerous tumor (See *malignant*)

beryllium — a metal that is hazardous to human health upon absorption. Dust is created by machining and it is frequently discharged from ceramic and propellant plants, and foundries.

beta particle — an electron, produced by electron tubes (e.g., television pictures tubes) and by radioactive decay. It may cause skin burns; it is stopped by a thin sheet of metal.

bias — an inadequacy in experimental design that leads to results or conclusions not representative of the population under study

bile — liquid secretion of the liver

bilirubin — a by-product formed by the lysis of hemoglobin during the destruction of RBCs

bioaccumulation — the retention and concentration of a substance by an organism

bioassay — test which determines the effect of a chemical on a living organism

biochemical — having to do with the chemistry of life

bioconcentration — the accumulation of a chemical in tissues of an organism to levels that are greater than the level in the medium in which the organism resides

biodegradable — capable of decomposing, in a relatively short time period, through the actions of microorganisms

biodegradation — decomposition of a substance into more elementary compounds by the action of microorganisms such as bacteria

biomagnification — the tendency of certain elements or chemicals to become concentrated as they move into and up the food chain; also bioconcentration and bioaccumulation

biosynthetic — a synthetic chemical action requiring energy from the body for reaction

biotransformation — conversion of a substance into other compounds by organisms; includes biodegradation

BLEVE — Boiling Liquid Expanding Vapor Explosion. A container (e.g., liquefied petroleum gas) boils when heated; a container failure releases gas to the atmosphere which, subsequently, releases energy in a rapid and violent explosion.

blood — circulatory fluid of humans; approximately one-half water with dissolved substances (plasma) and one-half a variety of cells and corpuscles (formed elements)

blood-brain barrier — a barrier composed of capillary endothelial cells and the processes of astrocytes that prevents or limits the free passage of substances in the blood from entrance to the CNS

blow down valve — a manually operated valve whose function it is to quickly reduce tank pressure to that of atmospheric pressure

boiling point — the temperature at which a liquid boils and changes rapidly to a vapor (gas) state at a given pressure (see *evaporation*); expressed in degrees Centigrade (°C) or degrees Fahrenheit (°F) at sea level pressure (760 mm Hg)

botulism — food poisoning caused by the toxic substances produced by the bacterium *Clostridium botulinum*

Bowman's Capsule — the enlarge proximal end of the nephron that along with the glomerulus makes up the renal corpuscle in the kidney

bradycardia — a slowness of the heart beat, as evidenced by a slowing of the pulse rate to less than 60 beats per minute

British Thermal Unit (BTU) — unit of measurement of heat necessary to raise one pound of water one degree Fahrenheit in temperature (from 63 to 64 °F)

broadcast application — in pesticide use, to spread a chemical or substance over an entire area

bronchiole — small distal branch of the bronchial tree, in which the diameter becomes progressively smaller and terminates in the alveolar duct; typically without cartilage support or cilia

bronchitis — inflammation of one or more bronchi, the larger air passages of the lungs

bronchopneumonia — pneumonia involving acute inflammation of the walls of the many relatively small areas of the lung adjacent to the smaller bronchi

Brucella abortus — a bacterium that causes brucellosis leading to abortion in cows, horses, and sheep as well as undulant fever in man

Brucella melitensis — a bacterium that causes brucellosis in man, abortion in goats, and a wasting disease in chickens; it may infect cows and hogs and be excreted in their milk

Brucella suis — a bacterium that may cause brucellosis in man; it may also infect horses, dogs, cows, monkeys, and goats

brucellosis — a widespread infectious disease affecting primarily cattle, pigs, and goats, but sometimes affecting other animals, including humans

bubo — an enlarged, inflamed lymph node, usually in the auxiliary and inguinal regions

Bubonic Plague — an acute infectious disease caused by the bacterium *Yersinia pestis,* which is transmitted to humans from the bite of the rat flea. The high mortality rate is due to pneumonia complications; was called the "black death" in the Middle Ages.

Building Associated Illness (BAI) — an illness related to indoor air pollution, of unknown etiology

Building Related Illness (BRI) — illness associated with indoor air pollution of known etiology

bulbourethral gland — paired glands associated with the male prostate gland; secretes an alkaline fluid designed to neutralize the acidic environment within the vagina

byssinosis — respiratory symptoms resulting from exposure to the dust of cotton, flax, and soft hemp. Symptoms range from acute dyspnea with cough and reversible breathlessness and chest tightness on one or more days of a workweek to permanent respiratory disability owing to irreversible obstruction of air passages.

calcium-binding Protein (CaBP) — a protein involved in the absorption of calcium from the small intestine; it can also bind with heavy metals such as cadmium and lead

Cal/OSHA — California Occupational Safety and Health Administration, a state agency in the Department of Industrial Relations which establishes and enforces worker health and safety regulations. It consists of the Division of Occupational Safety and Health (DOSH), the Consultation Service, the Standards Board, and the Appeals Board. (See *resources*)

cancer — a condition that occurs when a cell, or group of cells, proliferates in an unchecked, uncontrolled, or unregulated manner

cannula — a small tube made for insertion into a body cavity, duct, or vessel

carbaryl — a pesticide sold under the trade name Sevin; it is used to control fleas on cats and dogs and in mosquito-control programs

carbohydrate — a class of biomolecules as a multiple of $C\ H_2O$ which are termed sugars, saccarides, or polysaccharides

carbon — element with an atomic number of 6 and a nominal atomic weight of 12; forms the basis of organic compounds and life, as we know it

carbonate — a compound containing the carbonic acid (organic acid) group

carbon dioxide — a colorless, odorless, nonpoisonous asphyxiant, which is normally part of ambient air due to respiration or fossil fuel combustion

carbon monoxide — a colorless, odorless, poisonous gas produced by the incomplete combustion; turns the blood cherry red *carbonic acid* — a weak colorless acid formed by dissolving carbon dioxide in water; means of transporting carbon dioxide in the blood

carbon tetrachloride — a nonflammable, sweet-smelling, colorless liquid used as a degreasing or cleaning agent and a solvent

carboxyhemoglobin — a compound formed when carbon monoxide binds to hemoglobin

carcinogen — a chemical or physical agent capable of causing cancer. Such an agent is often described as *carcinogenic*. The ability to cause cancer is termed *carcinogenicity*. Words with similar meaning include *oncogenic* and *tumorigenic*

carcinogenesis — process leading to the development of a tumor

carcinoma — malignant tumor of epithelial tissue

cargo manifest — a shipping paper that lists all of the contents being carried by a transporting vehicle or vessel

case-control study — investigates the prior exposure of individuals with a particular health condition and those without it to infer why certain subjects got the disease and others did not; also known as the "Why me?" study

CAS Number — the Chemical Abstracts Service Registry Number is a numeric designation which is given to a specific chemical compound. This number may appear on the Material Safety Data Sheet. (See *MSDS*)

catabolic reaction — the process in which living cells break down substances into simpler substances

catalyst — a substance that changes the rate of a chemical reaction without undergoing a chemical change. Enzymes are organic catalysts.

cathartic — an orally administered substance that enhances the elimination, movement, of gastrointestinal contents

cathode — the negatively charged electrode in an electrochemical cell

cation — the positively charged ions in a solution; they migrate to the cathode

causality — that which brings about any condition or produces any effect

constitutional cause — one acting within the body that is not restricted to a specific site, but is systemic or has a genetic basis

exciting cause — one that leads directly to a specific condition

immediate cause — a cause that is operative at the beginning of the specific effect; called also the precipitating cause

local cause — one that is not general or constitutional, but is confined to the site where the effect is produced

precipitating cause — immediate cause

predisposing cause — anything that renders a person more liable to a specific condition without actually producing it

primary cause — the principal factor contributing to the production of a specific result

proximate cause — that which immediately precedes and produces an effect

remote cause — any cause that does not immediately precede and produce a specific condition; a predisposing, secondary, or ultimate cause

secondary cause — one that is supplemental to the primary cause

specific cause — one that produces a special or specific effect

ultimate cause — the earliest factor, in point of time, that has contributed to production of a specific result

caustic — something alkaline that strongly irritates, corrodes, or destroys living tissue (See *base*)

caustic soda — sodium hydroxide ($NaOH$), a strongly alkaline substance used as extensively in industrial chemical processing

CDC (Center for Disease Control) — federal centers for disease control and prevention; a branch of the U.S. Public Health Service under the Department of Health and Human Services. It is involved in the study and prevention of infectious diseases and in establishing standards, guidelines, and recommendations relative to toxic chemicals.

ceiling limit — the maximum concentration of a material in air that must never be exceeded, even for an instant

cell — the structured unit of life and of which tissues are made. There are many types of cells, e.g., epithelial cells, connective (blood) cells, nerve cells, and muscles cells. In higher animals each type of cell is specialized to perform a particular function.

Central Nervous System — CNS, consisting of the brain and spinal cord (See also PNS)

CNS Depressant — toxins or substances which effect the CNS by reducing capacity, diminishing sensations, or impairing function, e.g., ethanol or grain alcohol

CERCLA — Comprehensive Environmental Response, Compensation, and Liability Act of 1980. The act requires that the Coast Guard National response Center be notified in the event of a hazardous substance release. The act also provides for a fund (the Superfund) to be used for the cleanup of abandoned hazardous waste disposal sites.

cerebellum — the little brain involved in the control and coordination of voluntary skeletal muscle movements

cerebrum — the large anterior portion of the brain stem; the cerebral cortex is the site of cognitive thought

CFR — Code of Federal Regulations; a collection of the regulations that have been promulgated under U. S. law

chancre — a lesion (ulcer) associated with syphilis, usually appearing 2-3 weeks after infection

chelate — a chemical compound in which a metallic ion is sequestered and firmly bound into a ring with the chelating molecule; used in chemotherapeutic treatments for metal poisoning

chemical — a substance characterized by a definite molecular composition

chemical formula — a diagram that represents the composition of a substance using symbols to represent each element and subscript numbers showing the number of atoms of each element involved

chemical name — the scientific designation of a chemical in accordance with the nomenclature system; developed by the International Union of Pure and Applied Chemistry (IUPAC) of the Chemical Abstract Service's (CAS) rules of nomenclature

chemical reduction — a chemical reaction in which one or more electrons are transferred to the chemical being reduced from the chemical initiating the transfer (reducing agent)

chemoreceptor — a sensory nerve ending, like a taste bud, that is stimulated by and reacts to certain chemicals

CHEMTREC — CHEMical TRansportation Emergency Center; provides assistance during a hazardous materials emergency; source of information regarding specific chemicals and can contact manufacturer or other experts for additional information for onsite assistance

chloracne — a skin disease, resembling the acne of adolescence, caused by exposure to certain chlorinated aromatic hydrocarbons

chlorinated hydrocarbon — organic compound made up of atoms of carbon, chlorine, and usually hydrogen. Chlorinated hydrocarbons include DDT and PCBs; they tend to be very long-lived in the environment, to be toxic, and to accumulate in the food web

chloroform — a colorless, volatile liquid used as an anesthetic, solvent, and fumigant

cholera — an acute infectious disease caused by the bacterium Vibria cholera, characterized by profuse watery diarrhea and vomiting resulting in loss of fluid, cramps, and dehydration

choroid — the middle coat of the eye sandwiched between the retina and the sclera containing blood vessels and forming the lens apparatus

chrom-ulcer — an open sore or lesion of the skin or mucous membrane caused by exposure to chromium VI containing compounds

chromosome — the part of a cell that contains genetic material (See gene); rodlike structure in the nucleus of a cell that forms during mitosis; composed of DNA and protein

chronic — generally refers to sublethal exposures that occur consistently over extended periods of time

chronic effect — an adverse effect on a human or animal body, such as cancer, which can take months or years to develop after exposure

chronic exposure — long-term, low level exposure to a toxic chemical

cilia — small short motile projections from the cell surface which serve to pass material, e.g., mucous, over the surface of the cell or along a tube so lined

ciliary body — a smooth muscle portion of the choroid coat of the eye that serves to change the focal point of the lens

cirrhosis — a chronic disease of the liver characterized by replacement of normal liver tissue with connective tissue

clinical studies — controlled studies of humans exhibiting symptoms induced by chemical exposure

Clostridium botulinum — a bacterium that grows under anaerobic conditions of improperly processed food that produces a toxin causing botulism

clubfoot — a congenital abnormality associated with the structure of the foot

CN gas — a chemical warfare agent that causes lacrimation and severe irritation to the skin; additional symptoms of exposure may include tightness of the chest, nausea, and vomiting

CNS — abbreviation for Central Nervous System; refers to the brain and spinal cord; excludes the PNS Peripheral Nervous System, which are the nerves exiting the brain stem and traveling to various locations in the body

CNS depressant — a toxin or substance such as ethanol that affects the CNS by reducing capacity, diminishing sensations, or impairing function

codon — a sequence of 3 bases in a strand of DNA that provides the genetic code for a specific amino acid

cohort study — follows a group of healthy individuals who have different levels of exposure, and characterizes their health outcomes over time with respect to their level of exposure; also known as the "What will happen to me?" study

colic — acute abdominal pain

colitis — inflammation of the large intestine (colon)

collecting duct — a straight tubule that extends from the cortex of the kidney to the tip of the renal pyramid

combustible — able to catch on fire and burn. The National Fire Protection Association and the U.S. Department of Transportation generally define a combustible liquid as having a flash point of 37.8 °C (100 °F) or higher. (Also see *flash point*)

Class A — ordinary combustibles which leave a residue after burning

Class B — flammable liquids and gases

Class C — class A or B fires that occur in or near electrical equipment

Class D — combustible metals that are easily oxidized

comedo — a thickened secretion plugging a duct of the skin, especially of the sebaceous gland; it is usually associated with dermatitis or acne

competition — the ability of different molecules very similar in structure to combine with the same receptor

compound — a pure substance composed of two or more elements

compressed gas — any material or mixture having a container pressure exceeding 40 psi at 21.11 °C (70 °F) or having an absolute pressure exceeding 104 psi at 54.44 °C (130 °F)

concentration — the amount of a specific substance mixed into a given volume of air or liquid

conduction — heat transfer through the movement of atoms within a substance

confidence level — a statistical term that expresses how assured one can be that the results obtained from an experiment are meaningful; e.g., a 95% confidence level implies that there is a 95% probability that the observed results were due to the conditions of the experiment

confounding — finding an association for the wrong reason

confounding factors — variables other than chemical exposure level which can affect the incidence or degree of a parameter being measured

congenital — pertaining to a condition existing before or at birth

conjugation — the joining of two substances to form a single molecule; conjugation of toxic substances can increase water solubility and excretion

conjunctivitis — inflammation of the conjunctiva, the delicate mucous membrane that lines the eyelids and covers the exposed surface of the eye

connective — one of the four basic tissues; characterized by few cells and large amounts of interstitial material

contact dermatitis — a skin rash caused by direct contact with an irritating substance

contaminants — materials, chemicals, substances, or life forms located where they are not wanted

contraindication — an indication, symptom, or condition that makes inadvisable a particular treatment or procedure

controls — normal experimental subjects not treated or exposed

convection — heat transfer from one place to another by actual motion of the heated material

cornea — the clear transparent anterior portion of the eye

corrosive — a liquid or solid that causes visible destruction or irreversible alterations in human skin tissue at the place where it touches the skin

cost/benefit analysis — a quantitative evaluation of the costs which would be incurred versus the overall benefits to society of a proposed action such as the establishment of an acceptable dose of a toxic chemical

covalent bond — a bond involving the sharing of electrons between atoms

cross-sectional study — assesses a group's health status and exposure status simultaneously; also called the "Am I like my neighbors?" study

cubic meter — a metric unit of volume, commonly used in expressing concentrations of a chemical in a volume of air. One cubic meter equals 353 cubic feet or 1.3 cubic yards. One cubic meter also equals 1000 liters or one million cubic centimeters.

cumulative exposure — the summation of exposures of an organism to a chemical over a period of time

Curie — a measure of radioactivity, 37 billion disintegrations per second

cyanosis — a bluish discoloration, especially of skin and mucous membranes, owing to excessive concentration of reduced hemoglobin in the blood

cystic fibrosis — an inherited disease that affects primarily the respiratory system but also various glands of the body; usually begins during infancy and is characterized by chronic respiratory infections

DBCP (1,2-Dibromo-3-chlorpropane) — a nematocide soil fungicide

DDOH (2,2-Bis(p-chlorphenyl)ethanol) — a metabolite of DDT

DDT — the first chlorinated hydrocarbon insecticide (1,1,1- trichloro - 2,2 - bis (p-chloriphenyl) - ethane), dichlorodiphenyltrichloroethane. It has a shelf-life of 15 years and can collect in fatty tissues of certain animals. The EPA banned registration and interstate sale of DDT for all but emergency uses in the United States in 1972 because of its persistence in the environment and accumulation in the food chain.

deactivation — the process of making a substance inactive; rendering a toxic substance less toxic

deamination — the removal of an amine group from a molecule, usually by hydrolysis

decibel (dB) — a unit of sound measurement

decomposition — breakdown of chemical into simpler parts, compounds, or elements

degradation — chemical or biological breakdown of a complex compound into simpler compounds

DEHP (Di(2-ethylhexyl)phthalate) — a plasticizer for many resins and elastomers

dementia — a deterioration of intellectual faculties, such as memory, concentration, and judgement, often accompanied by emotional disturbances and personality changes

demography — the study of the characteristics of human populations such as size, growth, density, distribution and vital statistics

demyelinate — to destroy or remove the myelin sheath that surround high speed neural fibers

denaturation — the destruction of the usual nature of a substance, usually the change in the physical properties of proteins caused by heat or certain chemicals

dendrite — any of the usual branching processes involved in conducting information toward the nerve cell body

Dengue Fever — an acute mosquito-borne viral disease characterized by sudden onset, headache, fever, joint and muscle pain, and the appearance of a rash

dermal — referring to the skin

dermal exposure — contact between a chemical and the skin

dermatitis — inflammation of the skin causing redness (rash) and often swelling, pain, itching, and cracking. Dermatitis may be caused by an irritant or allergen.

dermis — layer of tissue located below the epidermis and above the subcutaneous fatty hypodermis of the skin

DES — diethylstilbestrol, a synthetic estrogen like compound

dialysis — the process of diffusing blood across a semipermeable membrane to remove toxic materials and maintain normal fluid, pH, and solute balance; artificial kidney

diaphragm — a muscular separation between the thoracic and abdominal cavities. Its contraction assists inhalation while its relaxation aids exhalation.

dichlorformoxime — a chemical warfare agent that is primarily a skin irritant. At high concentrations it may cause blistering and enter the bloodstream resulting in death.

dichotomy — a division into two mutually exclusive groups

dicumarol — an anticoagulant that inhibits the formation of prothrombin in the liver

dieldrin — an organochlorine insecticide

diethylstilbestrol — a synthetic estrogen that stimulates production of estrogen and progesterone in the placenta

diffusion — the movement of suspended or dissolved particles from a more concentrated to a less concentrated region as a result of the random movement of individual particles; the process tends to distribute them uniformly throughout the available volume

dioxin — common name for a family of chlorine-containing chemicals, some of which are supremely toxic. Dioxins are the by-products of combustion-based technologies involving chlorine and are known to cause certain types of cancers in humans.

diquat — a type of herbicide used to kill nuisance plants

disease — a non-healthy physiological state

distal convoluted tubule — the twisted latter portion of the nephron responsible for sodium, potassium, pH, and water balancing

distribution — the movement of a substance within an organism or the environment

DMPS (2,3-dimercapto-1-propanesulfonic Acid — a chelating agent used in the treatment of mercury and lead poisoning

dopamine –– a monoamine that occurs especially as a neurotransmitter in the brain and is a precursor of epinephrine and norepinephrine

dose — the amount of a chemical that enters or is absorbed by the body. Dose is usually expressed in milligrams of chemical per kilogram of body weight (mg/kg).

dose-response — a quantitative relationship between the dose of a chemical and an effect caused by the chemical

dose-response curve — a graphical presentation of the relationship between degree of exposure to chemical (dose) and observed biological effect or response

dosimeter — an instrument that measures exposure to radiation

down-regulation — a decrease in the total number of target-cell receptors for a given messenger in response to chronic high extracellular concentration of the messenger

d-tubocuraine — a drug used to treat poisoning from black widow spider bite, also used on tips of poison arrows. It can produce skeletal muscle relaxation; at high doses it may be lethal.

dust — small particulate material or matter; frequently suspended in air

dyspnea — difficult or labored breathing

dysrhythmia — disturbances of rhythm; e.g., speech, brain waves, or heart beat

ecology — the relationships of living things to one another and to their environment, or the study of such relationships

ecosystem — the interacting system of a biological community and its nonliving environment

ecotoxicological studies — measurement of effects of environmental toxicants on indigenous populations of organisms *eczema* — a superficial inflammatory process involving primarily the epidermis; characterized early by redness, itching, minute papules and vesicles, weeping of the skin, oozing, and crusting, and later by scaling, lichenification, and often pigmentation

edema — a swelling of body tissues due to water or fluid accumulation in tissues

ED_{50} — the effective dose for 50% of the tested subjects in a model system

efferent neurons — neuronal fibers extending from the CNS towards the PNS, motor, muscle, innervation

electrolyte — a liquid, most often a solution, that will conduct an electric current, generally, in an electrochemical cell

electromagnetic fields — the coupled energy fields, electric and magnetic) that are generated by an electric current passing through a conductor

elimination — the removal of a chemical or substance from the body by metabolism or excretion

embryo — the developing child during the first two months of human pregnancy. This stage is much shorter in other animal model systems.

emergency — a situation requiring immediate action (See *IDLH*)

emetic — an agent that induces vomiting

emphysema — literally, an inflation or puffing up; a condition of the lung characterized by an increase, beyond the normal, in the size of air spaces distal to the terminal bronchioles

empirical — based on experience and observation

emulsion — a mixture in which one liquid is suspended as tiny drops in another liquid, such as oil in water

encephalopathy — any degenerative disease of the brain

endangerment assessment — a site-specific risk assessment of the actual or potential danger to human health or welfare and the environment from the release of hazardous substances or waste. The endangerment assessment document is prepared in support of enforcement actions under CERCLA or RCRA.

endemic — pertaining to prevalence in a particular region

endoplasmic reticulum — intercellular membrane system continuous with the nuclear membrane associated with ribosomes and protein synthesis

endothermic reaction — a reaction in which heat is absorbed

endpoint — a biological effect used as an index of the effect of a chemical on an organism

ensemble — a group or set, frequently used in referring to a particular configuration of equipment designed or designated for dealing with a specific hazard or hazardous situation

enterocolitis — an inflammation of the intestine and colon

enterohepatic circulation — the recurrent cycle in which the bile salts and other substances excreted by the liver are reabsorbed through the intestinal mucosa, returned to the hepatic cells, and then excreted again

environmental fate — the destiny of a chemical after release to the environment; involves considerations such as transport through air, soil and water, bioconcentration, degradation, and ultimate disposition

Environmental Protection Agency (EPA) — the United States' Government Agency responsible for protection of the environment

EPA Registration Number — the number that appears on the pesticide label to identify the individual pesticide product, i.e., EPA Reg. No.

enzyme — a protein, synthesized by a cell, that acts as a catalyst in a specific chemical reaction

eosinophil — granular WBC which are found in elevated numbers in asthmatics

epidemiology — the study of the pattern of disease in a population of people — the study of the relationships of the various factors determining the frequency and distribution of diseases in a human community — the field of medicine concerned with the determination of the specific causes of localized outbreaks of infection (such as Legionnaire's Disease) of toxic disorders (such as lead poisoning) or any other disease of recognized etiology

epidemiological studies — investigation of elements contributing to disease or toxic effects in human populations

epidermis — the outer avascular layering of the skin which forms a protective barrier over the dermis; grows hair, ducts and glands *epigenesis* — the development of an organism from an undifferentiated cell, consisting in the successive formation and development of organs and parts that do not preexist in the fertilized egg

epigenetic — pertaining to non-genetic mechanisms

epilation — the loss of hair due to exposure to ionizing radiation

epilepsy — a disorder of the brain characterized by brief and sudden attacks of altered consciousness and motor activity

epinephrine — adrenaline, a neurotransmitter that is the principal blood-pressure raising hormone secreted by the adrenal medulla; serves medicinally as a heart stimulant, vasoconstrictor in controlling hemorrhages of the skin, and a muscle relaxant in bronchial asthma

epistemology — study of the methods and validity of knowledge

epithelium — one of the four basic tissues; characterized by many cells with little interstitial material

equilibrium — the state in which the movement of molecules, ions, atoms or energy in any one direction is equal to the movement of molecules or energy in the opposite direction

erethism — an excessive irritability or sensitivity to stimulation, particularly with reference to the sexual organs, but including any body parts; also a psychic disturbance marked by irritability, emotional instability, depression, shyness, and fatigue, which are observed in chronic mercury poisoning

erythremia — the redness of the skin produced by congestion of the capillaries

erythrocyte — red blood corpuscle which serves to carry oxygen from the lungs to the tissue; frequently mislabeled as a cell, which it is not

Escherichia coli (E. coli) — a bacterium normally found in the colon. The presence of this microorganism in milk or water is an indicator of fecal contamination. Food contaminated with E. coli may be lethal.

esophageal atresia — an abnormality characterized by lack of complete development of the esophagus in a way that there is no passageway to the stomach

esophagus — muscular tube extending from the larynx to the stomach

ester — an organic compound corresponding in structure to a salt in inorganic chemistry. Esters are derived from acids by the exchange of the replaceable hydrogen or an organic alkyl group.

estrogen — a steroid hormone secreted primarily by the ovaries. It is involved in the maintenance and development of the female reproductive organs.

ether — class of chemical with two moieties bonded to oxygen. Dimethyl ether is a volatile, colorless, highly inflammable liquid with an aromatic odor; it used as an anesthetic and solvent.

ethylene diaminetetraacetic acid (EDTA) — a chelating agent used primarily in the treatment of lead poisoning

etiology — the study of the causation of any disease

evaporation — the process by which a liquid is changed into a vapor and mixed into the surrounding air

evaporation rate — the rate at which a liquid is changed to a vapor under standard conditions, usually compared to the rate of another substance that evaporates very quickly

excretion — the removal of a substance from the body in urine, feces, sweat, or expired air

exhaust — that atmosphere removed from a particular site

exothermic reaction — reactions that produce heat

explosive — a material capable of burning, or detonating, suddenly and violently

Class A — a material or device that presents a maximum hazard through detonation

Class B — a material or device that presents a flammable hazard and functions by deflagration

Class C — a material or device that contains restricted quantities of either Class A or Class B explosives or both, but presents a minimum hazard

explosive limits — the range of concentrations (percent by volume in air) of a flammable gas or vapor that can result in an explosion from ignition in a confined space; usually given as Upper and Lower Explosive Limits (See *UEL* and *LEL*)

exposure — when an organism makes contacts with a hazard, exposure occurs

exposure assessment — the determination or estimation (qualitative or quantitative) of the magnitude, frequency, duration, route, and extent (number of people) of exposure to a chemical

exposure coefficient — term which combines information on the frequency, mode, and magnitude of contact with contaminated medium to yield a quantitative value of the amount of contaminated medium contacted per day

exposure level — the amount (concentration of a chemical) at the absorptive surfaces of an organism

exposure scenario — a set of conditions or assumptions about sources, exposure pathways, concentrations of toxic chemicals and populations (numbers, characteristics, and habits) which aid the investigator in evaluating the quantifying exposure in a given situation

extrapolation — estimation of unknown values by extending or projecting from known values

exudate — a fluid that has exuded out of a tissue or its capillaries due to injury or inflammation

facilitated diffusion — the movement of a substance by a carrier molecule across the cell membrane without expenditure of energy

false negative — obtaining a statistically nonsignificant result when an effect truly exists

false positive — obtaining a statistically significant result when there is no effect

fate — the transport and transformation of a pollutant or toxin

fenestrae — windows, the large pores in the endothelial lining of the of the capillaries and a variety of organs

ferrous — relating to or containing iron

Fetal Alcohol Syndrome — a syndrome characterized by developmental abnormalities in the fetus as a result of maternal consumption of alcohol during pregnancy

fetotoxic — toxic to the fetus

fetus — the developing child during the last seven months of human pregnancy

fiber — solid particle whose length is at least three times its width

fibrinogen — a globular protein in the blood plasma that forms filamentous strands of fiber to initiate blood clotting

fibrosis — a disease in which alveolar cell death occurs and the cells are replaced with fibrous connective tissue; it decreases elasticity of the lungs and does not participate in gas exchange

filtrate — liquid after passing through a filter

First Law of Thermodynamics (energy) — in any physical or chemical change, no detectable amount of energy is created or destroyed, but energy can be changed from one form to another

first pass effect — the phenomenon of removal of chemicals by the liver before entrance of the blood to the systemic circulation; it is associated with substances that are absorbed from the digestive tract

fissionable isotope — isotope, such as uranium 235, that can split apart when hit by a neutron at the right speed and thus undergo nuclear fission or splitting

flammable — catches on fire easily and burns rapidly. The National Fire Protection Agency and the U. S. Department of Transportation define a flammable liquid as having a flash point below 378 °C (100 °F). (Same as *inflammable*)

flammable gas — any compressed gas that will burn

flammable limits — the range of gas or vapor concentrations (percent by volume in air) that will burn or explode if an ignition source is present (See *LEL* and *UEL*)

flammable liquid — any liquid having a flash point below 37.78 °C (100 °F), as determined by tests prescribed in the Federal Regulations

flammable solid — any solid material, other than an explosive, which is liable to cause fires through friction, absorption of moisture, spontaneous chemical changes, retained heat from manufacturing or processing, or which can be ignited readily and when ignited burns vigorously and persistently

flash point — the lowest temperature at which a liquid gives off enough flammable vapor to ignite and produce a flame when an ignition source is present

flow — movement of mass in response to a force

force — that which accelerates mass or changes the velocity of an object

fluorosis — a disease characterized by the replacement of calcium in the bone with fluoride, which results in weakening of the bone

fly ash — noncombustible particles carried by flue gas; frequently, elementally concentrated due to the combustion process *food additive* — a natural or synthetic chemical deliberately added to processed foods

food web — food chain; the relationship of predators and their prey in natural ecosystems. Primary producers are eaten by small animals which, in turn, are eaten by large predators. Together, all of these relationships are called the food web.

footdrop — plantar flexion of the foot due to weakness or paralysis of the muscles of the anterior compartment of the lower leg

forensic toxicology — the medical aspects of the diagnosis and treatment of poisoning and the legal aspects of the relationships between exposure to and harmful effects of a chemical substance. It is concerned with both intentional and accidental exposures to chemicals.

fume — very fine solid particles formed from recondensed vaporized metals

fungicide — a pesticide that controls or inhibits fungus growth

gamma ray — photons of ionizing energy, typically, produced by nuclear reactions; also a high energy x ray

gas — third state of matter comprised of diffuse molecules, e.g., our atmosphere

gastric — pertaining to the stomach

gastroenteritis — an inflammation of the mucous membrane of the stomach and intestines

gastrointestinal — the digestive system

Gaylor's model — a linear graphical extrapolation from zero to the upper confidence level of the lower limit of a dose response curve with a vertical drop down to zero risk. This triangular region encompasses all possible risk and, therefore, eliminates discussion on how to model extrapolations to low dose.

Geiger counter — an electrical device that detects the presence of ionizing radiation

gene — the part of the chromosome that carries a particular inherited characteristic or trait

genome — the complete set of hereditary factors of an individual

genotoxic — damaging to genetic material

gestation — the duration of pregnancy; in humans nominally nine months

gingivitis — inflammation of the gums of the mouth

glial cells — population of nervous system cells responsible for the protection, care, and function of neurons

globin — the protein portion of the hemoglobin molecule of the RBC

globulin — any of a group of albuminous proteins; frequently of the immune system, i.e., gamma globulin

glomerular nephritis — an inflammation of the kidney that primarily affects of glomeruli of the nephron

glomerulus — the mass of capillary loops at the beginning of each nephron, nearly surrounded by Bowman's Capsule

glucosuria — the excretion of glucose in urine, especially in elevated amounts, an indication of diabetes

glucuronidation — the binding of glucuronic acid with a toxic compound or its metabolites to increase its excretion

glycogen — also known as animal starch, it is a polysaccharide carbohydrate reserve found in, e.g., muscle and liver tissue, that is polymerized glucose

glycoprotein — an organic molecule composed of a protein and a carbohydrate

gonads — repository for genetic material, the testes in the male and ovary in the female

gram (g) — a metric unit of mass. One U.S. ounce equals 28.4 grams; one pound equals 454 grams. There are 1000 milligrams (mg) in one gram. One kilogram equals 1000 grams and "weighs" 2.2 pounds.

granulocytopenia — a decrease in the number of granulocytes (neutrophils, basophils, and eosinophils)

granuloma — a term applied to a mass or nodule of chronically inflamed tissue with granulations that is usually associated with an infective process

Gray (Gr) — the international unit that is used as a measure of the amount of radiation absorbed per gram of tissue

half-life — the time taken for materials to lose one-half of their activity. For example, the half-life of DDT is 15 years; radium has a half-life of 1,580 years.

hard water — alkaline water containing dissolved mineral salts, which interfere with some industrial processes and prevent soap from lathering

hazard — that which may be hazardous to life upon exposure; typically toxic, flammable, explosive, and corrosive materials or unsafe living or working conditions

hazardous air pollutant — substances covered by Air Quality Criteria, which may cause or contribute to illness or death, e.g., asbestos, beryllium, mercury, and vinyl chloride

hazardous class — a group of materials designated by the Department of Transportation (DOT) that share a common major hazardous property, i.e., flammability, corrosivity, radioactivity

hazardous material — any chemical that is a physical hazard or a health hazard

hazardous waste — waste materials that are classified as toxic, corrosive, flammable, explosive, radioactive, or biological /infectious wastes

heart — major organ responsible for pumping blood throughout the body

heat of fusion — the quantity of heat that must be supplied to a material at its melting point to convert it completely to a liquid at the same temperature

heat of vaporization — the quantity of heat that must be supplied to a liquid at its boiling point to convert it completely to a gas at the same temperature

hematocrit — the percentage of the volume of a blood sample composed of formed elements (cells)

hematology — the study of blood and blood-forming tissues

hematopoiesis — the production of blood and blood cells; hemopoiesis

hematuria — blood in the urine. Color of the urine may be slightly smoky, reddish, or very red.

heme — the portion of the hemoglobin molecule that contains iron

hemocytoblast — a primordial cell capable of differentiating into any type of blood cell

hemodialysis — a procedure for removing metabolic waste products or toxic substances from the bloodstream by dialysis

hemoglobin — the red oxygen transporting substance contained in red blood corpuscles (RBC)

hemolysis — the rupture of RBCs membranes

hemolytic anemia — an anemia resulting from the abnormal destruction of RBCs, in response to certain toxic infectious agents and in certain inherited blood disorders

hemophilia — a hereditary blood disease characterized by a prolonged coagulation time

heparin — a glycoprotein that prevents blood from clotting

hepatic — pertaining to the liver

hepatitis — an inflammation and enlargement of the liver, usually caused by a virus

  a virus — blood borne virus that exists in the blood for only two weeks; infection occurs by contact with contaminated food or water, and is also sometimes contracted via blood transfusions

  b virus — also referred to as serum hepatitis; caused by a DNA virus that persists in the blood and is characterized by a long incubation period; usually transmitted as a result of contact with infected blood or other body fluids. The most obvious symptom is the development of jaundice.

  c virus — a form of hepatitis with clinical effects similar to those of hepatitis B, caused by a blood-borne retrovirus that may be of the hepatitis non-A, non-B type

  d virus — usually occurs with one of the other hepatitis viruses, their coexistence being necessary for HVD to survive. Cirrhosis of the liver often develops.

hepatocyte — a cell of the liver

hepatoma — a malignant tumor occurring in the liver

herbicide — a chemical used to kill plants

HESIS — Hazard Evaluation System and Information Service provides information to workers, employers, and health professionals about the health effects of toxic substances and how to use them safely; (510) 540-3014

histamine — an amine released from basophils, that causes dilation of capillaries, contraction of smooth muscle, stimulation of gastric acid secretion, and that promotes the inflammatory response. It is a neurotransmitter that is released during allergic reactions.

histology — the study of the structure of cells and tissues; usually involves microscopic examination of tissue slices

homeostasis — maintenance of a constant internal environment in an organism

hormone — a chemical substance secreted by an endocrine organ and transported throughout the body, to specific target sites, by the circulatory system

human equivalent dose — a dose which, when administered to humans, produces an effect equal to that produced by a dose in an animal model system

hydrocarbon — compounds found in fossil fuels that contain carbon and hydrogen, typically, organic compounds

hydrogen sulfide (H_2S) — a gas that smells like rotten eggs, emitted during organic decomposition, by-product of oil refining and burning; TWA 10 ppm, IDLH 300 ppm

hydrology — the study of the properties, distribution, behavior and effects of water on the earth's surface, in the soil, and underlying rocks and in the atmosphere

hydrolysis — the breakdown of a chemical into two parts concomitant with the addition of the elements of water (H^+ and OH^-) to the products

hydrophobic — antagonistic to water; any substance that is incapable of dissolving in water

hydroxyapatite — the complex crystal structure that makes up the lamellae of the bone

hygiene — the science of health and of its preservation

hyperkeratosis — the excessive growth of the horny tissue (outer layers) of the skin

hyperpigmentation — an excessive amount of pigment in a tissue or body part

hyperplasia — an abnormal increase in the number of cells

hypertrophy — excessive growth or enlargement of an organ or tissue

hypokinesis — abnormally decreased mobility; abnormally decreased motor function or activity

hypoxia — a deficiency of oxygen

IDLH — Immediately Dangerous to Life or Health; a term used to describe an environment which is very hazardous due to a high concentration of toxic chemicals or insufficient oxygen or both

igneous rock — rock formed when molten rock material (magma) wells up from the Earth's interior, cools, and solidifies into rock masses

ignition temperature — the lowest temperature at which a substance will catch on fire and continue to burn

incidence — an expression of the rate at which a certain event occurs, as the number of new cases of a specific disease occurring during a certain period (See *prevalence*)

incineration — the combustion (by burning) of organic matter or organic waste

incompatible — a term used to describe materials which could cause dangerous reactions from direct contact with one another

induced mutation — a change in the structure of a gene or chromosome of an organism as a result of exposure to an exogenous chemical or physical agent

industrial hygiene — that branch of preventive medicine concerned with the protection of health of the industrial population

infarct — an area of necrosis in a tissue caused by local lack of blood resulting from obstruction of circulation to the area

infectious — capable of being transmitted with or without contact

inflammable — Same as *flammable*

inflammation — a general nonspecific body response to tissue injury, characterized by swelling, redness, pain, heat, and the infiltration of the area by leukocytes

ingestion — taking in and swallowing a substance through the mouth

inhalation — breathing in a substance

initiation — viewed as the first step of carcinogenesis involving the induction of an irreversibly altered cell; a mutational event

inorganic compound — one of two major classes of chemical compounds; contains no carbon

insecticide — a chemical used to kill insects

intake — amount of material inhaled, ingested, or absorbed dermally during a specified period of time

integrated exposure assessment — a summation over time, in all media, of the magnitude of exposure to a toxic chemical

integrated pest management (IPM) — combined use of biological, chemical, or cultivation methods in proper sequence and timing to avoid economically unacceptable loss of a crop or livestock animals; seeks to minimize use of toxic chemicals (pesticides)

integument system — major organ system forming a barrier between the inside and outside of the body; consists of skin, hair, and nails

interstitial fluid — the fluid between cells or body parts

intradermally — into the skin, as with an injection

intramuscularly — into the muscle, as with an injection

intraperitoneally — into the peritoneal (body) cavity. This is the cavity that surrounds the digestive tract and its associated structures.

intravenously — into a vein, as with an injection

in vitro studies — studies of chemical effects conducted in tissues, cells or subcellular extracts from an organism model system, i.e., not in the living organism (in vivo)

in vivo studies — studies of chemical effects conducted in intact living model systems

ion — a molecule, atom, or radical, that has lost, or gained, one or more electrons and has, therefore, acquired a net electric charge. Positively charged ions are cations; negatively charged ions are anions. In general, an ion has entirely different properties from the element (atom) from which it was formed.

ionic bond — a bond formed by complete transfer of an electron from one atom to another, which results in ions that are oppositely charged and attract one another

iris — the circular, pigmented tissue of the eye located directly in front of the lens that gives color to the eye

irreversible effect — effect characterized by the inability of the body to partially or fully repair injury caused by a toxic agent

irritant — a substance that can cause an inflammatory response or reaction of the eye, skin, or respiratory system

ischemia — deficiency of blood owing to a functional constriction or actual obstruction of a blood vessel

isomer — two or more compounds that have the same molecular formula but have different orders of attachment of atoms and thus have different physical and chemical properties

isotope — atoms of the same element that differ in atomic weight

jaundice — a yellowish pigmentation of the skin, tissues, and body fluids caused by the deposition of bile pigments

Kaposi's Sarcoma — a neoplastic disease associated especially with AIDS, affecting the skin and mucous membranes, and characterized usually by the formation of pink to reddish-brown or bluish plaques, macules, papules, or nodules

keratin — an insoluble protein produced in the cells that comprise the outer layers of the skin

keratitis — an inflammation of the cornea

kidney — major organ responsible for filtering blood, regulations of electrolytes, and the excretion of waste via urine

kilogram (kg) — a metric unit of mass equals to 1000 grams, which equals 2.2 pounds

Kupffer Cells — phagocytic cells found in the liver; they are involved in the destruction and removal of foreign substances circulating in the blood; macrophages

lacrimation — the secretion and discharge of tears

laryngitis — inflammation of the larynx, a condition attended with dryness and soreness of the throat, hoarseness, cough, and dysphagia (difficulty in swallowing)

latency — the time between exposure and the first appearance of an effect

lavage — a washing, especially of a hollow organ such as the stomach, with repeated injections of water

leachate — materials in aqueous solution that occur as waste seeps through (e.g., solid waste) and have the potential of polluting water supplies

leaching — a process in which various chemicals in upper layers of soil or other solid materials are dissolved and carried into the subsoil and ground-water

lead — a useful, easily workable, toxic heavy metal element widely used by and in our society. In humans it is known to cause neurological, reproductive, and growth disorders.

Legionnaire's Disease — a severe, often fatal disease characterized by pneumonia and dry cough; as the disease progresses it can lead to cardiovascular collapse. It is caused by the bacterium *Legionella pneumophila*.

LEL — Lower Explosive Limit (See *explosive limits*)

lens — the transparent structure of the eye between the cornea and the retina

lethal — that which has the ability to kill life

Lethal Concentration - 50% (LC_{50}) — a concentration of chemical in air that will kill 50 percent of the test animals inhaling it

Lethal Dose - 50% (LD_{50}) — the dose of a chemical that will kill 50 percent of the test animals receiving it. The chemical may be given by mouth (oral), applied to the skin (dermal), or injected (peritoneal). A given chemical will generally show different LD_{50} values depending on how it is given to the animals. It is a rough measure of acute toxicity.

leukemia — a disease of the bloodstream that results in an increase in the number of white blood cells

leukocyte — a white blood cell; it is part of the immune system that circulates in the blood and lymph and participates in reactions against invading microorganisms or foreign particles

ligand — the molecules attached to the central atom(s) by coordinate covalent bonds

lipid — any of various substances that are soluble in nonpolar organic solvents, such as chloroform and ether. Along with proteins and carbohydrates, lipids constitute the principal structural components of living cells.

lipophilic — substances that have a strong affinity for, or are soluble in, lipids

lipoprotein — a protein bound with a lipid such as cholesterol, phospholipid, or triglyceride

liter — a metric unit of volume. One U.S. quart is a little less than one liter. One liter equals 1000 cubic centimeters.

litmus — a visual indicator of the acidity or alkalinity of various substances

local effects — those effects which tend to be limited to point of contact, e.g., poison oak and corrosives (acids and bases)

logistic regression — a statistical method for calculating odds ratios for individual risk factors where a variety of risk factors may be contributing to the occurrence of the disease

Lowest Observable (Adverse) Effect Level - LOEL (LOAEL) — the lowest dose which produces an observable adverse effect

lumen — the cavity of a tubular organ, such as a blood vessel

lungs — paired organs responsible for the exchange of gases between the blood and the atmosphere

Lyme Disease — a recurrent inflammatory disease characterized by the appearance of red splotches on the skin and development of fever, fatigue, headache, and swelling of the joints; there may also be associated effects on the cardiovascular and nervous systems. It is caused by a spirochete, *Borrelia burgdorferi*.

lymph — a clear or yellowish fluid derived from interstitial fluid and found in lymph vessels

lymph node — vascularized mass of lymphoid tissue which produces lymphocytes that fight bacteria and filter the lymph. They are located throughout the body except the CNS.

lymphocyte — a type of WBC characterized by a large nucleus with little cytoplasm; which grows in size when there are infections; involved with specific immune response

lymphoma — malignancy of the lymphoid tissues

macrophage — any large, mononuclear, phagocytic cell

makeup air — in workplace ventilation, air introduced into an area to replace the air that has been removed

malaise — a vague feeling of discomfort or illness

malignant — injurious or deadly, as in cancer

materials balance — an accounting of the mass flow of a substance from sources of production, through distribution and use, to disposal or distribution, and including any releases to the environment

Material Safety Data Sheet (MSDS) — a document produced by chemical manufacturers that describes physical and chemical properties of chemicals, first aid information, personal protective equipment to use, and suggested means of disposal

medium — the environmental vehicle by which a substance (i.e., a pollutant) is carried to the site, e.g., surface water, soil, or groundwater

megakaryocyte — a large cell associated with bone marrow which is the source of the blood platelets

melanosis — an unusual deposit of black pigment in different parts of the body

melting point — the temperature at which a solid substance changes to the liquid state

mercurialism — a chronic poisoning by mercury as a result of working with metallic mercury; symptoms include soreness of gums and teeth, increased salivation, and diarrhea

messenger RNA (mRNA) — a type of ribonucleic acid involved in the transcription of DNA in the nucleus and the transport of that information into the cytoplasm for ribosomes to utilize for the sequencing of amino acids in proteins

metabolism — the sum of the chemical reactions occurring within a cell or a whole organism; includes the energy-releasing breakdown of molecules, catabolism, and the synthesis of new molecules, anabolism

metabolite — any product of metabolism, especially a transformed chemical

metal fume fever — a syndrome resembling influenza produced by inhalation of metallic oxide fumes such as zinc oxide. Symptoms include chills, weakness, sweating, and anorexia.

metamorphic rock — rock produced when a preexisting rock is subjected to high temperatures, high pressures, chemically active fluids, or a combination of these agents

metastasis — the spread of a disease (cancer) to another part of the body

meter — a measure of length based on the spectrographic color line of the element krypton; 1 meter = 39.37 inches

methemoglobinemia — a condition in which more than one percent of the iron in hemoglobin has been oxidized from Fe^{+2} to Fe^{+3} and is unable to carry oxygen

methyl isocyanate — a chemical used in the production of pesticides; exposure causes irritation of mucous membranes, coughing, and increased saliva production; it can be acutely toxic when inhaled

methylation — a Phase II metabolic reaction that plays a minor role in the transformation of toxic substances to a different toxic substance, which is retained in the body longer

micro — prefix meaning one one-millionth

milligram (mg) — a metric unit of mass. One gram equals 1000 mg. One ounce equals 28375 mg.

mg/kg — a way of expressing dose: milligrams of a substance (mg) per kilogram (kg) of body weight (See *dose*)

mg/m³ — a measure of concentration: weight of substance (mg) in a volume of air (m^3), often used to express PELs and TLVs (See *exposure limits*)

mm Hg — a unit of measurement for pressure, millimeters (mm) of the metal mercury (Hg). At sea level, the earth's atmosphere exerts 760 mm Hg of pressure.

Minamata disease — a neurological disease named after an afflicted village in Japan caused by the consumption of fish contaminated with organic methylmercury industrial wastes

mist — liquid particles of various sizes suspended in the atmosphere

mitochondria — a specialized subcellular organelle that manufactures ATP

mixture — a combination of substances held together by physical rather than chemical means

modeling — use of mathematical equations to simulate and predict real events and processes

model system — biological system used to predictively emulate effects on humans

molecular weight — the sum of the atomic weights of the atoms in a molecule. Each proton and each neutron is given the value of 1 atomic mass unit, called a Dalton in molecular biology.

molecule — combination of two or more atoms of one or more chemical elements held together by chemical bonds; smallest particle of a compound capable of having the properties of the compound

monitoring — measuring concentrations of substances in environmental media, humans, or model systems

monocyte — a large WBC that transforms into a macrophage in response to tissue damage

monomer — See *polymerization*

morbidity — the relative incidence of a disease in a population

mortality — the relative incidence of deaths in a population

MSDS — Material Safety Data Sheet; a form listing the properties and hazards of a product or a substance

MSHA — Mine Safety and Health Administration, an agency in the U. S. Department of Labor which regulates safety and health in the mining industry. This agency also tests and certifies respirators. (See *NIOSH*)

MTD — Maximum Tolerated Dose; the highest dose of a chemical that does not alter the life span or severely affect the health of an animal

mucociliary escalator — a mechanism found in the respiratory tract, it is comprised of ciliated epithelium with numerous mucus cells. The upward beating of the cilia facilitates the removal of particulate matter caught in the mucous from the respiratory system.

mucous — thick gooey secretion produced by the goblet cells in the mucus membrane

mucus membrane — the moist, soft lining of the nose, mouth, throat, bronchus, and eyes

muscarinic — a type of postganglionic receptor that interacts with the substance muscarine. Effects of stimulation are similar to stimulation produced by acetylcholine. Nicotinic receptors are not affected.

muscle — one of the four basic tissue types; characterized by a fundamental ability to contract

mutagen — a chemical or physical agent able to change the genetic material in cells

mutagenesis — process leading to the alteration of DNA

mutagenicity — the capacity of a chemical or physical agent to cause permanent alteration of the genetic material within living cells

mutation — an inheritable change in the kind, structure, sequence, or number of component parts of a cell's DNA

myelin — a lipoprotein sheath created by Schwann or oligodendrocyte cell membrane wrapped around a neural process and allowing high speed neural impulse propagation

myelotoxin — a cytotoxin that causes destruction of the bone marrow cells

myocardium — the middle layer of the walls of the heart, composed of cardiac muscle

myometrium — the middle layer of the wall of the uterus, composed of smooth muscle

myopathy — any disease of muscle tissue

NAAQS — National Ambient Air Quality Standards; maximum allowable level, averaged over a specific period of time, for a certain pollutant in outdoor air

NAD (Nicotinamide Adenine Dinucleotide) — energy transport molecule used in the cell to transfer energy from, e.g., glucolysis, to ATP production by the cytochrome transport system of the mitochondria

narcosis — the state of stupor or unconsciousness produced by a chemical

nausea — stomach upset accompanied by a feeling that one is about to vomit

necropsy — the examination of a dead body; autopsy. The term necropsy is often used with respect to examination of animals other than humans.

necrosis — the death of tissue, usually as individual cells, groups of cells, or in small localized areas

neoplasm — a tumor, new growth, results from a more rapid than normal division of one or more cells

nephritis — inflammation of the kidney; a focal or diffuse proliferate or destructive process, which may involve the glomerulus, tubule, or interstitial renal tissue

nephron — the functional unit of the kidney, consisting of the renal corpuscle, the proximal convoluted tubule, the Loop of Henle, and the distal convoluted tubule

nervous — one of the four basic tissues; characterized by the ability to conduct impulses

neuroglia — nurse or helper cells of the nervous system which maintain the neurons

neuron — specialized cells of the nervous system designed to communicate information throughout the body

neurotransmitter — a chemical substance released from the end of an axon resulting in a stimulatory or inhibitory effect on the target

neutralization — the chemical addition of either an acid or base to a solution to adjust the pH to 70

neutron — a neutrally charge particle that may be emitted from the nuclei of atoms

neutrophil — a granular WBC which wanders throughout the body acting as a phagocyte

NFPA — National Fire Protection Association. NFPA has developed a scale for rating the severity of fire, reactivity, and health hazards of substances. References to these ratings frequently appear on MSDSs.

nicotinic receptor — a subset of acetylcholine receptors that bind nicotine, nonmuscarinic receptors

NIOSH — National Institute for Occupational Safety and Health, a federal agency which conducts research on occupational safety and health questions and recommends new standards to federal OSHA. NIOSH, along with MSHA, tests and certifies respirators.

noise pollution — any unwanted, disturbing, or harmful sound that impairs or interferes with hearing, causes stress, hampers concentration and work efficiency, or causes accidents

nondegradable pollutant — material (such as lead) that is not broken down by natural process

nonpoint source — source of pollution from human activities that occurs on large or dispersed land areas such as crop fields, streets, and lawns. These sources discharge pollutants into the environment over a large area.

nonrenewable resource — resource that exists in a fixed amount (stock) in the Earth's crust and has the potential for renewal only by geological, physical, and chemical processes taking place over millions to billions of years

nonroutine task — a predictable task that occurs infrequently

No Observable (Adverse) Effect Level NOEL (NOAEL) — maximum dose at which no adverse effect is observable

norepinephrine — noradrenaline, a neurotransmitter utilized by the adenergic sympathetic nervous system

nucleic acid — a molecule composed of many nucleotides chemically bound together, such ad DNA and RNA

nucleotide — any of the various compounds consisting of the bases, Adenine, Thymine, Guanine, Cytosine, or Urine, and a sugar, deoxyribose or ribose, and phosphate groups, forming the basic constituent of DNA and RNA

nucleus — the center of an atom, making up most of the atom's mass. The nucleus contains one or more positively charge protons and one or more neutrons with no electrical charge.

nutrient — any food or element that an organism requires to live, grow, or reproduce

observation bias — a misidentification of an individual based upon incomplete information, resulting in an individual being placed in the wrong study group; this may lead to identifying an association between the causative agent and the measured effect that may be either weaker or stronger than what is really there

occupational study — studies in which subjects are chosen from the workplace

odds ratio — the comparison between the odds of exposure among cases to the odds of exposure among controls

odor threshold — the lowest concentration of a substance in air that can be smelled. For a given chemical, different people usually have very different odor thresholds.

olestra — a fat substitute approved by the FDA that can be used in certain food types such as potato chips; it has been shown to affect normal gastrointestinal functioning in such ways as diarrhea

oligodendrocytes — neuroglial cells within the CNS that form myelin sheaths around neurons

oncogenic — a substance that causes tumors, whether benign or malignant

oncology — study of cancer

opacity — the measure or indicator of how much light can pass through a substance *oral* — of the mouth, through or by the mouth

organic compound — one of two major classes of chemical compounds; contains carbon

ortho-chlorobenzylidene malononitril — CS gas

OSHA, Federal — Occupational Safety and Health Administration, an agency in the U. S. Department of Labor which establishes workplace safety and health regulations. Many states, including California, have their own OSHA programs. State OSHA programs are monitored by federal OSHA to ensure they are "at least as effective as" the federal OSHA program.

osmosis — the passage of water through a selective semipermeable membrane from an area of less concentration to a region of higher concentration, diluting the more concentrated solute

osteocyte — a mature bone cell that is surrounded by bone matrix; bone cell

Osteomalacia — a disease that results in adults from a deficiency of vitamin D or calcium; it is characterized by a softening of the bones with accompanying pain and weakness

Osteoporosis — a bone disorder in which bone and cartilage become thinned, hardened, and brittle

oxidant — a substance that supplies oxygen for chemical reactions; oxidation is the opposite of reduction

oxidation-reduction — the process of substances combining with oxygen or passing electrons

oxidizer — a chemical other than a blasting agent or explosive that initiates or promotes combustion in other materials, thereby causing fire

ozone layer — layer of gaseous ozone (O_3) in the stratosphere, 18 to 48 km (11 to 30 miles) above the surface of the Earth, that filters out harmful ultraviolet radiation from the sun

p-value (probability value) — the probability that an index of effect is as extreme or more extreme than that observed even if no effect exists (i.e., if the null hypothesis is false)

palliative — a substance that seems to relieve or alleviate symptoms without curing

pallor — the lack of color especially in the face; paleness

pancytopenia — a decrease in the number of all WBCs

pandemic — epidemic disease spread over many regions (See *endemic*)

papule — a small, circumscribed, solid elevation on the skin

paranoia — mental disorder characterized by delusion

Parasympathetic Nervous System (PNS) — subdivision of the autonomic nervous system; it is involved in involuntary functions

parenteral — the introduction of a substance into the body other than by the digestive system, such as by subcutaneous or intravenous injection

paresthesia — a skin sensation, such as burning, prickling, itching, or tingling, with no apparent cause

Parkinson's Syndrome — a neurological disorder characterized by tremors

particulates — fine liquid or solid particles such as dust, smoke, mist, fumes, or smog, found in the air or point source emissions

passive transport — the movement of molecules, ions, or atoms across the cell membrane without the expenditure of energy. Protein carrier molecules may be involved in the transport.

pathogen — any disease-causing agent, usually applied to living agents

pathogenic — causing or capable of causing disease

pathology — the study of diseased tissue

pathway — the flow of a substance through a series reactions or operations

PCB — polychlorinated biphenyl; a group of toxic, persistent chemicals used in transformers and capacitorsz. Its sale or use was banned in the United states in 1979.

PEL — Permissible Exposure Level, a maximum allowable exposure level under OSHA regulations (See *exposure limits*)

penetration — entrance of a substance through the pores of a barrier

Peripheral Nervous System (PNS) — distal portion of the nervous system from the spinal cord to the muscle, motor, and skin, sensory

permeation — diffusion of chemicals through the material of a barrier *permissible dose* — the dose of a chemical that may be received by an individual without the expectation of significantly harmful results

persistent pesticides — pesticides that do not break down chemically and remain in the environment after a growing season

pesticide — a chemical used to kill pests; includes herbicides, insecticides, and rodenticides

petechiae — tiny nonraised, perfectly round, purplish red spots caused by intradermal or submucosal hemorrhaging

petrochemicals — chemicals obtained by refining (distilling) crude petroleum (oil) and used as raw materials in the manufacture of most industrial chemicals, fertilizers, pesticides, plastics, synthetic fibers, paints, medicines, and other products

pH — expresses how acidic or basic a solution or chemical is, using a scale of 1 to 14. For example, a pH of 1 indicates a strongly acidic solution, a pH of 7 indicates a neutral solution, and a pH of 14 indicates a strongly alkaline solution.

phagocytosis — the process by which a cell engulfs solid particles; characterized by the formation of cellular extensions called pseudopodia that surround and facilitate the endocytosis of the particles

pharmacodynamics — the interaction between the molecules of toxic substance and the specific sites of action, the receptor sites

pharmacokinetics — the behavior of chemicals inside biological model systems; including absorption, uptake, distribution, metabolism, biotransformation, concentration, accumulation, and excretion

phocomelia — the developmental abnormality characterized by an abnormal shortening of the long bones of the arms and legs

phosgene — a colorless gas formed by the reaction of carbon dioxide and chlorine; it is used in the synthesis of isocyanates, polyurethane resins, carbamates and dyes; used also as a chemical warfare agent that causes lung irritation

phospholipid — a lipid containing phosphorus, fatty acids, and a nitrogen-containing substance. It is the main component of the lipid bilayer that forms the structural backbone of all cell membranes.

photon — a quantitative unit of electromagnetic energy with varying wavelengths, generally regarded as a discrete particle having zero rest mass, no electrical charge, and an indefinitely long lifetime

photophobia — abnormal sensitivity, usually of the eyes, to light

phylogenetic — pertaining to the evolutionary relationship among organisms

physical hazard — a chemical for which there is scientifically valid evidence that it is a combustible liquid, a compressed gas, an explosive or flammable substance, or organic peroxide, an oxidizer, or a pyrophoric, unstable (reaction), or water-reactive substance

physiology — the study of body function

pica — an eating disorder characterized by a craving to ingest any material not fit for food, including starch, clay, crayons, grass, metals, or plaster

picrotoxin — a toxic substance obtained from the seed of a shrub, *Animirta cocculas* which can produce uncontrolled muscle spasms

pinocytosis — the engulfing and absorption of droplets of liquids by cells

pitot tube — sensor designed to measure gas flow rates or air velocity

placenta — the organ that forms the bridge between the fetal and the maternal bloodstreams

platelet — the minute fragments of megakaryocytes, that play a critical role in blood clotting

plenum — a large main tube or duct which accommodates mass flow

pneumoconiosis — a condition of respiratory tract due to inhalation of dust particles

pneumonitis — inflammation of the lungs

point mutation — a change in a single base in the DNA structure

point source — a stationary location where discharge occurs, e.g., the end of a pipe

poison — a chemical or substance that, in relatively small amounts, is able to produce injury by chemical or biochemical actions, when it comes into contact with a susceptible cell or tissue

Class A — a poisonous gas or liquid of such a nature that a very small amount of the gas, or vapor, when mixed with air, is dangerous to life and health

Class B — any substance known to be so toxic to humans that it poses a severe health hazard during transportation

pollution — substance introduced into air, water, soil, or food that is not normally present (or not normally present in such high concentrations) and that can adversely affect the health, survival, or activities of humans or other living organisms

Polychlorinated Biphenyl (PCB) — a group of toxic, persistent, chemicals used in transformers and capacitors; sale or use was banned in the United States in 1979

polymerization — a chemical reaction in which small molecules (monomers) combine to form much larger molecules (polymers). A hazardous polymerization is a reaction that occurs at a fast rate, and releases large amounts of energy. Many monomers are hazardous in the liquid and vapor states, but form much less hazardous polymers. An example is vinyl chloride monomer, which causes cancer but forms the relatively nontoxic polyvinyl chloride (PVC) plastic.

polyneuritis — an inflammation of several nerves at one time, marked by paralysis, pain, and muscle atrophy

polyuria — the passage of a large volume of urine in a given period, a characteristic of diabetes

population at risk — a population subgroup that is more likely to be exposed to a chemical, or is more sensitive to a chemical, than is the general population

potentiation — situation whereby the effects of two substances are greater than the summation of their individual effects

ppb — parts per billion, a measure of concentration, such as parts of a chemical per billion parts of air or water. One thousand ppb equals one part per million (ppm).

ppm — parts per million, a measure of concentration, such as parts of a substance per million parts of air or water. PELs and TLVs are often expressed in ppm. (See also *ppb*)

precipitate — insoluble material produced by chemical reaction in a solution

precursor — a chemical from which another chemical is made

presynaptic terminal — the end of an axon that releases a neurotransmitter which diffuses across the synaptic gap to the post synaptic terminal

prevalence — the total number of cases of a disease in existence at a certain time in a designated area (See *incidence*)

prevalence study — an epidemiological study which examines the relationships between diseases and exposures as they exist in a defined population at a particular point in time

Preven — hormonal cocktail successful in the immediate prevention of ovulation

priority pollutant — the Clean Water Act amendments of 1977 listed 126 "priority pollutants" based on criteria of toxicity, resistance, and potential for exposure of living things

procarcinogen — chemicals that require metabolism to another, more reactive or toxic chemical form before their carcinogenic action can be expressed

prognosis — a forecast as to the probable outcome of an illness

progression — the process whereby a benign tumor becomes malignant due to additional heritable changes to the initiated cells; irreversible event

progressor agent — a chemical or substance that causes the transition of cells from the promotion stage to the progression stage of cell growth

promotion — viewed as the second step of carcinogenesis; the experimentally defined process by which the initiated cell clonally expands into a visible tumor, often a benign lesion; reversible event

proportional mortality ratio — the proportion of total deaths represented by a particular cause of death in the occupational cohort measured against the same cause in the reference population

prospective study — an epidemiological study which examines the development of disease in a group of persons determined to be presently free of the disease

prostration — an extreme weakness or exhaustion

protein — a molecule consisting of one or more polypeptide chains. Each chain is composed of a large number of amino acids

proteinurea — excessive amounts of protein in the urine

prothrombin — a plasma protein needed for blood clotting

protocol — the rules and outline of an experiment

proton — positively charged particle in the nucleus of every atom. Each proton has a relative mass of 1 and carries a single positive charge.

proximal convoluted tubule — the convoluted portion of the nephron of the kidney that extends from Bowman's capsule to the Loop of Henle

pseudopodia — a temporary infolding or extension of the cell membrane; usually associated with phagocytosis or in cell locomotion

psi — pounds per square inch; a unit of pressure. At sea level, Earth's atmosphere exerts 14.7 psi, which equals 760 mm Hg

psoriasis — a chronic, hereditary, recurrent, papulosquamous dermatitis, the distinctive lesion of which is a vivid red mascula, papule, or plaque covered almost to its edge by livery lamellated scales. It usually involves the scalp and extensor surfaces of the limbs, especially the elbows, knees, and shins

psychoneurosis — a mental or behavioral disorder of mild or moderate severity. It leads to disturbances in thought, feelings, and attitudes

pulmonary edema — filling of the lungs with fluid, which produces coughing and difficulty breathing and may cause drowning or death by suffocation

pupillary light reflex — a constriction of the pupil upon stimulation of the retina by light

Purkinje fiber — a specialized cardiac muscle fiber that conducts action potentials through cardiac muscle; part of the conduction system of the heart

pustule — a small inflamed skin swelling that is filled with pus; a small swelling similar to a blister or pimple

pyrophoric — capable of igniting spontaneously when exposed to dry or moist air at or below 54.44 °C (130 °F)

qualitative — descriptive of kind, type, or direction, as opposed to size, magnitude, or degree

quantitative — descriptive of size, magnitude or degree

rad — a unit of measurement of any kind of radiation absorbed by tissue

radiation — the emission or transfer of energy by photons or particles; photons or particles released by nuclear reactions

radioactive — a substance capable of giving off atomic emissions in the form of particles or photons

radionuclide — a nuclide (atom) that is of either artificial or natural origin and exhibits radioactivity

reaction — a chemical transformation or change

reactivity — the ability of a substance to undergo a chemical reaction (such as combining with another substance). Substances with high reactivity are often quite hazardous.

receptor

   in biochemistry — a specialized molecule in a cell that binds a specific chemical with high specificity and high affinity

   in cellular — a specific protein in either the plasma membrane or interior of a target cell with which a chemical messenger combines to exert its effects

   in exposure assessment — an organism that receives, may receive, or has received environmental exposure to a chemical

reduction — the gaining of electrons by a molecule; a reaction in which hydrogen is combined with or oxygen is removed from a compound

Reference Dose/Concentration (RfD/RfC) — estimates of a daily exposure to an agent that is assumed to be without an adverse health impact on the human population

relative risk — the measure of risk for those exposed compared to those who are not exposed

rem — the amount of ionizing radiation required to produce the same biologic effect as one rad of x rays

renal — pertaining to the kidney

reproduction — production of offspring by parents

reservoir — a tissue depot in an organism, or a place in the environment, where a toxic chemical, toxon, accumulates, from which it may be released at a later time

respirator — a device worn to prevent inhalation of hazardous substances

resting membrane potential — the voltage (mV) built up across a cell membrane due to normal transport phenomena, typically negative in excitable cells

reticulocyte — the RBC immediately after its nucleus has been expelled and it has excited the bone marrow, persists for a day or so before becoming a RBC, typically found in concentrations of 1% of the RBC populations, can be used to gage RBC production rates

retrospective study — an epidemiological study which compares diseased persons with non-diseased persons and works back in time to determine exposure

reversible effect — an effect which is not permanent, especially adverse effects which diminish when exposure to a toxic substance is terminated

rhinitis — an inflammation of the lining of the nasal cavity; in acute cases it increases the secretion of mucus

ribonucleic acid (RNA) — single stranded nucleic acid molecule

ribosome — an intracellular structure responsible for protein synthesis

risk — the potential for realization of unwanted negative consequences or events

risk analysis — identifying hazards and evaluating the nature and severity of risks (risk assessment) to make decisions about allowing reducing, increasing, or eliminating risks (risk management) and communicating information about risks to decision makers or the public (risk communication)

risk assessment — a qualitative or quantitative evaluation of the environmental and/or health risk resulting from exposure to a chemical or physical agent (pollutant); combines exposure assessment results with toxicity assessment results to estimate risk; the scientific activity of evaluating the harmful, toxic properties of a substance or condition and the circumstances of human exposure to these events, with the objective of determining the probability that exposure will cause adverse effects

risk-benefit analysis — estimate of short and long-term risks and benefits of using a particular product or technology

risk communication — communicating information or perspectives about risks to decision makers or the public

risk estimate — a description of the probability that organism exposed to a specified dose of chemical will develop an adverse response, e.g., cancer

risk factor — characteristics (such as race, sex, age, obesity) or variables (e.g., smoking, occupational exposure level) associated with increased probability of a toxic effect

risk management — the process of applying a risk assessment to the conditions that exist in society, so as to balance exposures to toxic agents against needs for products and processes that may be inherently hazardous

risk specific dose — the dose associated with a specified risk level

rodenticide — a type of pesticide designed to kill rodents

route of entry — the means by which material may gain access to the body, for example, inhalation, ingestion, and eye or skin contact

route of exposure — the avenue by which a chemical comes into contact with an organism, e.g., inhalation, ingestion, dermal contact, and injection

RU486 — drug cocktail that functions as an abortifacient

safe exposure level — the level of exposure that will not result in a health hazard

safety factor — a factor that presumably reflects the uncertainties inherent in the process of extrapolating data about toxic exposures, i.e., intraspecies and interspecies variations. With this approach, an allowable human exposure to a compound can be determined by dividing the NOEL established in chronic animal toxicity studies by some safety factor.

salt — the compound formed with the hydrogen of an acid is replaced by metal, or its equivalent, e.g., ammonium, NH_4. Generally, the reaction of an acid and a base yields a salt and water; salts tend to ionize, dissolve, in water solutions.

sarcomas — malignant tumors of the connective tissue

sarin — an extremely toxic nerve gas composed of the isopropyl ester of methylphosphonofloride acid

saturation — the degree to which receptors are occupied by a messenger. If all are occupied, the receptors are fully saturated; if half are occupied, the saturation is 50%, and so on.

saturnine gout — a disease with gout-like symptoms produced by lead poisoning

SCBA — self-contained breathing apparatus; a respiratory protection device that consists of a supply of oxygen, or oxygen-generating material, carried by the wearer

Schwann cell — a neuroglial cell forming the myelin sheaths around peripheral nerve cell processes

sclera — outer tunic of the eye forming the cornea anteriorly and the white of the eye posteriorly

sebaceous gland — a gland in the integument formed by the epidermis, invaginating into the dermis, usually in conjunction with a hair follicle, that secretes sebum (oil)

secondary pollutant — a harmful chemical formed in the atmosphere when a primary air pollutant reacts with normal air components or with other air pollutants

Second Law of Thermodynamics — in any conversion of heat energy to useful work, some of the initial energy input is always degraded to lower quality, more dispersed (higher entropy), less useful energy, usually low-temperature heat that flows into the environment

secretion — the passage of material formed by a cell from the inside to the outside

sedimentary rock — a rock that forms from deposition of eroded materials and in some cases from the compacted shells, skeletons, or other remains of dead organisms

selection bias — occurs when individuals assigned to a group are not representative of the group being studied; therefore, there may be a weaker or stronger association identified than actually exists

selectively permeable — the property of cell membranes that selectively allows certain molecules to pass while others cannot

sensitization — a condition of being made sensitive to a specific substance such as protein or pollen, antigen or hapten, usually as a result of repeated exposure; activation of an immune response

sevin — an insecticide (1-naphthyl--N-methyl carbamate) used to control fleas on cats and dogs; also used in mosquito control programs

shale oil — a slow-flowing, dark brown, heavy oil obtained when kerogen (solid, waxy mixture of hydrocarbons) in oil shale (a rock formation) is vaporized at high temperatures and then condensed

sharps — the medical articles that may cause punctures or cuts; includes all broken medical glassware, syringes, needles, scalpel blades, suture needles, and disposal razors

Sick building syndrome (SBS) — an illness caused by the emission of odors, vapors, or gases from the materials used to construct buildings or its furnishings; symptoms include irritation of eyes, throat, and nose; dry skin; dizziness and nausea; lack of concentration and fatigue

Sievert — an international unit that is used for measuring the biologic dose of ionizing radiation. Like the rem, the sievert takes into account the relative biological effectiveness of each form of ionizing radiation on living tissue. A sievert is roughly equivalent in RBE to 1 Gray or 100 rads of gamma radiation. One millisievert equals 10 ergs of gamma radiation energy transferred to 1 gram of living tissue.

silicosis — a lung disease caused by inhalation of dust-containing silicon dioxide. Damage to the alveoli results in cell death; dead alveoli are replaced with connective tissue in localized areas, nodules, unable to exchange gases.

sink — a place in the environment where a compound or material collects or flows to or towards; opposite of source

sinoatrial node — a mass of specialized cardiac muscle cells located in the right atrium. Self-excitable, they act as the "pacemaker" of the cardiac conduction system.

skin absorption — the ability of some hazardous chemicals to pass directly through the skin and enter the bloodstream

smog — air pollution associated with smoke and fog; excess criteria air pollutants in the atmosphere

smoke — particles suspended in air after incomplete combustion of materials containing carbon

sodium-potassium pump — an active transport mechanism associate with the cell membrane that is responsible for maintaining potassium concentrations inside of the cell while decreasing the internal sodium concentration

soil — complex mixture of inorganic minerals (clay, silt, pebbles, and sand), decaying organic matter, water, air, and living organisms

solubility — the degree to which a chemical can dissolve in a solvent, forming a solution

solute — a dissolved substance

solution — a mixture in which the components are uniformly dispersed. All solutions consist of some kind of a solvent (such as water or other liquid) which dissolves another substance, usually a solid.

solvent — a substance, usually a liquid, into which another substance (solute) is dissolved

somatic — pertaining to body cells, as opposed to reproductive cells

sorption — a surface phenomenon which may be either absorption or adsorption, or a combination of the two; often used when the specific mechanism is unknown

source — the origin of a chemical or substance or where it flows from or away

species — a group of organisms that resemble one another in appearance, behavior, chemical makeup and processes, and genetic structure

specific gravity — the ratio of the weight of a volume of the product to the weight of an equal volume of water

specificity — selectivity; the ability of a receptor to react with only one type or a limited number of structurally related types of molecule

stability — an expression of the ability of a material to remain unchanged under expected and reasonable conditions of storage and use

standardized incidence ratio — the rate of incidence of disease in the worker group compared to that rate in the reference group

standardized mortality ratio — the rate of mortality in the worker group due to a specified cause compared to the rate for the same in the reference group

statistical power — the probability that one can detect an effect if there really is one

statistical significance — the probability of obtaining a result as extreme or more extreme as that observed even if the null hypothesis is true

STEL — Short-Term Exposure Limit; a term used by ACGIH to indicate the maximum average concentration allowed for a continuous 15 minute exposure period

stereoselectivity — when a molecule has a pair of isomers (mirror-image molecules that are alike in many ways, such as melting and boiling points and solubility, yet are different) that are present in the mixture in equal amounts (a racemic mixture), then metabolism occurs at different rates for each isomer, while one may appear largely unchanged and still be foreign when found in the urine, the other will appear as a fully transformed metabolite

stochastic — based on the assumption that the actions of a chemical substance results from probabilistic events

stomatitis — an inflammation of the mucous membrane of the mouth

stratification — the division of a population into subpopulations for sampling purposes; the separation of environmental media into layers, as in lakes

subatomic particle — extremely small particles (electrons, protons, and neutrons) that make up the internal structure of atoms

subchronic — of intermediate duration, usually used to describe studies or levels of exposure between five and 90 days

sublethal — pertaining to a dose level that is less than an amount necessary to cause death

sublimation — going directly from the solid to the gaseous state, e.g., dry ice

substrate — a chemical that serves of the substance acted upon by an enzyme

sulfation — a Phase II metabolic reaction in which a sulphate group is reacted with a toxic molecule; the end product is highly water soluble

Superfund sites — contaminated sites, the cleanup of which was authorized by CERCLA

supersensitivity — the increased responsiveness of a target cell to a given messenger, resulting from up-regulation

survival analysis — charts the survival of the group with the risk factor and the group without it to determine if the survivals differ

synapse — the junction between a nerve cell and another neuron, motor, or neural effector cell

synergistic effect — a biological response to dosage of multiple substances in which the net effects are greater than the sum of the effects of the individual agents

synthetic chemicals — chemicals that people make, rather than those that nature makes; they tend to be long-lived in the environment

Synthetic reaction — a chemical action in which larger molecules are formed from simpler ones

systemic effects — those effects which are due to absorption and distribution of an agent into and throughout the body. They impact one or more of the ten basic systems of which humans are composed.

Tabes Dorsalis — a disease of the spinal cord associated with syphilis; symptoms include staggering walk, pain, and paresthesia

tabum — an extremely toxic nerve gas containing the ethyl ester of dimehtylphysphoramidocyanidic acid

tachycardia — an excessively increased heartbeat, greater than 140 bpm

tachypnea — an excessively rapid respiration rate

tamoxifen — drug used for treating breast and uterine cancers by blocking estrogen receptors. Side effects include blood clotting, hot flashes, weight gain, vaginal discharge to dryness, and the promotion of bone growth.

target tissue — the tissue in which a toxon exerts its effects

technical pesticide — highly concentrated pesticide, which is to combine with other materials to formulate pesticide products

teratogen — a chemical or physical agent which can lead to malformations in the fetus and birth defects in children (liveborn offspring). Such an agent is called teratogenic.

teratogenesis — the induction of structural or functional development abnormalities by exogenous factors acting during gestation; interference with normal embryonic development

teratogenicity — the ability of a physical or chemical agent to cause birth defects, nonhereditary congenital malformations, in offspring

tetraethyl pyrophosphate — an organophosphate pesticide used for aphids, mites, and as a rodenticide

tetrodotoxin — a nerve poison found in eggs of the California newt and certain puffer fish

thalidomide — a drug initially used to treat morning sickness, the nausea associated with pregnancy. It causes severe birth defects, especially associated with limb formation.

therapeutic index — the ratio of the dose required to produce toxic or lethal effects to dose required to produce nonadverse or therapeutic responses

threshold — the lowest dose of a chemical at which a specific measurable (observable) effect occurs

tight building syndrome — an illness caused by reduced ventilation in well insulated buildings; symptoms similar to those associated with sick building syndrome

time series — a set of statistics that are compiled and reported over a period of time

Time-Weighted Average (TWA) — the average value of a parameter (e.g., concentration of a chemical in air) that varies over time

tissue — a group of cells similar in structure, function, location, and embryonic origin

TLV — Threshold Limit Value, an exposure limit recommended by the ACGIH (See *exposure limits*)

toxaphene — a waxy white chlorinated camphene used as an organochlorine pesticide

toxemia — a general intoxication sometimes due to the absorption of bacterial (toxic) products formed at a local source of infection

toxicant — a harmful substance or agent that may injure an exposed organism

toxicity — the quality or degree of being poisonous or harmful to plant, animal, or human life; the degree of danger posed by a substance to life

toxicity assessment — characterization of the toxicological properties and effects of a chemical, including all aspects of its absorption, metabolism, excretion and mechanism of action, with special emphasis on establishment of dose-response characteristics *toxicology* — the study of the adverse effects of substances on biological systems; the sum of what is known regarding poisons; the scientific study of poisons, their actions, their detection, and the treatment of the conditions produced by them

toxicosis — any disease condition due to poisoning

toxic shock syndrome (TSS) — a rare and sometimes fatal disease caused by a toxin or toxins produced by certain strains of the bacterium *Staphylococcus aureus*

toxic substance — any substance that can cause acute or chronic injury to the human body or that is suspected of being able to cause diseases or injury under some conditions

toxin — from the Greek (*of or for the bow*) a poison; a substance toxic (harmful or injurious, capable of killing) to life; frequently used to refer specifically to a protein produced by some higher plants, certain animals, and pathogenic bacteria, which is highly toxic for other living organisms. Such substances are differentiated from the simple chemical poisons and the vegetable alkaloids by their high molecular weight and antigenicity.

toxinemia — poisoning of the blood

toxon — a toxic substance within a living organism which will react with or at some specific target (site, enzyme, or cell). This reaction induces a toxic effect.

toxoplasmosis — a disease due to infection with the protozoa *Toxoplasm gondii*; symptom include muscle pain, fever, pneumonitis, hepatitis, and encephalitis; sometimes referred to as "Cat Scratch Fever"

trace element — an element essential to plant and animal nutrition in trace amounts, 1,000 ppm or less

trade name — the trademark name or commercial name given to material by the manufacturer or supplier

trade secret — any confidential formula pattern, process, device, information, or compilation of information that is used in an employer's business, and that gives the employer an opportunity to obtain an advantage over competitors who do not know about or use it

transfer RNA — a type of ribonucleic acid that attaches a specific amino acid and transports it to the ribosome to incorporate into a chain of amino acids during the process of protein synthesis

transferrin — a beta globulin in blood plasma capable of transporting dietary iron to the liver, spleen, and bone marrow

transformation — the chemical alteration of a compound by breakdown into component elements, conversion into other molecules, or reactions with other compounds; acquisition by a cell of the property of uncontrolled growth

transport — movement of a substance along a pathway

Tri-Ortho-Cresyl Phosphate — used as a plasticizer for polyvinyl rubber, vinyl plastics, and polystyrene; targets tissues of the nervous system

tumor — an abnormal mass of tissue, the growth of which exceeds and is uncoordinated with that of normal tissue; initially benign, but which may progress to malignancy

TWA — Time Weighted Average; the average concentration of a chemical in air over the total exposure time, usually an 8-hour work day

UEL — Upper Explosive Limit (See *explosive limits*)

ulcer — an injury of the skin or of a mucous membrane, such as the lining of the stomach, that is accompanied by formation of pus and necrosis of surrounding tissue

ultraviolet radiation — electromagnetic radiation with a wavelength between 10 and 390 nm; a nonionizing form of radiation used for microscopy and sterilization; portion of sun's spectrum responsible for sun burn and promotion of skin cancer

uncertainty factor — a number (equal to or greater than one) used to divide NOAEL or LOAEL values derived from measurements in animals or small groups of humans, in order to estimate a NOAEL value for the whole human population

unction — an ointment; the application of an ointment or salve

unstable — materials or substances which are capable of undergoing rapid chemical change or decomposition *up-regulation* — an increase in the total number of target-cell receptors for a given messenger in response to a chronic low extracellular concentration of the messenger

uric acid — a crystallizable acid occurring as an end product of purine metabolism; normally excreted by the body. Decreased excretion is associated with kidney disease and lead poisoning; cause of gouty arthritis.

vapor — gaseous form of a substance that is primarily a liquid or solid at standard pressure and temperature

vapor pressure — a measure of the tendency of a liquid to evaporate and become a gas; the pressure exerted by a saturated vapor above its own liquid in a closed container at given conditions of temperature and pressure, usually expressed in mm Hg; the higher the vapor pressure, the greater the tendency of the substance to evaporate (See *evaporation*, *mm Hg*, and *volatility*)

vector — an object, organism, or thing that transmits disease from one host to another

ventricular fibrillation — arrhythmia characterized by fibrillary contractions of the ventricular muscle owing to rapid repetitive excitation of myocardial fibers without coordinated contraction of the ventricle

vertigo — a sensation of irregular or whirling motion, either of oneself or of external objects

vesicant — an agent that induces blistering, also called blister gas, e.g., mustard gas and lewisite

vesicle — a small sac inside of the cell containing a liquid or gas

vital capacity — the volume of air that can be expelled from the lungs following full inhalation

vitreous humor — a transparent, jellylike substance located between the lens and the retina of the eye; aids the eye in maintaining its shape

volatility — a measure of how quickly a substance forms vapors at ordinary temperatures. The more volatile the substance is, the faster it evaporates, and the higher the concentrations of vapor (gas) in the air.

waste disposal methods — proper disposal methods for contaminated material, recovered liquids or solids, and their containers

water pollution — the introduction into water of a substance that is not normally present (or not normally present in such a high concentration) and that is potentially toxic or otherwise undesirable

water reactive — a chemical that reacts with water to release a gas that is either flammable or presents a health hazard; also denotes danger when wet

wettable powder — a finely ground pesticide dust that will mix with water to form a suspension for application. Although this formulation may not burn, it may release toxic fumes upon exposure to elevated temperatures, i.e., fire.

wood preservative — pesticide used to treat any wood product to prevent damage by pests or dry rot

wristdrop — a condition in which the hand is flexed at the wrist and cannot be extended due to median nerve injury or paralysis of the extensor muscles of wrist and hand; frequently due to impact at the base of the palm while the hand is hyperextended

xenobiotic — a chemical foreign to the biologic system; i.e., chemicals that are not normal endogenous compounds for the biologic system

x ray — extremely high-energy, short wavelength, <12 nm, electromagnetic radiation emitted as the result of the electron transitions in the inner orbits of atoms due to electron bombardment in x-ray tubes or by the shaking of electrons in synchrotrons and particle accelerators

Sample Problem Answers

(See Chapter 12)

Example Six

First find the number of hours for each sampling period, leaving the minutes for later.

1. For example, 8:01 am to 11:20 am is 3 hours (8:00 am to 11:00 am).
   There are 7 hours between 8:10 am and 3:45 pm (8:00 am to 3:00 pm).

2. Multiply the number of hours by 60 to convert to minutes.

3. Find the difference in minutes by subtracting the start minutes from the stop minutes. For example, 8:01 am to 11:20 am is 20 minutes, minus 1 minute or + 19 minutes. Then, 8:10 am to 3:45 pm is 45 minutes, minus 10 minutes, or +35 minutes.

4. Finally, add the totals (hours and minutes) to get the total time in minutes.

Sample Period 1	8:00 am to 11:00 am	=
	3 hours × 60	= 180 minutes
	20 min. - 1 min.	= +19 minutes
		= 199 minutes
Sample Period 2	12:00 pm to 4:00 pm	=
	4 hours × 60	= 240 minutes
	34 min. – 15 min.	= +19 minutes
		= 259 minutes

Total Sampling Time = Sample Period 1 + Sampling Period 2
= 199 minutes + 259 minutes
= 458 minutes

Sample Period 1

7:45 am	Start pump
3:05 pm	Stop pump, end sample

7:00 am to 3:00 pm	=	8 hours × 60
	=	480 minutes
05 min. – 45 min.	=	– 40 minutes
	=	+440 minutes

Note that the minus sign is OK; remember to subtract the start minutes from the stop minutes.

Example Seven

2.0 lpm × 480 min. = 960.0 liters min.

Example Ten

Calculate the air volume for each sample by multiplying the flow rate by the time.

Air Volume (m³) 0.194 0.234 0.228 0.202

Calculate the concentration by dividing the weight of the sample by the air volume for each sampling period.

Concentration (mg/m³) 12.33 18.64 21.56 12.32

Determine the time-weighted average (TWA) or equivalent exposure:

$$TWA = \frac{(C1 \times T1) + (C2 \times T2) + (C3 \times T3) + (C4 \times T4)}{Total\ Time}$$

$$= \frac{(12.33 \times 97) + (18.64 \times 117) + (21.56 \times 114) + (12.32 \times 101)}{429}$$

$$= \frac{(1196.01 + 2180.88 + 2457.84 + 1244.32)}{429}$$

TWA = 14.75 mg/m³

Example Twelve

a. What is the respirable dust concentration?

Volume = 1.7 lpm x 480 min. = 816 l = 0.816 m³

$$Concentration = \frac{1.602\ mg}{0.816\ m^3} = 1.96\ mg/m^3$$

b. What is the PEL for respirable dust from this sample?

$$PEL = \frac{(10\ mg/m^3)}{\%\ quartz + 2\ (\%\ cristobalite) + 2(\%\ tridymite)}$$

$$= \frac{10\ m/m^3}{17 + 2(8) + 2(0)} = \frac{10\ m/m^3}{17 + 16}$$

$$= \frac{10\ mg/m^3}{33}$$

PEL = 0.303 mg/m³

The actual concentration of respirable dust (1.96 mg/m³) is above the PEL for respirable dust containing 17% quartz and 8% cristobalite.

Example Thirteen

a. What is the time-weighted average concentration, assuming zero exposure during the unsampled time?

Sample Period 1

1.7 lpm × 220 min. = 374 l = 0.374 m³

$$C\ 1 = \frac{2.198\ mg}{0.374\ m^3} = 5.87\ mg/m^3$$

Sample Period 2

1.7 lpm × 241 min. = 410 l = 0.410 m³

$$C\ 2 = \frac{2.404\ mg}{0.410\ m^3} = 5.86\ mg/m^3$$

Sample Period 3

Time: 19 minutes

Concentrations: 0

$$TWA = \frac{(C1 \times T1) + (C2 \times T2) + (C3 \times T3)}{T1 + T2 + T3}$$

$$TWA = \frac{(5.87\ 220) + (5.86\ 241) + (0\ 19)}{220 + 241 + 19} = \frac{1291.4 + 1413.1}{480}$$

$$TWA = \frac{2704\ mg/m^3}{480\ min.}$$

$$TWA = 5.63\ mg/m^3$$

b. What is the PEL for respirable dust containing quartz and cristobalite for this sample?

Filter No:	1	2
	2.198 mg	2.404 mg
	× 0.14	× 0.20
Weight of quartz:	0.308 mg	0.481 mg
	2.198 mg	2.404 mg
	× 0.05	× 0.11
Weight of cristobalite:	0.110 mg	0.264 mg

Percent Quartz in entire sample

$$= \frac{0.308\ mg + 0.481\ mg}{2.198\ mg + 2.404\ mg} = \frac{0.789\ mg}{4.602\ mg} = 0.171$$

$$= 17.1\ \%$$

Percent Cristobalite in entire sample

$$= \frac{0.110 \text{ mg} + 0.264 \text{ mg}}{2.198 \text{ mg} + 2.404 \text{ mg}} = \frac{0.374 \text{ mg}}{4.602 \text{ mg}} = 0.081$$

$$= 8.1\%$$

$$\text{PEL} = \frac{10 \text{ mg/m}^3}{17.1\% + 2(8.1\%) + 2(0\%)} = \frac{10 \text{ mg/m}^3}{33.3\%} = 30.03 \text{ mg/m}^3$$

Bibliography

Amdur, Mary O. *Casarett and Doull's Toxicology: The Basic Science of Poison.* New York: McGraw-Hill, 1991.

Ariens, E.J. and A.M. Simonis. *Introduction to General Toxicology.* New York: Academic Press, 1978.

Baselt, Randall C. *Disposition of Toxic Drugs and Chemicals in Man.* St. Louis, MO: Mosby-Year Book, 1989.

Berkow, Robert, editor. *Merck Manual. General Medicine.* Vol. 1. 16th edition. Rahway, NJ: Merck & Company, Inc. 1992

Budavari, Susan, editor. *Merck Index of Chemicals and Biologicals.* 11th edition. Rahway, NJ: Merck & Company, 1989.

Cohen, Gerald M. *Target Organ Toxicity.* Boca Raton, FL: CRC Press, 1986.

Colton, C.E., L.R. Birkner, and L.M. Brosseau. *Respiratory Protection: A Manual and Guideline.* 2nd edition. Fairfax, VA: American Industrial Hygiene Association, 1991.

Dorland, Newman W. *Dorland Illustrated Medical Dictionary.* Philadelphia, PA: W.B. Saunders Company, 1994.

Dreisbach, Robert H. and William O. Robertson. *Handbook of Poisoning.* East Norwalk, CT: Appleton and Lange, 1987.

Duffus, J. H. and Edward Arnold. *Environmental Toxicology.* 1980.

Eckert, Charles. *Emergency Room Care.* 4th edition. Boston, MA: Little, Brown & Company, 1981.

Emergency Response Guidebook, 1990: Guidebook for First Response to Hazardous Materials Incidents. Washington, DC: United States Printing Office, 1990.

Environmental Protection Agency Staff. *Toxicology Handbook.* Rockville, MD: Government Institutes, Inc., 1986.

Friedman, Gary D. *Primer of Epidemiology.* 4th edition. New York: McGraw-Hill, 1988.

Albert P. Li, editor. *Genetic Toxicology.* Boca Raton, FL: CRC Press, 1991.

Gilman, Alfred G. *Goodman and Gilman's The Pharmacological Basis of Therapeutics.* 8th edition. New York: McGraw-Hill, 1990.

Hallerbeck, William H. *Quantitative Risk Assessment for Environmental and Occupational Health.* Boca Raton, FL: Lewis Publishers, 1986.

Hawkins, N., S. Norwood, and J.C. Rock, editors. *A Strategy for Occupational Exposure Assessment*. Fairfax, VA: American Industrial Hygiene Association, 1991.

Hayes, A. Wallace, editor. *Principles and Methods of Toxicology*. New York: Raven Press, Limited, 1994.

Karvonen, M. and M.L. Mikheev, editors. *Epidemiology of Occupational Health*. Albany, NY: World Health Organization, 1986.

Kirk, H. Ray. *OSHA Compliance Guide, 1992-1994: General Safety Standards, Sample Programs and Forms*. Washington, DC: United States Government Printing Office, 1994.

Lu, Frank C. *Basic Toxicology: Fundamentals, Target Organs, and Risk Assessment*. 2nd edition. Bristol, PA: Hemisphere Publishing Corporation, 1990.

NIOSH Pocket Guide to Chemical Hazards. Washington, DC: United States Government Printing Office, 1993.

Clayton, George D. and Florence E. Clayton, editors. *Patty's Industrial Hygiene and Toxicology*. 4th edition. New York: John Wiley & Sons, Inc., 1994.

Physician's Desk Reference. 49th edition. Montvale, NJ: Medical Economics Data, Inc., 1995.

Plog, Barbara A., editor. *Fundamentals of Industrial Hygiene*. 3rd edition. Itasca, IL: National Safety Council, 1988.

Raw, G. J. *Sick Building Syndrome: A Review of the Evidence on Causes and Solutions 1992*. Lanham, MD: UNIPUB, 1992.

Scialli, Anthony R. *A Clinical Guide to Reproductive and Developmental Toxicology*. Boca Raton, FL: CRC Press, Inc., 1991.

Siemiatycki, Jack. *Risk Factors for Cancer in the Workplace*. Boca Raton, FL: CRC Press, Inc., 1991.

Sittig, Marshall. *Handbook of Toxic and Hazardous Chemicals and Carcinogens*. 3rd edition. Park Ridge, NJ: Noyes Press, 1992.

Spiro, T.G. and W. M. Stigliana. *Environmental Issues in Chemical Perspective*. State University of New York, 1980.

Talty, John T., editor. *Industrial Hygiene Engineering: Recognition, Measurement, Evaluation and Control*. 2nd edition. Park Ridge, NJ: Noyes Data Corporation, 1989.

Turnbull, G. J. *Occupational Hazards of Pesticide Use*. Washington, DC: Taylor & Francis, Inc., 1985.

Williams, Phillip L. and James L. Burson, editors. *Industrial Toxicology*. New York: Van Nostrand Reinhold, 1985.

Wyngaarden, James. *Cecil Textbook of Medicine*. 19th edition. Philadelphia, PA: W. B. Saunders Company, 1991.

Index

B

Bacillus, 116, **238**
Background, 239
Bacteria, 239,
Bactericidal, 177
Baghouse, 192, **239**
BAI, 240
BAL, 239
Barbecues, 98
Barbiturate, 33, 72, 81
Barium, 79
Baroreceptor, 239
Basal cell carcinoma, 239
Base, 239
Baseline, 200
Base excision repair, 239
Base-pair, 88
Basophil, 239
Batrachotoxin, 71
Battery, 98,141, 165
Battery-powered, 165
Baygon, 105, 210
Beard, 170
Beaver, 112
Becquerel, 111
Bee, 63
BEI, 4, 51, 202, 203, 205, **239**
Belladonna, 72
Bellows-type, 138
Benign, 29, 88, 89, 113, **239,**
Benzanthracene, 89
Benzene, 4, 9, 19, 32, 43, 75, 91,
    93, 102, 185, 208, 225
Benzidine, 90
Benzoates, 82
Benzpyrene, 87
Bergamot, 64
Bernadina, 3
Bernard, 3
Beryllium, 65, 75, 76, 97, 98, 209,
    239
Beta particle, 239
BHA, 82
Bhopal, 4, 220
BHT, 82
Bias, 239
Bilayer, 15, 27, 30
Bile, 18, **239**
Biliary, 90
Bilirubin, 31, 61, 77, 211, **239**
Bioaccumulation, 29, 35, 37, 92,
    239
Bioactivation, 31
Bioassay, 239

Biochemical, 6, 58, 59, 60, 89, **239**
Biocides, 104, 105
Bioconcentration, 239
Biodegradable, 239
Biodegradation, 239
Biohazardous, 115, 117, 118
Biomagnification, 239
Biopsy, 73
Biosynthetic, 239
Biotechnology, 117
Biotransformation, 14, 31, 32, **239**
Biphenyl, 4, 35, 65, 211
Bismuth, 97
Bithionol, 64
Bitter-almonds, 100
Bladder, 16, 18, 89, 90, 105
Blancophores, 64
Blastgate, 193, 194
BLEVE, 239
Blindness, 31, 59
Blister,64, 65
Block, 23, 58, 64, 71, 72, 80, 114,
    211
Blood, 5, 239
Blood-brain, 29, 71, **238, 239**
Blood-forming, 30, 209, 212
Blood-gas, 22, 32
Blotchy, 64
Blow down valve, 239
Blower, 165, 171, 172, 189
Blueskying, 129
Boiling point, 239
Bolus, 82
Borgia, 3
Botulinum, 62, 71, 116, **240**
Bowman's Capsule, 240
Bq, 111
Bradycardia, 240
Brain, 8, 29, 71, 73, 99, 101, 105,
    209
Breast, 105
Breathable, 42, 160, 164, 171
British Thermal Unit (BTU), 240
Broadcast applications, 240
Bromacetone, 65
Bromacetophenone, 65
Bromide, 8, 209
Bromine, 24
Bromobenzene, 65, 75, 81
Bromoform, 102
Bromosulfophthalein, 77
Bronchi, 22, 65, 101, 104, **240**
Bronchiole, 240
Bronchitis, 65, 104, **240**
Bronchogenic, 89

Bronchopneumonia, 104, **240**
Brownian movement, 22, 23
Brucella, 240
Brucellosis, 117, **240**
BTU, 240
Bubo, 240
Bubonic, 240
Building illness, 240
Bulbourethral, 240
BUN, 211
Burns, 11, 18, 64, 138, 195, 201
Butadiene, 102
Butane, 101
Butazone, 75
Butylamine, 104
Butylparabens, 82
Butyric, 72
Byssinosis, 240

C

CaBP, 240
Cadmium, 79, 81, 89, 93, 97, 98,
    209, 210
Caffeine, 10, 66, 72
Calcium, 18, 29, 30, 35, 72, 81,
    101, 209
Calcium-binding, 240
Calibration, 111, 130, 138, 139,
    140, 141, 146
Calomel, 99
CAL-OSHA, 51, 184, **240**
Cancer, 240
Cannula, 240
Capillary, 29, 204
Carbamate, 72, 94, 105, 210
Carbaryl, 72, **240**
Carbohydrate, 240
Carbon dioxide, 240
Carbon monoxide, 240
Carbon tetrachloride, 241
Carbonate, 240
Carbonic acid, 241
Carbonyl, 48, 65, 86
Carbon-black, 145
Carboxyhemoglobin, 60, 115, 209,
    212, **241**
Carboxyl, 48, 49, 86
Carboxylic, 104
Carcinogenic, 9, 18, 42, 43, 44, 54,
    55, 89, 90, 91, 92, 104, 112,
    211, 218, 219, 222, **241**
Carcinomas, 89, 90, 91, **241**
Cardiopulmonary, 209

Dinoflagellates, 71
Dinucleotide, 59
Diolefins, 102
Dioxin, 105, 106, 106, 210, **245**
Diquat, 59, **245**
Disease, 245
Disposal, 14, 110, 119, 120, 121, 135, 192, 193
Disprove, 225, 228
Distal convoluted tubule, 245
Distribution, 14, 18, 27, 28, 48, **245**
Disulfide, 73, 100
Disulfiram, 59
Ditches, 126
Dithiocarbamate, 59
Dithiocarbamates, 59
Diuresis, 79
Diuretic, 33, 34
Dizziness, 50, 60, 100, 101, 103, 115, 210, 211
DMPS, 245
DNA, 6, 9, 49, 58, 62, 63, 85, 86, 87, 88, 89
Dogs, 8, 54
Dopamine, 245
Doping, 33
Dosages, 64, 66
Dose, 245
Dose-response, 41, 43, 45, 89, 94, 203, 216, 218, 219, 228, **245**
Dose-effect, 91, 223
Dose-time, 41
DOSH, 51
Dosimeter, 245
DOT, 118
Double-blind, 227
Downdraft, 191
Downstream, 167, 190
Downwind, 136
Down-regulation, 245
Drowning, 24, 100
Drowsiness, 55
Drunk, 26, 101
Ductwork, 188, 189, 190, 191, 192, 194
Dullness, 73
Dust, 245
Dyscrasias, 75
Dyspnea, 98, 100, 241, **245**
Dysrhythmia, 245
D-tubocuraine, 245

E

Ebers, 2
Ecology, 7, 235 , **245**
Economic, 189, 207, 217, 220
Ecosystem, 6, 245
Ecotoxicological, 6 , **245**
Ectodermal, 89
Eczema, 63, 64, **246**
Eczematous, 98
Edema, 65, 81, 98, 99, 100, 101, 102, 103, 104, **246**
Effective dose, 246
EDTA, 247
EEGLs, 53
Effector, 57, 58, 61, 66, 69
Efferent neurons, 246
Eggs, 8, 92, 100
EHA, 53
Eight-hour, 38, 142
EKG, 238
Elastic, 161, 201
Electrocardiogram, 205
Electrochemical, 138
Electroencephalograms, 57
Electrolytes, 18, 32, 33, 65, 246
Electromagnetic, 109, 126, 246
Electromyograms, 57
Electron, 111
Electroplating, 98
Electrostatic, 111, 140, 192
Elimination, 16, 30, 31, 50, 114, 127, 185, 203, 204, 246
Embryo, 2, 54, 94, **246**
Embryotoxicity, 88
Emergencies, 1, 53, 139, 168, 169
Emergency, 246
Emetic, 246
Emphysema, 65, 98, 106, 246
Emphysematous, 100
Empirical, 246
Emulsificant, 15
Emulsion, 246
Encapsulate, 8, 161, 179, 180
Encephalopathy, 246
Endangerment assessment, 246
Endemic, 246
Endocrine, 61, 62, 94
Endoplasmic, 76, **246**
Endothermic, 246
Endpoint, 246
Endrin, 105, 210
Enlargement, 209
Ensemble, 179, 180, 206, **246**
Enterocolitis, 246

Enterohepatic, 246
EPA Registration number, 246
Enzyme, 246
Eosinophils, 246
EPA, 53, 105, 106, 112, 114, 118, 132, 218, **246**
Ependymal, 71
Epichlorohydrin, 103
Epidemic, 99, 127
Epidemiology, 3, 127, 215, 225, 229, **246**
Epidermal, 91
Epidermis, 20, **246**
Epigenesis, 247
Epigenetic, 89, **247**
Epilation, 247
Epilepsy, 73, **247**
Epinephrine, 66, **247**
Epistemology, 247
Epithelium, 89, 100, 112, **247**
Epoxy, 89, 103
Equilibrium, 16, 30, 34, 210, **247**
Erethism, 247
Ergonomic, 126
Ergs, 111
ERPG, 53
Erythremia, 247
Erythrocyte, 60, 61, 73, 74, 204, 212, **247**
Erythropoiesis, 74
Escape-impairing, 53
Escherichia, 247
Esophageal, 247
Esophagus, 16, **247**
ESP, 192
Ester, 89, 104, **247**
Esterase, 71
Estrogen, 61, 62, 89, 113, **247**
Esu, 111
Ethane, 48, 86, 101, 102, 210
Ethanol, 29, 31, 59, 59, 86, 89, 102, 103,
Ether, 32, 61, 88, 103, **247**
Ethylamine, 104
Ethylbenzene, 102
Ethylene, 9, 73, 81, 102, 103, 209
Ethylene diaminetetraacetic acid (EDTA), 247
Ethyleneamine, 63
Etiology, 109, 112, **247**
Evaporation, 205, **247**
Evidence, 54, 115, 121, 129, 130, 173, 199, 206, 225
Excitability, 72

mV, 70
Mycobacterium, 117
Myelin, 71, 73, **255**
Myelotoxin, 255
Myocardium, 78, 79, 80, **255**
Myoglobin, 82
Myometrium, 255
Myopathy, 255

N

NAAQS, 255
NAD, 255
NADP, 59
NADPH, 59
Nail, 34, 35, 112, 157, 204
Nanocurie, 111
NaOH, 101
Naphthalene, 74, 102
Narcosis, 52, 61, 102, 103, 104, **255**
Narcotic, 61, 62, 99, 102
NAS, 53
Nasal, 99, 115, 184
Nausea, 50, 59, 60, 99, 100, 103, 105, 106, 115, 211, **255**
Neck, 161
Necropsy, 255
Necrosis, 14, 65, 73, 80, 81, **255**
Necrotic, 65, 76
Needles, 116, 117
Neighborhood, 226
Neighboring, 220
Nematocide, 104
Neonates, 60
Neoplasm, 54, 89, **255**
Neoplasms, 89
Neoplastic, 1
Neostigmine, 72
Nephritis, 82, 98, **255**
Nephron, 32, 81, 82, 240, 246, **255**
Nephrotoxic, 82
Nephrotoxicity, 81
Nephrotoxins, 81
Nero, 3
Nervous, 69, **256**
Neuritis, 73
Neuroglia, 29, 69, **256**
Neurohumoral, 61, 62, 71, 72
Neurological, 94, 113, 209
Neuromuscular, 61, 62, 71, 72, 105, 211
Neuron, 58, 61, 69, 70, 71, 72, **256**
Neuropathy, 98, 104

Neuropsychiatric, 98
Neurotoxic, 58
Neurotoxicity, 211
Neurotoxins, 71
Neurotransmission, 61, 62
Neurotransmitter, 58, 71, 72, **256**
Neuro-stimulation, 58
Neutralization, 256
Neutron, 110, **256**
Neutrophil, 256
NFPA (National Fire Protection Association), 53, **256**
NiCad, 98
Nickel, 79, 89, 93, 97
Nickel-cadmium, 98
Nickel-chromium, 189
Nicotine, 10, 31, 33, 59, 61
Nicotinic receptors, 61, **256**
NIOSH, 1, 8, 51, 53, 115, 132, 161, 167, 168, 171, **256**
Nitrates, 74, 80, 82
Nitric acid, 65, 99
Nitrites, 60, 74, 80
Nitrofurantoin, 60
Nitrogen, 23, 24, 99, 101, 188, 211
Nitrogen-containing, 60,
Nitrosamines, 18, 76, 89
Nitrosoureas, 89
Nitrous, 23, 65, 99
NOAEL (No Observable Adverse Effect Level), 41, **256**
Node, 76, 78, 79, 90
NOEL, 41, 219, 223, **256**
Noise, 2, 8, 126, 144
Noncarcinogenic, 219
Nonchemical, 109
Nondegradable pollutant, 256
Nondestructive, 140
Nonemergency, 200
Nonencapsulating, 179
Nonexposure, 144
Nongenotoxic, 113
Nonhazardous, 205
Nonhuman, 44
Nonincapacitating, 53
Noninfectious, 115, 117
Nonionizing, 109
Nonlethal, 14, 63
Nonmuscarinic, 259
Nonoccupational, 113, 202, 203
Nonorganic, 174
Nonpoint source, 256
Nonpolar, 32, 61
Nonprescription, 201
Nonrecorded, 201

Nonrenewable resource, 256
Nonreversible, 14
Nonroutine task, 256
Nonself, 63
Nonsmokers, 92, 112, 115, 227, 228
Nonspecific, 23, 54
Nonsteroid, 31
Nonsteroidal, 64
Nontechnical, 219
Nontoxic, 33, 90, 164, 193
Nonuniformly, 189
Nonverifiable, 141
Non-dusty, 192
Non-ionized, 32
Non-normal, 23
Noradrenaline (norepinephrine), 61, 66
Norepinephrine, 61, 66, **256**
Nose, 11, 22, 65, 130, 132, 161, 184
Nostrils, 161
Notebook, 140
Notes, 129, 130, 141
Nozzle, 188
No-Adverse-Response Level, 223
No-Observable-Effect Level, 223
NRC, 53, 118
NRDC, 224
NSAID (Non Steroid Anti Inflammatory Drug), 31, 64
Nuclear, 110, 118, 220
Nucleic acid, 6, 18, 58, 62, 85, 87, **256**
Nucleotide, 256
Nucleus, 62, 76, 78, 110, **256**
Nuisance, 140, 173
Numbness, 210, 211
Numerical, 216, 224
Nurse, 69, 117, 125, 127, 229
Nutrient, 15, 18, **256**
Nutritional, 18, 31
Nuts, 62

O

Obesity, 59
Objectives, 127, 128
Observation, 73, 127, 201, 223,
Observation bias, 256
Obstetrical, 116
Obstruction, 23, 81, 172, 195, 211
Occupation, 200, 201, 207
Occupational study, 256

Perforated, 161
Perfume, 64
Perfusion-limited, 34
Pericardium, 112
Peripheral, 61, 71, 112, 209
Peripheral Nervous System, 71, 257
Periphery, 71
Periportal, 75
Peritoneum, 112
Periwinkle, 82
Permeability, 71, 72, 79, 101
Permeate, 179
Permeation, 257
Permissible Exposure Levels, 257
Permits, 120
Permitting, 9, 117
Peroxide, 61
Persistent pesticide, 258
Person-hours, 178
Perspiration, 19
Pesticide, 93, 105, 105, 140, 216, 224, **258**
Pet, 105
Petechiae, 258
Petrochemicals, 258
Petroleum, 91, 101, 102
pH, 17, 18, 30, 32, 33, 33, 81, 212, **258**
Phagocyte, 16, 22, 24, 74, **258**
Pharmaceutical, 25, 47, 208
Pharmacodynamics, 48, 57, 203, **258**
Pharmacokinetics, 14, 25, 26, 30, 48, 203, **258**
Pharmacology, 1, 33, 42, 44, 101
Pharynx, 22
Phenazone, 60
Phenobarbital, 34
Phenol, 103
Phenotypic, 89
Phenoxybenzamine, 72
Phentolamine, 72
Phenylalanine, 238
Phenylbutazone, 91
Phenylhydrazine, 74
Phenyls, 65
Phillippus, 3
Philosybin, 62
Phocomelia, 94, **258**
Phorbol, 89
Phosgene, 24, 65, 100, 184, **258**
Phosphatase, 77, 211
Phosphate, 59, 104,
Phosphates, 49, 81, 86

Phosphide, 106
Phospholipid, 15, **258**
Phosphorous, 53, 76, 106
Phosphorylate, 106
Photoactivation, 91
Photoallergies, 64
Photochemical, 64
Photoionization, 139
Photon , 110, **258**
Photophobia, 258
Photoreaction, 64
Photosensitivity, 6, 64
Phototoxic, 64
Phthalic, 104
Phylogenetic, 258
Physiology, 6, **258**
Physostigmine, 72
Pica, 258
Picocurie, 111, 114
Picrotoxin, 72, **258**
Picture, 135
PID (Photo Ionization Detector, 139
Pigmentation, 64
Pinocytosis, 258
Pipettes, 117
Pitot tube, 194, **258**
Pitted, 221
pKa (ionization constant), 32, 81
Placebo, 227
Placenta, 31, 246, **258**
Plan, 118, 119, 129, 135, 136, 187, 200, 202
Planning, 53, 136
Plaque, 113
Plasma, 28, 29, 30, 32, 33, 71, **73,** 75, 81, 100, 204, 211
Plastic, 115, 118, 161
Plasticizers, 104
Platelet, 28, 74, 75, 208, 212, **258**
Plating, 98, 125
Platinum, 138
Pleated, 163
Plenum, 191, 193, **258**
Pleura, 112, 113
Plotting, 42
Plume, 136
Pneumoconiosis, 65, 89, 163, **258**
Pneumonia-like, 117
Pneumonitis, 98, 100, 101, 102, **258**
PNS, 71, 73
Poietins, 73
Point source, 258
Point-of-operation, 190

Poison, 2, 3, 8, 14, 19, 42, 61, 114, 126, 138, **258**
Polar, 15, 30, 140
Polarization, 71, 72
Pollutants, 7, 50, 51, 100, 115, 203
Pollution, 259
Polychlorinated Biphenyl (PCB), 4, 35, 65, 211, **259**
Polycyclic, 90, 91, 102
Polyethylene, 207
Polymer, 115
Polymerase, 62
Polymerization, 259
Polymers, 140
Polyneuritis, 259
Polynuclear, 102
Polyphenol, 102
Polypropylene, 103
Polytetrafluoroethylene, 184
Polyurethanes, 64
Polyuria, 81, **259**
Polyvalent, 62
Polyvinyl, 140
Pontus, 2
Population at risk, 259
Pore, 27, 140, 141, 163, 164, 179, 193
Porous polymers, 140
Portable, 139, 191
Portal, 75, 76
Positive pressure, 173
Postganglionic (See ANS, 238)
Postsynaptic, 71, 72
Potassium, 16, 32, 65, 70, 71, 72, 79, 81, 101, 164
Potency, 3, 43, 66
Potent, 43, 62, 66, 87, 103, 224
Potentials, 57, 79, 139
Potentiation, 62, 66, **259**
Pott, 4
Powder, 49, 183, 187
Powered respirators, 165, 166
ppb, 99, **259**
PPE, 8, 14, 20, 41, 48, 131, 136, 138, 157, 158, 164, 172, 177, 179, 185, 188, 202, 205, 206
ppm, 8, 10, 60, 99, 100, 100, 101, 106, 115, 128, 137, 170, 177, 212, **259**
Precautions. 64, 116, 124
Precipitate, 103, **259**
Precipitators, 192
Precursor, 259
Predesigned, 194
Pregnancy, 8, 9, 54, 94, 99, 203

Regurgitation, 18
Rejection, 82
Relative risk, 260
Relatives, 217
Relief, 206
rem, 111, **260**
Remediation, 47, 114
Remedy, 3, 64
Removal, 30, 59, 110, 113, 114, 177, 192, 201
Renal, 33, 34, 74, 81, 82, 102, 103, 116, **260**
Repair, 168, 176, 178
Replication, 87
Report, 4, 113, 124, 138, 139, 199, 201, 219
Reporters, 221
Reproduction, 115, 111, **260**
Reproductive, 6, 54, 55, 92, 93, 94
Reptiles, 2, 58
Rescues, 166
Reservoir, 260
Resin-impregnated, 163
Resistance, 53, 63, 98, 127, 162, 163, 172, 179, 192
Resorbed, 34, 81
Resorcinol, 103
Resorption, 33, 81
Respirable, 140, 144, 148, 149
Respiration, 60, 61, 72, 100, 106
Respirator, 260
Respiratory, 21
Respirator-inlet, 166
Responsibility, 124, 126, 127, 130, 170, 178
Resting potential, 261
Restroom, 144
Rests, 161
Retardation, 88, 99, 113
Reticulocyte, 212, **261**
Reticulum, 76
Retina, 20
Retroperitoneally, 80
Retrospective study, 261
Reversible effect, 261
Re-evaluated, 128
Re-uptake, 32, 62, 71
RfC, 260
RfD, 223, 260
Rhubarb, 81
Ribonucleic acid, 261
Ribose, 85
Ribosome, 62, **261**
Rifampicin, 62

Right-to-know, 53
Rinse, 177, 212, 213
Risk, 261
Risk analysis, 261
Risk assessment, 261
Risk-benefit analysis, 222, **261**
Risk estimate, 261
Risk-factor, 225, **261**
Risk management, 261
Risk specific dose, 261
RMV (Respiratory Minute Volume), 172
RNA, 49, 58, 62, 63, 85, 86, 94
Rodenticide, 63, 104, 106, **261**
Rodlike, 112
Rods, 20
ROE (Routs of Entry), 157
Roentgen, 111, 139
Roman, 2
Roughage, 34
Route of Exposure, 261
Routes, 2, 9, 34, 103, 104, 113, 157, 158, 218
RTECS (Registry of Toxic Effects of Chemical Substances), 132
RU 486, 261
Rubber, 7, 59, 161, 177
Rusty, 157

S

Safe exposure level, 261
Safeguards, 175
Safety factor, 261
Salicylanilides, 73
Salicylic acid, 33, 65
Saliva, 16, 105, 117, 211
Salmonella, 117
Salmonellosis, 117
Salt, 261
Sampling, 125
Sandalwood, 64
Sanding, 183
Sanitary, 226
Sanitation, 128, 188
Sanitized, 118
SARA, 53
Sarcomas, 89, **261**
Sarin, 261
Saturation, 28, 37, 49, 60, **262**
Saturnine gout, 262
Saunas, 34
Saxitoxin, 71
Scar, 22

SCBA, 160, 166, 167, 168, 171, 172, 176, 179, **262**
Scent, 174
Schizoid molecules, 30
Schwann, 71, 261
Scintillation, 139
Sclera, 262
Sclerosis, 73
Screening, 200, 205, 208
Scrotal, 4, 91, 94
Scrubber, 192
Seafood, 99
Sebaceous, 262
Secondary pollutant, 262
Second Law of Thermodynamics, 262
Secretion, 20, 209, **262**
Sediment, 212
Sedimentary rock, 262
Sedimentation, 22, 23
Seizures, 211
Selection bias, 262
Selectively permeable, 262
Selectivity, 28, 164
Selenium, 97
Self, 63
Self-contained, 160, 166, 167, 172
Self-determination, 206
Self-Test, 185
Seminoma, 94
Semipermeable, 16
Sensation, 11, 73, 100, 101, 164, 224
Senses, 126, 129, 137, 194
Sensitive, 54, 61, 64, 81, 82, 94, 99, 137, 138, 139
Sensitization, 102, **262**
Sensitize, 6, 14, 82 , 103, 115
Sensory, 20, 69, 73, 98, 100
Sensory-motor, 99
Separator, 190, 191, 192
Sequencing, 187
Sequester, 35
Sequestration, 65
Serum, 77, 204, 211
Sevin, 72, 105, **262**
Sex, 2, 92, 93, 203, 217, 225, 229
Sexual, 92, 217
Sexuality, 93
SF (Safety Factor), 223
SGOT (test), 211
SGPT (test), 211
SH (sulfhydryl), 58
Shaking, 192
Shale oil, 262

Succinylcholine, 72
Sugars, 10, 16, 18, 28, 29, 32, 48, 73, 81, 85, 86
Suit, 8, 161, 167, 171, 172, 179, 180
Sulfate, 89, 105
Sulfation, 263
Sulfhemoglobin, 60, 61
Sulfide, 8, 60, 72, 100, 184
Sulfites, 82
Sulfonamide, 31, 60, 81
Sulfur, 23, 65, 100, 101, 105, 184
Sulfuric, 101
Sulfur-containing, 60, 61, 100
Sunblocks, 64
Sunburn, 20, 64
Sunlight, 64
Sunscreen, 64
Sunstroke, 65
Suntan, 64
Superfund, 53, **263**
Supersensitivity, 263
Supine, 212
Supplied-air, 171, 172
Surfactants, 30, 61, 74
Survival analysis, 264
Swallowed, 10, 16, 22
Sweat, 18, 19, 34, 105, 106, 205, 208, 211
Swelling, 65, 209
Swollen, 64
Sympathetic nervous system, 71
Synapse, 70, 71, 72, **264**
Synergism, 9, 64, 82
Synergistic effect, 8, 62, 64, 66, 92, 208, **264**
Synovial, 30
Synthetic chemicals, 264
Synthetic reaction, 264
Syringes, 116, 117, 118
Systemic effects, 264
Systemic, 11, 13, 14, 15, 32, 33, 34

T

Tabes Dorsalis, 264
Tabum, 264
Tachycardia, 79, 106, **264**
Tachypnea, 264
Talc, 184
Tamoxifen, 264
Tampon, 225
Tanning, 98

Target, 2, 14, 25, 29, 30, 37, 42, 57, 66, **264**
Taste, 101, 164
Taster, 3
TB, 117
TBB (Total Body Burden), 25, 33, 35, 38, 39
TCDD, 210
Te, 60, 146
Tea, 33
Tear, 41, 65
Technical pesticide, 264
Teeth, 16, 161
Teflon, 184
TEM, 63
Temperature, 18, 45, 49, 99, 103, 126, 138, 164, 172, 177, 179, 180, 184, 185, 186, 203, 206, 208
Terata, 87, 94
Teratogen, 54, **264**
Teratogenesis, 54, **264**
Teratogenic, 6, 7, 9, 86, 94, 218, 264
Teratogenicity, 264
Teratogens, 54, 87, 94
Teratology, 9
Test, 3, 7, 41, 44, 53, 54, 55, 73, 116, 118, 140, 161, 170, 173, 174, 175, 176, 187, 195, 202, 205, 208, 209, 210, 211, 227, 228
Testicular, 93
Testis, 31
Tetanus, 201
Tetrachloride, 19, 65, 75, 76, 81, 102, 209
Tetrachlorodibenzo-p-dioxin, 210
Tetrachloroethane, 209
Tetrachloroethylene, 81, 89, 209
Tetrachlorsalicylanilide, 64
Tetracycline, 62
Tetraethyl, 73, **264**
Tetrodotoxin, 61, 71, **264**
Tetroxide, 99
Thalassemia, 60
Thalidomide, 4, 9, 94, **264**
Thallium, 73, 97
THC, 62
Theobromine, 72
Theophrastus, 3
Theophyline, 72
Therapeutic index, 264
Thermophilic, 63
Thermoregulatory, 45

Thiazides, 33
Thioacetamide, 76
Thiocyanates, 80
Thioguanine, 82
Thioridazine, 76
Thorium, 65
Threshold, 264
Threshold limit Value (TLV), 264
Throat, 22, 65, 99, 101, 164, 184
Throbbing, 60
Thrombocytes, 28, 73, 74
Thrombocytopenia, 73
Thymus, 82
Tidal volume, 23
Tightening, 173, 174
Tightness, 60, 177, 211
Tight building syndrome, 264
Tight-fitting, 161
Time-concentration, 26
Time-consuming, 176
Time-dose, 50
Time-lapsed, 136
Time-out, 131
Time series, 264
Time-weighted average, 51, 52, 149, 164, 187, **264**
Tin, 97
Tissue, 264
Titanium, 64
TLV, 4, 5, 48, 51, 52, 55, 60, 115, 128, 158, 170, 203, 205, 206, 212, **264**
TLV-C, 52
TLV-STEL, 52
TLV-TWA, 51, 52
Tn, 147
Tobacco, 24, 89, 91, 94, 200
Toenails, 98
Tolazoline, 72
Tolerance, 207
Toluene, 43, 63, 93, 102, 185, 208
Tongue, 16
Topical, 45
Toxaphene, 264
Toxemia, 264
Toxicant, 15, 43, **264**
Toxicity, 264
Toxicity assessment, 264
Toxicology, 265
Toxicosis, 265
Toxic shock syndrome (TSS), 265
Toxic substance, 265
Toxikos, 2
Toxin, 265
Toxinemia, 265

Government Institutes Mini-Catalog

PC #	ENVIRONMENTAL TITLES	Pub Date	Price
629	ABCs of Environmental Regulation: Understanding the Fed Regs	1998	$49
627	ABCs of Environmental Science	1998	$39
585	Book of Lists for Regulated Hazardous Substances, 8th Edition	1997	$79
579	Brownfields Redevelopment	1998	$79
4088	CFR Chemical Lists on CD ROM, 1997 Edition	1997	$125
4089	Chemical Data for Workplace Sampling & Analysis, Single User Disk	1997	$125
512	Clean Water Handbook, 2nd Edition	1996	$89
581	EH&S Auditing Made Easy	1997	$79
587	E H & S CFR Training Requirements, 3rd Edition	1997	$89
4082	EMMI-Envl Monitoring Methods Index for Windows-Network	1997	$537
4082	EMMI-Envl Monitoring Methods Index for Windows-Single User	1997	$179
525	Environmental Audits, 7th Edition	1996	$79
548	Environmental Engineering and Science: An Introduction	1997	$79
643	Environmental Guide to the Internet, 4rd Edition	1998	$59
560	Environmental Law Handbook, 14th Edition	1997	$79
353	Environmental Regulatory Glossary, 6th Edition	1993	$79
625	Environmental Statutes, 1998 Edition	1998	$69
4098	Environmental Statutes Book/CD-ROM, 1998 Edition	1997	$208
4994	Environmental Statutes on Disk for Windows-Network	1997	$405
4994	Environmental Statutes on Disk for Windows-Single User	1997	$139
570	Environmentalism at the Crossroads	1995	$39
536	ESAs Made Easy	1996	$59
515	Industrial Environmental Management: A Practical Approach	1996	$79
510	ISO 14000: Understanding Environmental Standards	1996	$69
551	ISO 14001: An Executive Report	1996	$55
588	International Environmental Auditing	1998	$149
518	Lead Regulation Handbook	1996	$79
478	Principles of EH&S Management	1995	$69
554	Property Rights: Understanding Government Takings	1997	$79
582	Recycling & Waste Mgmt Guide to the Internet	1997	$49
603	Superfund Manual, 6th Edition	1997	$115
566	TSCA Handbook, 3rd Edition	1997	$95
534	Wetland Mitigation: Mitigation Banking and Other Strategies	1997	$75

PC #	SAFETY and HEALTH TITLES	Pub Date	Price
547	Construction Safety Handbook	1996	$79
553	Cumulative Trauma Disorders	1997	$59
559	Forklift Safety	1997	$65
539	Fundamentals of Occupational Safety & Health	1996	$49
612	HAZWOPER Incident Command	1998	$59
535	Making Sense of OSHA Compliance	1997	$59
589	Managing Fatigue in Transportation, *ATA Conference*	1997	$75
558	PPE Made Easy	1998	$79
598	Project Mgmt for E H & S Professionals	1997	$59
552	Safety & Health in Agriculture, Forestry and Fisheries	1997	$125
613	Safety & Health on the Internet, 2nd Edition	1998	$49
597	Safety Is A People Business	1997	$49
463	Safety Made Easy	1995	$49
590	Your Company Safety and Health Manual	1997	$79

Government Institutes

4 Research Place, Suite 200 • Rockville, MD 20850-3226
Tel. (301) 921-2323 • FAX (301) 921-0264
Email: giinfo@govinst.com • Internet: http://www.govinst.com

Please call our customer service department at (301) 921-2323 for a free publications catalog.

CFRs now available online. Call (301) 921-2355 for info.

GOVERNMENT INSTITUTES ORDER FORM

4 Research Place, Suite 200 • Rockville, MD 20850-3226
Tel (301) 921-2323 • Fax (301) 921-0264
Internet: http://www.govinst.com • E-mail: giinfo@govinst.com

3 EASY WAYS TO ORDER

1. **Phone:** **(301) 921-2323**
   Have your credit card ready when you call.

2. **Fax:** **(301) 921-0264**
   Fax this completed order form with your company purchase order or credit card information.

3. **Mail:** **Government Institutes**
   4 Research Place, Suite 200
   Rockville, MD 20850-3226 USA
   Mail this completed order form with a check, company purchase order, or credit card information.

PAYMENT OPTIONS

❑ **Check** *(payable to Government Institutes in US dollars)*

❑ **Purchase Order** *(This order form must be attached to your company P.O. Note: All International orders must be prepaid.)*

❑ **Credit Card** ❑ VISA ❑ Master Card ❑ AMERICAN EXPRESS

Exp.____/____

Credit Card No. _____

Signature _____

(Government Institutes' Federal I.D.# is 13-2695912)

CUSTOMER   INFORMATION

Ship To: *(Please attach your purchase order)*

Name: _____

GI Account # *(7 digits on mailing label)*: _____

Company/Institution: _____

Address: _____
(Please supply street address for UPS shipping)
_____

City: _____ State/Province: _____

Zip/Postal Code: _____ Country: _____

Tel: (____) _____

Fax: (____) _____

Email Address: _____

Bill To: *(if different from ship-to address)*

Name: _____

Title/Position: _____

Company/Institution: _____

Address: _____
(Please supply street address for UPS shipping)
_____

City: _____ State/Province: _____

Zip/Postal Code: _____ Country: _____

Tel: (____) _____

Fax: (____) _____

Email Address: _____

Qty.	Product Code	Title	Price

❑ **New Edition No Obligation Standing Order Program**

Please enroll me in this program for the products I have ordered. Government Institutes will notify me of new editions by sending me an invoice. I understand that there is no obligation to purchase the product. This invoice is simply my reminder that a new edition has been released.

15 DAY MONEY-BACK GUARANTEE

If you're not completely satisfied with any product, return it undamaged within 15 days for a full and immediate refund on the price of the product.

Subtotal _____
MD Residents add 5% Sales Tax _____
Shipping and Handling (see box below) _____
Total Payment Enclosed _____

Within U.S:	**Outside U.S:**
1-4 products: $6/product	Add $15 for each item (Airmail)
5 or more: $3/product	Add $10 for each item (Surface)

SOURCE CODE: BP01